Numerical Methods for Inverse Problems

Numerical Methods for Inverse Problems

Edited by:

Olga Moreira

ARCLER PRESS

www.arclerpress.com

Numerical Methods for Inverse Problems

Olga Moreira

Arcler Press
224 Shoreacres Road
Burlington, ON L7L 2H2
Canada
www.arclerpress.com
Email: orders@arclereducation.com

ISBN: 978-1-77407-707-8 (Hardcover)

DECLARATION

ABOUT THE EDITOR

Olga Moreira is a Ph.D. in Astrophysics and B.Sc. in Physics and Applied Mathematics. She is an experienced technical writer and researcher which former fellowships include postgraduate positions at two of the most renown European institutions in the fields of Astrophysics and Space Science (the European Southern Observatory, and the European Space Agency). Presently, she is an independent scientist working on projects involving machine learning and neural networks research as well as peer-reviewing and edition of academic books.

TABLE OF CONTENTS

LIST OF CONTRIBUTORS

Shangqin He
School of Mathematics and Statistics, NingXia University, Yinchuan 750021, China
College of Mathematics and Information Science and Technology, Hebei Normal University of Science and Technology, Qinhuangdao 066004, China

Xiufang Feng
School of Mathematics and Statistics, NingXia University, Yinchuan 750021, China

Dan Qu
College of Mathematics and Statistics, Hanshan Normal University, Chaozhou, P.R. China

Yan-Bo Ma
College of Mathematics and Statistics, Hanshan Normal University, Chaozhou, P.R. China

Baiyu Wang
College of Mathematics and Econometrics, Hunan University, Changsha, Hunan 410082, China

Fatma Kanca
Department of Information Technologies, Kadir Has University, Istanbul, 34083, Turkey

Songshu Liu
School of Mathematics and Statistics, Northeastern University at Qinhuangdao, Qinhuangdao 066004, China

Lixin Feng
School of Mathematical Sciences, Heilongjiang University, Harbin 150080, China

Chunlong Sun
School of Science, Shandong University of Technology, Zibo 255049, China
Department of Mathematics, Southeast University, Nanjing 210096, China

Gongsheng Li
School of Science, Shandong University of Technology, Zibo 255049, China

Xianzheng Jia
School of Science, Shandong University of Technology, Zibo 255049, China

Upeksha Perera
Department of Mathematics, University of Kelaniya, Kelaniya 11600, Sri Lanka
Institut für Mathematik, Universität Potsdam, 14476 Potsdam, Germany

Christine Böckmann
Institut für Mathematik, Universität Potsdam, 14476 Potsdam, Germany

Ratinan Boonklurb
Department of Mathematics and Computer Science, Faculty of Science, Chulalongkorn University, Bangkok 10330, Thailand

Ampol Duangpan
Department of Mathematics and Computer Science, Faculty of Science, Chulalongkorn University, Bangkok 10330, Thailand

Phansphitcha Gugaew
Department of Mathematics and Computer Science, Faculty of Science, Chulalongkorn University, Bangkok 10330, Thailand

I. Khélifa
Department of Mathematics, LMA, Badji Mokhtar University, Annaba, Algeria

L. Chorfi
Department of Mathematics, LMA, Badji Mokhtar University, Annaba, Algeria

Florian Bruckner
Christian Doppler Laboratory of Advanced Magnetic Sensing and Materials, Institute of Solid State Physics, Vienna University of Technology, Austria

ClaasAbert
Christian Doppler Laboratory of Advanced Magnetic Sensing and Materials, Institute of Solid State Physics, Vienna University of Technology, Austria

GregorWautischer
Christian Doppler Laboratory of Advanced Magnetic Sensing and Materials, Institute of Solid State Physics, Vienna University of Technology, Austria

Christian Huber
Christian Doppler Laboratory of Advanced Magnetic Sensing and Materials, Institute of Solid State Physics, Vienna University of Technology, Austria

ChristophVogler
Institute of Solid State Physics, Vienna University of Technology, Austria

Michael Hinze
Department of Mathematics, University of Hamburg, Germany.

Dieter Suess
Christian Doppler Laboratory of Advanced Magnetic Sensing and Materials, Institute of Solid State Physics, Vienna University of Technology, Austria

Marcin Lawnik
Faculty of Applied Mathematics, Silesian University of Technology, Kaszubska 23, 44-100 Gliwice, Poland

S. L. Han
Institute of Industrial Science, The University of Tokyo, 4-6-1 Komaba, Meguro-ku, Tokyo 153-8505, Japan

Takeshi Kinoshita
Institute of Industrial Science, The University of Tokyo, 4-6-1 Komaba, Meguro-ku, Tokyo 153-8505, Japan

Ngoc-Hien Nguyen
Singapore-MIT Alliance, National University of Singapore, Singapore 117576, Singapore

Boo Cheong Khoo
Department of Mechanical Engineering, National University of Singapore, Singapore 117576, Singapore

Karen Willcox
Massachusetts Institute of Technology, Cambridge, MA 02139-4307, USA.

A. Pascarella
Dipartimento di Neuroscienze, Universitá di Parma, Italy

A. Sorrentino
Department of Statistics, University of Warwick, United Kingdom

LIST OF ABBREVIATIONS

ADM	Adomian Decomposition Method
BPES	Boubaker Polynomials Expansion Scheme
CDM	Chebyshev Differentiation Matrices
CP	Chebyshev Polynomials
CSCM	Chebyshev Spectral Collocation Method
CG	Conjugate Gradient
DQM	Differential Quadrature Method
dSPM	Dynamic Statistical Parametric Mapping
EVP	Eigenvalue Problems
EEG	Electroencephalography
FDM	Finite Difference Method
FEM	Finite Element Method
FIM-SCP	finite Integration Method based on Shifted Chebyshev Polynomials
FIM	Finite Integration Method
FSLP	Fourth-Order Sturm–Liouville Problem
GE	Gaussian Elimination Method
HAM	Homotopy Analysis Method
HPM	Homotopy Perturbation Method
IDE	Integro-Differential Equation
IS	Iterative Solution
LGM	Lie-Group Methods
MEG	Magnetoencephalography
MI	Magnus Integrators
MCMC	Markov Chain Monte Carlo
MM	Matrix Methods
ML	Maximum Likelihood
MNM	Modified Numerov's Method
NDE	Nondestructive Evaluation

NM	Numerov's Method
PDEs	Partial Differential Equations
POD	Proper orthogonal decomposition
RV	Random Variable
SPPS	Spectral parameter Power Series
SLP	Sturm–Liouville Problem
SQMR	Symmetric Quasi-Minimal Residual
TFDE	Time-Fractional Diffusion Equation
VIM	Variational Iteration Method

PREFACE

Given a model, forward (a.k.a. direct) problems consist of predicting the observable data. In a nutshell, given an input the forward problem computes its outputs. On the contrary, inverse problems consist of inferring the unknown physical parameters, or constraining the theoretical model in some way given a set of measured data (i.e. given its outputs, the inverse problem attempts to find what the inputs were). Although particular inverse problems have been theorized as early as the Babylonian Period, this subject is considered to be a relatively new branch of mathematics as the first formulations only emerged after Albert Tarantola introducing an application of inversion problem to seismology in the early 2000's.

A problem is well-posed if all the following three properties are held true: (1) **existence** - a solution exists; (2) **uniqueness** - the solution is unique; and (3) **stability** - the solution's behavior changes continuously with the initial conditions. Inverse problems are often ill-posed as they often do not have a unique solution in the sense that there are more than one model leading to the same observational effects. Additionally, some inverse problems such that of the heat equation are known suffer from instability problems, small changes in the data can produce large changes in the solution.

Ill-posed inverse problems need to be re-formulated by adding assumptions about the data or by using numerical treatments for minimizing approximation errors and regularizing unstable solutions. Nowadays, inverse problems are extensively used in a wide range of scientific and engineering settings. A significant number of numerical methods have been proposed to address the issues present in ill-posed inverse problems. This book includes 14 open-access articles featuring several numerical methods for treating inverse problems.

The first part of the book focuses on Cauchy-like (i.e. partially differential equations with specific boundary conditions) inverse problems. This class of inverse problems has many applications across all science and engineering domains. The inverse problems included in the second part of the book are dedicated to specific examples of their applications in physics, medical imaging and machine learning.

Chapter 1 and 2 are centered on the inverse problem of the Laplacian equation, i.e. the inversion of elliptic partial differential equations. This is an ill-posed problem that produces serious instabilities as a tiny deviation in the data can cause large errors in the solution. Developing numerical treatments for the Laplacian equation inverse problem is rather important as it appears in many engineering and physical settings. Chapter 1 considers the 3-D Laplace equation with non-homogeneous Neumann boundary conditions. To deal with the ill-posed problem, the authors proposes mollification method with bivariate de la Vallée Poussin kernel. Stable estimates are obtained under priori bound assumptions and an appropriate choice of the regularization parameter. Chapter 2 is focused on the Robin boundary value problem for the Laplace's equation and its related inverse problem. The inverse problem is solved using a preconditioned Krylov subspace method based on a block matrix decomposition. By transforming this inverse problem into an equivalent boundary integral equation, the authors succeed in developing a fast algorithm for the discretization of the linear Laplacian inverse equations' system which reduces time complexity of the computations.

Chapter 3 focuses on the inversion of parabolic equations with a time-dependent source term, an ill-posed inverse problem which appears in a wide variety of physical and engineering settings such as elasticity, plasticity, hydrology, material sciences, heat transfer, combustion systems, medical imaging, transport problems, and control theory. To solve this ill-posed problem, a meshless method based on the moving least squares is proposed. Using an appropriate length of time, the authors are able to find a numerically stable and highly efficient solution as demonstrated in the experiments included in this paper.

Chapters 4 to 6 focus on the inversion of fractional differential equations. Chapter 4 considers the inverse problem of finding the time-dependent coefficients of a heat equation with periodic boundary and integral over-determination conditions. This ill-posed problem is treated using a Crank-Nicolson finite-difference scheme combined with an iteration method. Chapter 5 considers a 2-D time-fractional inverse diffusion problem which is severely ill-posed; i.e. the inverse problem's solution does not depend continuously on the data. To treat this ill-posed problem, a modified kernel method is used for finding an approximated solution and convergence estimates are obtained based on both priori and posteriori choices of the regularization parameters. Chapter 6 considers the inverse problem of determining the multiple fractional orders in a multi-term time-fractional diffusion equation (TFDE). A regularization algorithm consisting of a combination of the homotopy method with an optimal perturbation technique is applied to the inversion problem resulting in a stable solution.

Chapter 7 considers the direct and inverse eigenvalue problems of linear differential operators which play an important role in all vibration problems in engineering and physics. The authors develop a universal method for solving a Sturm-Liouville problem of any order with arbitrary boundary conditions. To solve the inverse problem, the authors propose a Lie group method in combination with Magnus expansion.

Chapter 8 to 11 feature domain-specific inverse problem applications. Chapter 8 considers the inversion of a 1-D time-dependent Volterra integro-differential equation (TVIDE) involving two integration terms with respect to time and space. This inversion has important applications in RLC circuit analysis and neuroscience. The inverse problem is solved using a numerical algorithm based on a finite integration method with shifted Chebyshev polynomials. Chapter 9 considers the inverse problem of determining the geometric shape of a perturbed strip with boundary conditions. This is a specific inverse boundary value problem associated with the two-dimensional harmonic equation which has important applications in electrostatic or thermal imaging methods such as the detection of a corrosion surface in non-destructive testing. The authors use a Gauss-Newton method to reduce the inverse problem to a non-linear equation system, and least squares method to regularize the solution. Chapter 10 considers the inverse magnetostatic problem which can be understood as the constrained optimization of a partial differential equation system inversion. This inverse problem is solved by means of an adjoint approach based on the Fredkin-Koehler method for the solution of the forward problem. Chapter 11 considers the inverse Lyapunov exponent problem. The Lyapunov exponent is often used to generate chaotic functions which play an important role in many fields of science and technology (e.g. cryptography, chemical reactors models, and weather forecasts). The author proposes inverse method based on the construction of a piece-wise linear model.

Chapters 12 and 13 consider the statistical inverse problem and Bayesian inference based methods. Chapter 12 considers the stochastic inverse problem for the identification of the non-linear damping of present in harmonic oscillators such that of a ship roll motion. The inverse problem is solved using a hierarchical Bayesian model. Chapter 13 considers the stochastic inverse problem in the application of water quality management in reservoir systems. This statistical inverse problem is solved using a combination of a reduced-order model and a Bayesian inference formulation.

Last chapter 14 considers the Magnetoencephalography (MEG) inverse problem. The paper introduces an inversion method based on Bayesian filtering equations known as the Highly Automated Dipole EStimation (HADES) method. It also overviews the several methods for solving the linear and non-linear MEG inverse problem commonly used in this research field.

The majority of the selected papers include a description of the forward and inverse problems and respective methodology used for approximating and regularizing the solution, as well as several numerical examples.

CHAPTER 1

A Numerical Approximation Method for the Inverse Problem of the Three-Dimensional Laplace Equation

Shangqin He [1,2] and Xiufang Feng [1]

[1]School of Mathematics and Statistics, NingXia University, Yinchuan 750021, China

[2]College of Mathematics and Information Science and Technology, Hebei Normal University of Science and Technology, Qinhuangdao 066004, China

ABSTRACT

In this article, an inverse problem with regards to the Laplace equation with non-homogeneous Neumann boundary conditions in a three-dimensional case is investigated. To deal with this problem, a regularization method (mollification method) with the bivariate de la Vallée Poussin kernel is proposed. Stable estimates are obtained under a priori bound assumptions

Citation: He, S., & Feng, X. (2019). A Numerical Approximation Method for the Inverse Problem of the Three-Dimensional Laplace Equation. Mathematics, 7(6), 487. (13 pages) DOI: https://doi.org/10.3390/math7060487

and an appropriate choice of the regularization parameter. The error estimates indicate that the solution of the approximation continuously depends on the noisy data. Two experiments are presented, in order to validate the proposed method in terms of accuracy, convergence, stability, and efficiency.

Keywords: three-dimensional Laplace equation; ill-posed; de la Vallée Poussin kernel; mollification method; regular parameter; error estimate

INTRODUCTION

The inverse problem of the Laplace equation appears in many engineering and physical areas, such as geophysics, cardiology, seismology, and so on [**1**,**2**,**3**]. It has been widely recognized that the inverse problem for the Laplace equation has a central position in all Cauchy problems of elliptic partial differential equations. The inverse problem of the Laplace equation is seriously ill-posed, where a tiny deviation in the data can cause a large error in the solution [**4**]. It is difficult to develop numerical solutions with conventional methods. Some different methods have been researched, such as the quasi-reversibility [**5**], Tikhonov regularization [**6**], wavelet [**7**], conjugate gradient [**8**], central difference [**9**], Fourier regularization [**10**], and mollification [**11**,**12**,**13**] methods.

The main procedure of the mollification method is using the kernel function to construct a mollification operator by convolution with the measurement data. Manselli, Miller [14], and Murio [15,16] constructed mollification operators by using the Weierstrass kernel to solve some inverse heat conduction problems (IHCP). There have been reports on using the Gaussian kernel to solve the Cauchy problem of elliptic equations [17,18,19,20,21]. Hào [22,23,24] adopted the Dirichlet kernel and de la Vallée Poussin kernel to solve some kinds of two-dimensional equations; including the two-dimensional Laplace equation However, the three-dimensional case was not considered, Moreover, the analysis method used for error estimate was does not generalize to the three-dimensional case well.

Our primary interest is to solve the inverse problem of the three-dimensional Laplace equation with non-homogeneous Neumann boundary conditions. In order to guarantee solvability for the inverse problem provided, a regularization method using the bivariate de la Vallée Poussin kernel is presented.

This paper is organized as follows: In **Section 2**, the mathematical problem for the three-dimensional Laplace equation and its ill-posedness are illustrated. In **Section 3**, we introduce the bivariate de la Vallée Poussin kernel and its properties, following which our mollification regularization method is proposed. In **Section 4**, some stability estimate results are given, in the interior 0<z<d and at the boundary z=d, under a priori assumptions. The numerical aspect of our proposed method is showed in **Section 5**. Concluding remarks are given in **Section 6**.

MATHEMATICAL PROBLEM AND THE ILL-POSEDNESS ANALYSIS

We give thought to the following inverse problem of the three-dimensional Laplace equation with non-homogeneous Neumann boundary conditions:

$$\begin{cases} \Delta u(x,y,z) = 0,\ (x,y) \in \mathbb{R}^2,\ \ 0 < z < d, \\ u(x,y,0) = g(x,y),\ (x,y) \in \mathbb{R}^2, \\ u_z(x,y,0) = h(x,y),\ (x,y) \in \mathbb{R}^2, \end{cases} \tag{1}$$

where $\Delta = \frac{\partial^2}{\partial x^2} + \frac{\partial^2}{\partial y^2} + \frac{\partial^2}{\partial z^2}$ is three-dimensional Laplace operator; and g(x,y), h(x,y) are given vectors in $L^2(R^2)$. The solution u(x,y,z) will be determined by the noisy data $g^\delta(x,y)$ and $h^\delta(x,y)$ in $L^2(R^2)$ that satisfy:

$$\|g - g^\delta\|_{L^2(\mathbb{R}^2)} + \|h - h^\delta\|_{L^2(\mathbb{R}^2)} \le \delta, \tag{2}$$

were δ>0 denotes the error level, and $\|\cdot\|_{L2(R2)}$ denotes the L^2– norm [**1**].

Note that the solution of the problem (**1**) is the sum $u=u_1+u_2$ of the solutions for the following two problems:

$$\begin{cases} \Delta u_1(x,y,z) = 0,\ (x,y) \in \mathbb{R}^2,\ \ 0 < z < d, \\ u_1(x,y,0) = g(x,y),\ (x,y) \in \mathbb{R}^2, \\ (u_1)_z(x,y,0) = 0,\ (x,y) \in \mathbb{R}^2, \end{cases} \tag{3}$$

and

$$\begin{cases} \Delta u_2(x,y,z) = 0,\ (x,y) \in \mathbb{R}^2,\ \ 0 < z < d, \\ u_2(x,y,0) = 0,\ (x,y) \in \mathbb{R}^2, \\ (u_2)_z(x,y,0) = h(x,y),\ (x,y) \in \mathbb{R}^2. \end{cases} \tag{4}$$

Therefore, in order to simplify the process of the Cauchy problem (**1**), we only need to solve problems (**3**) and (**4**), respectively.

For $\varphi \in L^2(R^2)$, the Fourier transform for a variable $r=(x,y) \in R^2$ is defined by

$$\hat{\varphi}(\xi,z)=\frac{1}{2\pi}\int_{\mathbb{R}^2}e^{-i\xi\cdot\mathbf{r}}\varphi(\mathbf{r},z)dr,$$

where $\xi=(\omega,\eta)\in R^2$ and $\xi\cdot r=\omega x+\eta y$.

The inverse Fourier transform for a variable $\xi=(\omega,\eta)\in R^2$ is defined by

$$\varphi(\mathbf{r},z)=\frac{1}{2\pi}\int_{\mathbb{R}^2}e^{i\xi\cdot\mathbf{r}}\hat{\varphi}(\xi,z)d\xi.$$

The Parseval equality [**16**] is as follows:

$$\|\varphi\|_{L^2(\mathbb{R}^2)}=\|\hat{\varphi}\|_{L^2(\mathbb{R}^2)}. \tag{5}$$

Adopting the Fourier transform for the variable $r=(x,y)\in R^2$ to problems (**3**) and (**4**), we obtain

$$\begin{cases} \widehat{(u_1)}_{zz}=(\omega^2+\eta^2)\widehat{u_1},\ (\omega,\eta)\in\mathbb{R}^2, 0<z<d, \\ \widehat{(u_1)}(\omega,\eta,0)=\widehat{g}(\omega,\eta),\ (\omega,\eta)\in\mathbb{R}^2, \\ (\hat{u_1})_z(\omega,\eta,0)=0,\ (\omega,\eta)\in\mathbb{R}^2, \end{cases} \tag{6}$$

and

$$\begin{cases} \widehat{(u_2)}_{zz}=(\omega^2+\eta^2)\widehat{u_2},\ (\omega,\eta)\in\mathbb{R}^2, 0<z<d, \\ \widehat{(u_2)}(\omega,\eta,0)=0,\ (\omega,\eta)\in\mathbb{R}^2, \\ (\hat{u_2})_z(\omega,\eta,0)=\widehat{h}(\omega,\eta),\ (\omega,\eta)\in\mathbb{R}^2. \end{cases} \tag{7}$$

The solution of problem (**6**) is

$$\hat{u}_1(\omega,\eta,z)=\hat{g}(\omega,\eta)\cosh(z\sqrt{\omega^2+\eta^2}). \tag{8}$$

Or

$$u_1(x,y,z)=\frac{1}{2\pi}\int_{\mathbb{R}^2}\hat{g}(\omega,\eta)\cosh(z\sqrt{\omega^2+\eta^2})e^{i\xi\cdot\mathbf{r}}d\xi.$$

The solution of problem (**7**) is

$$\hat{u}_2(\omega,\eta,z)=\hat{h}(\omega,\eta)\frac{\sinh(z\sqrt{\omega^2+\eta^2})}{\sqrt{\omega^2+\eta^2}}. \tag{9}$$

Or

$$u_2(x,y,z)=\frac{1}{2\pi}\int_{\mathbb{R}^2}\hat{h}(\omega,\eta)\frac{\sinh(z\sqrt{\omega^2+\eta^2})}{\sqrt{\omega^2+\eta^2}}e^{i\xi\cdot\mathbf{r}}d\xi.$$

Note that $\cosh(z\sqrt{\omega^2+\eta^2})$ and $\frac{\sinh(z\sqrt{\omega^2+\eta^2})}{\sqrt{\omega^2+\eta^2}}$ are unbounded with respect to the variable (ω,η); a small perturbation in the measured data g^(ω,η) and h^(ω,η) may result in a huge deviation in the solution $u_Q(x,y,z)$ and $u_2(x,y,z)$. Therefore, the problems (**3**) and (**4**) are severely ill-posed.

MOLLIFICATION METHOD AND REGULARIZATION SOLUTION

Mollification Operator

The bivariate de la Vallée Poussin kernel [**22**] function is defined by:

$$V_\alpha(x,y) = \frac{(\cos(\alpha x) - \cos(2\alpha x))(\cos(\alpha y) - \cos(2\alpha y))}{\alpha^2 x^2 y^2} \qquad (\alpha > 0),$$

where α>0 is called the mollification radius (or mollification parameter). $V_Q(x,y)$ has the following properties [**22**]:

(1) $V_\alpha(z)$ is an entire function of exponential type of degree α belong to $L^p(R^2)(1<p\leq\infty)$;

(2) $1/\pi^2\int_{R2} V_\alpha(x,y)dxdy=1$;

(3) $1/\pi^2\int_{R2}|V_\alpha|dxdy<12(\alpha\geq1)$; and

(4) $V\hat{}_\alpha$ is the Fourier transform of Vα(x,y), satisfying:

$$\frac{2}{\pi}\hat{V}_\alpha(\omega,\eta) = 1,\ (\omega,\eta) \in \Delta_\alpha,\ \Delta_\alpha := \{|\omega| \leq \alpha, |\eta| \leq \alpha\},$$

where

$$\hat{V}_\alpha(\omega,\eta) = v(\omega)v(\eta)$$

and

$$v(\omega) = \sqrt{\frac{\pi}{2}}\begin{cases} 1, & |\omega| < \alpha, \\ 2 - \frac{\omega}{\alpha}, & \alpha < |\omega| \leq 2\alpha, \\ 0, & 2\alpha < |\omega|, \end{cases}$$

$$v(\eta) = \sqrt{\frac{\pi}{2}}\begin{cases} 1, & |\eta| < \alpha, \\ 2 - \frac{\eta}{\alpha}, & \alpha < |\eta| \leq 2\alpha, \\ 0, & 2\alpha < |\eta|. \end{cases}$$

For any function $D_\alpha \in L^1(R^2)$ and $f \in L^p(R^2)$, $1 \le p \le \infty$, we define two-dimensional convolution [**22**] by

$$(D_\alpha * f)(x,y) = \int\int_{\mathbb{R}^2} D_\alpha(x',y') f(x-x',y-y')dx'dy'.$$

It is well-known that [**22**]

$$(D_\alpha \hat{*} f)(\omega,\eta) = 2\pi \hat{D}_\alpha(\omega,\eta)\hat{f}(\omega,\eta) \tag{10}$$

and

$$\|D_\alpha * f\|_{L^p(\mathbb{R}^2)} \le \|D_\alpha\|_{L^1(\mathbb{R}^2)}\|f\|_{L^p(\mathbb{R}^2)}. \tag{11}$$

We define the mollification operator T_α by $T_\alpha: R^2 \to R^2$

$$(T_\alpha g^\delta)(x,y) = \frac{1}{2\pi}(V_\alpha * g^\delta)(x,y),\ (T_\alpha h^\delta)(x,y) = \frac{1}{2\pi}(V_\alpha * h^\delta)(x,y).$$

From (**10**), we have

$$(T_\alpha \hat{g}^\delta)(\omega,\eta) = \frac{2}{\pi}\hat{V}_\alpha(\omega,\eta)\hat{g}^\delta(\omega,\eta),\ (T_\alpha \hat{h}^\delta)(\omega,\eta) = \frac{2}{\pi}\hat{V}_\alpha(\omega,\eta)\hat{h}^\delta(\omega,\eta).$$

Regularization Approximation Solution

Instead of solving the problems (**3**) and (**4**) with the data $g^\delta(x,y)$ and $h^\delta(x,y)$, we attempt to re-construct the noisy data $g^\delta(x,y)$ and $h^\delta(x,y)$ by $(T_\alpha g^\delta)(x,y)$ and $(T_\alpha h^\delta)(x,y)$, respectively. We obtain the problems, with the re-constructed data, as follows:

$$\begin{cases} (u_1)_{xx}^{\alpha,\delta} + (u_1)_{yy}^{\alpha,\delta} + (u_1)_{zz}^{\alpha,\delta} = 0, & (x,y,z) \in \Omega, \\ u_1^{\alpha,\delta}(x,y,0) = (T_\alpha g^\delta)(x,y), & (x,y) \in \mathbb{R}^2, \\ (u_1)_z^{\alpha,\delta}(x,y,0) = 0, & (x,y) \in \mathbb{R}^2, \end{cases} \tag{12}$$

and

$$\begin{cases} (u_2)_{xx}^{\alpha,\delta} + (u_2)_{yy}^{\alpha,\delta} + (u_2)_{zz}^{\alpha,\delta} = 0, & (x,y,z) \in \Omega, \\ u_2^{\alpha,\delta}(x,y,0) = 0, & (x,y) \in \mathbb{R}^2, \\ (u_2)_z^{\alpha,\delta}(x,y,0) = (T_\alpha h^\delta)(x,y), & (x,y) \in \mathbb{R}^2. \end{cases} \tag{13}$$

The solution for problem (**12**) is

$$\hat{u}_1^{\alpha,\delta}(\omega,\eta,z) = (T_\alpha \hat{g}^\delta)(\omega,\eta)\cosh(z\sqrt{\omega^2+\eta^2}),$$

or

$$u_1^{\alpha,\delta}(x,y,z) = \frac{1}{2\pi}\int_{\mathbb{R}^2} (\hat{T_\alpha g^\delta})(\omega,\eta)\cosh(z\sqrt{\omega^2+\eta^2})d\omega d\eta.$$

The solution to problem (**13**) is

$$\hat{u}_2^{\alpha,\delta}(\omega,\eta,z) = (\hat{T_\alpha h^\delta})(\omega,\eta)\frac{\sinh(x\sqrt{\omega^2+\eta^2})}{\sqrt{\omega^2+\eta^2}},$$

or

$$u_2^{\alpha,\delta}(x,y,z) = \frac{1}{2\pi}\int_{\mathbb{R}^2} (\hat{T_\alpha h^\delta})(\omega,\eta)\frac{\sinh(x\sqrt{\omega^2+\eta^2})}{\sqrt{\omega^2+\eta^2}}d\omega d\eta.$$

According to (**11**) and the properties (2) and (3) of the kernel V_α, we have the following conclusion:

Remark 1.

If $\|g - g^\delta\|_{L^2(\mathbb{R}^2)} \le \delta$ *and* $\|h - h^\delta\|_{L^2(\mathbb{R}^2)} \le \delta$ *hold, then*

$$\|T_\alpha g^\delta - g\|_{L^2(\mathbb{R}^2)} \le 12\delta,\ \|T_\alpha h^\delta - h\|_{L^2(\mathbb{R}^2)} \le 12\delta.$$

PARAMETER SELECTION AND ERROR ESTIMATES

Lemma 1. *For* z∈(0,d], *the following inequalities hold*

(1) $\frac{\cosh(z\sqrt{\omega^2+\eta^2})}{\cosh(d\sqrt{\omega^2+\eta^2})} \le 2e^{-(d-z)\sqrt{\omega^2+\eta^2}}$,

(2) $\cosh(z\sqrt{\omega^2+\eta^2}) < e^{z\sqrt{\omega^2+\eta^2}}$,

(3) $\frac{\sinh(z\sqrt{\omega^2+\eta^2})}{\sinh(d\sqrt{\omega^2+\eta^2})} \le e^{-(d-z)\sqrt{\omega^2+\eta^2}}$; *and*

(4) $\frac{\sinh(z\sqrt{\omega^2+\eta^2})}{\sqrt{\omega^2+\eta^2}} < de^{z\sqrt{\omega^2+\eta^2}}$.

Proof. Inequalities (1) and (2) are easy to obtain. From the inequalities

$$\frac{\sinh(z\sqrt{\omega^2+\eta^2})}{\sinh(d\sqrt{\omega^2+\eta^2})} = \frac{e^{z\sqrt{\omega^2+\eta^2}}}{e^{d\sqrt{\omega^2+\eta^2}}}\cdot\frac{1-e^{-2z\sqrt{\omega^2+\eta^2}}}{1-e^{-2d\sqrt{\omega^2+\eta^2}}} \le \frac{e^{z\sqrt{\omega^2+\eta^2}}}{e^{d\sqrt{\omega^2+\eta^2}}},$$

and the Taylor expansion

$$\frac{\sinh(z\sqrt{\omega^2+\eta^2})}{\sqrt{\omega^2+\eta^2}} = \sum_{n=0}^{\infty} \frac{(\sqrt{\omega^2+\eta^2})^{2n} z^{2n+1}}{(2n+1)!}$$
$$\leq d \sum_{n=0}^{\infty} \frac{(z\sqrt{\omega^2+\eta^2})^{2n}}{(2n)!}$$
$$= d\cosh(z\sqrt{\omega^2+\eta^2}),$$

we can arrive at (3) and (4).

In the next content, we give stability convergence estimates between the exact solution for problems (**3**) and (**4**) and the regularization approximate solution of problems (**12**) and (**13**) in $0<z<d$ and at the boundary $z=d$, respectively. Convergence estimates will be obtained when we choose a suitable regularization parameter α.

Error Estimates in the Interior

The convergence estimates for the proposed regularization method, in the case of $0<z<d$, will be given in this section, and we obtain the approximation results as following:

Theorem 1. *Let* $u_1(x,y,z)$ *and* $u^{\alpha,\delta}_1(x,y,z)$ *be the exact solution and the approximation solution for problem (1) with the exact input data* $g(x,y)$ *and mollified data, respectively. Assume the a priori bounds* $\|u_1(\cdot,\cdot,d)\|_{L2(R2)} \leq E$ *and* $\|g-g^\delta\|_{L2(R2)} \leq \delta$ *hold. We have the following estimate:*

$$\|u_1 - u_1^{\alpha,\delta}\|_{L^2(\mathbb{R}^2)} < 118Ee^{-(d-z)\alpha} + 36\delta e^{4z\alpha}. \tag{14}$$

If α *is selected as*

$$\alpha = \frac{1}{4d}\ln(E/\delta), \tag{15}$$

we have

$$\|u_1 - u_1^{\alpha,\delta}\|_{L^2(\mathbb{R}^2)} < 118E^{(3d+z)/4d}\delta^{(d-z)/4d} + 36E^{z/d}\delta^{1-z/d}, \tag{16}$$

where E is a finite positive constant.

Proof. From Parseval's equality (**5**) and the properties of double integrals, we have

$$
\begin{aligned}
\|u_1 - u_1^{\alpha,\delta}\|^2_{L^2(\mathbb{R}^2)} &= \|\hat{u}_1 - \hat{u}_1^{\alpha,\delta}\|^2_{L^2(\mathbb{R}^2)} = \sum_{k=1}^{25} \int_{D_k} |\hat{u}_1 - \hat{u}_1^{\alpha,\delta}|^2 d\omega d\eta \\
&= \sum_{k=1}^{16} \int_{D_k} |\hat{u}_1|^2 d\omega d\eta + \sum_{k=17}^{25} \int_{D_k} |\hat{u}_1 - \hat{u}_1^{\alpha,\delta}|^2 d\omega d\eta \\
&\le E^2 \sum_{k=1}^{16} (\sup_{D_k} |A(\omega,\eta)|)^2 + \sum_{k=17}^{25} \int_{D_k} (\hat{g} - (\widehat{T_\alpha g^\delta}))B(\omega,\eta)|^2 d\omega d\eta.
\end{aligned}
$$

Here, $\mathbb{R}^2 = \bigcup_{k=1}^{25} D_k$ (see **Figure 1**) and

$$A(\omega,\eta) = \frac{\cosh(z\sqrt{\omega^2+\eta^2})}{\cosh(d\sqrt{\omega^2+\eta^2})}, \quad B(\omega,\eta) = \cosh(z\sqrt{\omega^2+\eta^2}), \quad (\omega,\eta \in \mathbb{R}).$$

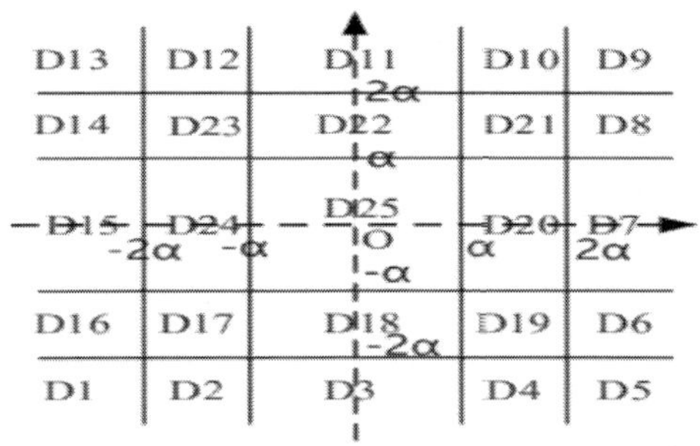

Figure 1. For every Dk(k=1,2,···,25)

Furthermore,

$$
\begin{aligned}
&\sum_{k=17}^{25} \int_{D_k} |(\hat{g} - (\widehat{T_\alpha g^\delta}))B(\omega,\eta)|^2 d\omega d\eta = \\
&\int_{D_{17}} |(\hat{g} - (2-\tfrac{\omega}{\alpha})(2-\tfrac{\eta}{\alpha})\hat{g}^\delta)B(\omega,\eta)|^2 d\omega d\eta + \int_{D_{18}} |(\hat{g} - (2-\tfrac{\eta}{\alpha})\hat{g}^\delta)B(\omega,\eta)|^2 d\omega d\eta + \\
&\int_{D_{19}} |(\hat{g} - (2-\tfrac{\omega}{\alpha})(2-\tfrac{\eta}{\alpha})\hat{g}^\delta)B(\omega,\eta)|^2 d\omega d\eta + \int_{D_{20}} |(\hat{g} - (2-\tfrac{\omega}{\alpha})\hat{g}^\delta)B(\omega,\eta)|^2 d\omega d\eta + \\
&\int_{D_{21}} |(\hat{g} - (2-\tfrac{\omega}{\alpha})(2-\tfrac{\eta}{\alpha})\hat{g}^\delta)B(\omega,\eta)|^2 d\omega d\eta + \int_{D_{22}} |(\hat{g} - (2-\tfrac{\eta}{\alpha})\hat{g}^\delta)B(\omega,\eta)|^2 d\omega d\eta + \\
&\int_{D_{23}} |(\hat{g} - (2-\tfrac{\omega}{\alpha})(2-\tfrac{\eta}{\alpha})\hat{g}^\delta)B(\omega,\eta)|^2 d\omega d\eta + \int_{D_{24}} |(\hat{g} - (2-\tfrac{\omega}{\alpha})\hat{g}^\delta)B(\omega,\eta)|^2 d\omega d\eta + \\
&\int_{D_{25}} |(\hat{g} - \hat{g}^\delta)B(\omega,\eta)|^2 d\omega d\eta.
\end{aligned}
$$

It is easy to verify that $9 < (2-\frac{\omega}{\alpha})(2-\frac{\eta}{\alpha}) < 16,\ (\omega,\eta) \in D_{17}$,

$D_{17}=(-2\alpha,-\alpha)\times(-2\alpha,-\alpha)$.

From Minkowski's inequality, we have

$$
\begin{aligned}
&\left(\int_{D_{17}} |(\hat{g} - (2 - \frac{\omega}{\alpha})(2 - \frac{\eta}{\alpha})\hat{g}^{\delta})B(\omega,\eta)|^2 d\omega d\eta\right)^{1/2} \\
&= \|(\hat{g} - (2 - \frac{\omega}{\alpha})(2 - \frac{\eta}{\alpha})\hat{g}^{\delta})B(\omega,\eta)\|_{L^2(D_{17})} \\
&\leq \|\hat{g}B(\omega,\eta)\|_{L^2(D_{17})} + 16\|(\hat{g}^{\delta} - \hat{g} + \hat{g})B(\omega,\eta)\|_{L^2(D_{17})} \\
&\leq 17\|\hat{g}B(\omega,\eta)\|_{L^2(D_{17})} + 16\|(\hat{g}^{\delta} - \hat{g})B(\omega,\eta)\|_{L^2(D_{17})} \\
&\leq 17E \sup_{D_{17}} |A(\omega,\eta)| + 16\delta \sup_{D_{17}} |B(\omega,\eta)|.
\end{aligned}
$$

Using a similar analysis, we can obtain the integral estimates of the other D_k(k=18,···,24).

Utilizing the inequality $\sqrt{A+B+C} < \sqrt{A} + \sqrt{B} + \sqrt{C}$(A>0,B>0,C>0), we have

$$
\begin{aligned}
&\|u_1 - u_1^{\alpha,\delta}\|_{L^2(\mathbb{R}^2)} = \|\hat{u}_1 - \hat{u}_1^{\alpha,\delta}\|_{L^2(\mathbb{R}^2)} \\
&\leq E\sum_{k=1}^{16} \sup_{D_k} |A(\omega,\eta)| + \sum_{k=17}^{25} \left(\int_{D_k} |(\hat{g} - (T_\alpha \hat{g^{\delta}}))B(\omega,\eta)|^2 d\omega d\eta\right)^{1/2}.
\end{aligned}
$$

According to (1) and (2) of Lemma 1, we obtain

$$
\begin{aligned}
&\|u_1 - u_1^{\alpha,\delta}\|_{L^2(\mathbb{R}^2)} \leq 2E(4e^{-2\sqrt{2}(d-z)\alpha} + 8e^{-\sqrt{5}(d-z)\alpha} + 4e^{-2(d-z)\alpha}) + \\
&(34Ee^{-(d-z)\sqrt{2}\alpha} + 16\delta e^{2\sqrt{2}z\alpha}) + (10Ee^{-(d-z)\alpha} + 4\delta e^{\sqrt{5}z\alpha}) + (10Ee^{-(d-z)\sqrt{2}\alpha} + 4\delta e^{2\sqrt{2}z\alpha}) + \\
&(4Ee^{-(d-z)\alpha} + \delta e^{\sqrt{5}z\alpha}) + (4Ee^{-(d-z)\sqrt{2}\alpha} + \delta e^{2\sqrt{2}z\alpha}) + (4Ee^{-(d-z)\alpha} + \delta e^{\sqrt{5}z\alpha}) + \\
&(10Ee^{-(d-z)\sqrt{2}\alpha} + 4\delta e^{2\sqrt{2}z\alpha}) + (10Ee^{-(d-z)\alpha} + 4\delta e^{\sqrt{5}z\alpha}) + \delta e^{\sqrt{2}z\alpha} \\
&\leq 118Ee^{-(d-z)\alpha} + 36\delta e^{2\sqrt{2}z\alpha} \\
&< 118Ee^{-(d-z)\alpha} + 36\delta e^{4z\alpha}.
\end{aligned}
$$

Taking parameter α to be α=1/4 dln(E/δ), we arrive at (**16**).

Similarly, the error estimate for problem (**4**) can be obtained in the following way.

Theorem 2. *Let* u_2(x,y,z) *and* $u^{\alpha,\delta}_2$(x,y,z) *be the exact solution and approximation solution for problem (**4**) with the exact input data* h(x,y) *and mollified data, respectively. Assume that the a priori bounds* $\|u_2(\cdot,\cdot,d)\|L^2(R^2) \leq E$ *and* $\|h-h^\delta\|L^2(R^2) \leq \delta$ *hold, we have*

$$\|u_2 - u_2^{\alpha,\delta}\|_{L^2(\mathbb{R}^2)} < 59Ee^{-(d-z)\alpha} + 36d\delta e^{4z\alpha}. \tag{17}$$

*If α is chosen as in (**15**), then we have*

$$\|u_2 - u_2^{\alpha,\delta}\|_{L^2(\mathbb{R}^2)} < 59E^{(3d+z)/4d}\delta^{(d-z)/4d} + 36dE^{z/d}\delta^{1-z/d}. \tag{18}$$

As for our problem (**1**), combining the results of Theorems 1 and 2 and the Minkowski inequality, we have the error estimate, as follows:

Theorem 3. *Let* u(x,y,z) *and* uα,δ(x,y,z) *be the exact solution and regular solution for problem (1). Assume that condition (2) and* $\max\{\|u_1(\cdot,\cdot,d)\|L^2(R^2), \|u_2(\cdot,\cdot,d)\|L^2(R^2)\} \le E$ *hold. Then, we have*

$$\|u - u^{\alpha,\delta}\|_{L^2(\mathbb{R}^2)} < 177Ee^{-(d-z)\alpha} + 36(d+1)\delta e^{4z\alpha}. \tag{19}$$

*If α is chosen as in (**15**), then we have*

$$\|u - u^{\alpha,\delta}\|_{L^2(\mathbb{R}^2)} < 177E^{(3d+z)/4d}\delta^{(d-z)/4d} + 36(d+1)E^{z/d}\delta^{1-z/d}. \tag{20}$$

Error Estimates at the Boundary

The estimates (**14**), (**17**), and (**19**) give no information about the error estimates at z=d, as the constraints $\|u_1(\cdot,\cdot,d)\|L^2(R^2) \le E$ and $\|u_2(\cdot,\cdot,d)\|L^2(R^2) \le E$ are too weak for this purpose. Therefore, to ensure stability of the solution $u_1(x,y,z)$, $u_2(x,y,z)$ at z=d, we need the Sobolev space $Hp(R^2)(p \ge 0)$ [**1**],

$$H^p(\mathbb{R}^2) = \{\varphi(x,y) \in L^2(\mathbb{R}^2) : \|\varphi\|_{H^p(\mathbb{R}^2)} < \infty\},$$

where $\|\cdot\|Hp(R2)$ is defined by

$$\|\varphi(\cdot,\cdot,d)\|_{H^p(\mathbb{R}^2)} = \|(1+\omega^2+\eta^2)^{p/2}\hat{\varphi}(\omega,\eta,d)\|_{L^2(\mathbb{R}^2)}.$$

If p=0, then $H^0 = L^2(R^2)$.

Theorem 4. *Let* $u_1(x,y,d)$ *and* $u^{\alpha,\delta}{}_1(x,y,d)$ *be the exact and regularization solutions, respectively, of problem (3) at* z=d. *Suppose that the a priori bounds* $\|u_1(\cdot,\cdot,d)\|_{Hp(R2)} \le Ep(p>0)$ *and* $\|g - g^\delta\|_{L2(R2)} \le \delta$ *hold. Then, we have the following inequality*

$$\|u_1(\cdot,\cdot,d) - u_1^{\alpha,\delta}(\cdot,\cdot,d)\|_{L^2(\mathbb{R}^2)} < \frac{59E_p}{(1+\alpha^2)^{p/2}} + 36\delta e^{4d\alpha}. \tag{21}$$

If the regular parameter α is selected as

$$\alpha = \frac{1}{8d}\ln(E_p/\delta), \tag{22}$$

then we have

$$\|u_1(\cdot,\cdot,d) - u_1^{\alpha,\delta}(\cdot,\cdot,d)\|_{L^2(\mathbb{R}^2)} < 59E_p\left(\frac{8d}{\ln(E_p/\delta)}\right)^p + 36E_p^{1/2}\delta^{1/2}, \tag{23}$$

where E_p *is a positive constant only depending on p.*

Proof. From Parseval's equality (**5**), we have

$$\|u_1(\cdot,\cdot,d)-u_1^{\alpha,\delta}(\cdot,\cdot,d)\|^2_{L^2(\mathbb{R}^2)}=\|\hat{u}_1(\cdot,\cdot,d)-\hat{u}_1{}^{\alpha,\delta}(\cdot,\cdot,d)\|^2_{L^2(\mathbb{R}^2)}$$

$$=\sum_{k=1}^{25}\int\int_{D_k}|\hat{u}_1(\cdot,\cdot,d)-\hat{u}_1{}^{\alpha,\delta}(\cdot,\cdot,d)|^2 d\omega d\eta$$

$$=\sum_{k=1}^{16}\int\int_{D_k}|\hat{u}(\cdot,\cdot,d)|^2 d\omega d\eta+\sum_{k=17}^{25}\int\int_{D_k}|\hat{u}_1(\cdot,\cdot,d)-\hat{u}_1{}^{\alpha,\delta}(\cdot,\cdot,d)|^2 d\omega d\eta,$$

where D_k(k=1,2,···,25) are same as in Theorem 1. Let

$$R(\omega,\eta)=\frac{1}{(1+\omega^2+\eta^2)^{p/2}}\quad(\omega,\eta\in\mathbb{R}).$$

Using the properties of the double integral, and the inequality $\sqrt{A+B+C}<\sqrt{A}+\sqrt{B}+\sqrt{C}$(A>0,B>0,C>0), we have

$$\|u_1(\cdot,\cdot,d)-u_1^{\alpha,\delta}(\cdot,\cdot,d)\|_{L^2(\mathbb{R}^2)}$$

$$\le E\sum_{k=1}^{16}\sup_{D_k}|R(\omega,\eta)|+\sum_{k=17}^{25}\left(\int\int_{D_k}|\hat{u}_1(\cdot,\cdot,d)-\hat{u}_1{}^{\alpha,\delta}(\cdot,\cdot,d)|^2 d\omega d\eta\right)^{1/2}.$$

Using a similar method as in Theorem 1 and the monotonicity of the function R(ω,η), we obtain

$$\|u_1(\cdot,\cdot,d)-u_1^{\alpha,\delta}(\cdot,\cdot,d)\|_{L^2(\mathbb{R}^2)}<\frac{59E_p}{(1+\alpha^2)^{p/2}}+36\delta e^{4d\alpha}.$$

If we chose α as α=1/8 dln(Ep/δ) and utilize inequality $1+\alpha^2>\alpha^2$, then (**23**) can be obtained.

Similar to Theorem 4, the error estimate for problem (**4**) can be obtained as follows.

Theorem 5. *Let* $u_2(x,y,d)$ *and* $u^{\alpha,\delta}_2(x,y,d)$ *be the exact and regularization solutions, respectively, for problem* (**4**) *at* z=d. *Suppose that the a priori bounds* $\|u_2(\cdot,\cdot,d)\|_{Hp(R2)}\le E_p$ *and* $\|h-h^\delta\|_{L2(R2)}\le\delta$ *hold. Then, we have the following inequality*

$$\|u_2(\cdot,\cdot,d)-u_2^{\alpha,\delta}(\cdot,\cdot,d)\|_{L^2(\mathbb{R}^2)}<\frac{59E_p}{(1+\alpha^2)^{p/2}}+36d\delta e^{4d\alpha}. \tag{24}$$

If the regularization parameter α is chosen as in (**22**), *then*

$$\|u_2(\cdot,\cdot,d)-u_2^{\alpha,\delta}(\cdot,\cdot,d)\|_{L^2(\mathbb{R}^2)}<59E_p\left(\frac{8d}{\ln(E_p/\delta)}\right)^p+36dE_p^{1/2}\delta^{1/2}. \tag{25}$$

Thus, as for problem (**1**), using the results of Theorems 4 and 5 and the Minkowski inequality, we have the stable error estimate, as follows:

Theorem 6. *Let* $u(x,y,d)$ *and* $u^{\alpha,\delta}(x,y,d)$ *be the exact and regularization solutions, respectively, for problem (1) at* $z=d$. *Suppose that condition (2) and* $\max\{\|u_1(\cdot,\cdot,d)\|_{H^p(R^2)},\|u_2(\cdot,\cdot,d)\|_{H^p(R^2)}\}\le E_p (p>0)$ *hold. We have the convergence estimate, as follows:*

$$\|u(\cdot,\cdot,d)-u^{\alpha,\delta}(\cdot,\cdot,d)\|_{L^2(\mathbb{R}^2)} < \frac{118E_p}{(1+\alpha^2)^{p/2}} + 3e(d+1)\delta e^{d/\alpha}. \tag{26}$$

*If the regularization parameter α is selected as in (**22**), then*

$$\|u(\cdot,\cdot,d)-u^{\alpha,\delta}(\cdot,\cdot,d)\|_{L^2(\mathbb{R}^2)} < 118E_p\left(\frac{8d}{\ln(E_p/\delta)}\right)^p + 3e(d+1)E_p^{1/2}\delta^{1/2}. \tag{27}$$

Remark 2. *In this part, we consider the stable error estimates in the cases* $0<z<d$ *and* $z=d$, *respectively. In the interior,* $0<z<d$, *the a priori bound for* $\|u(\cdot,\cdot,d)\|_{L^2(R^2)}$ *is sufficient, and the convergence estimate converges quickly to zero as* $\delta\to0+$. *However, for the case* $z=d$, *although a stronger a priori bound for* $\|u(\cdot,\cdot,d)\|_{H^p(R^2)} (p>0)$ *is imposed, the error estimate is only of logarithmic type, with order* $(\ln(E_p/\delta))^{-p} (p>0)$.

NUMERICAL EXAMPLES

We performed two numerical examples to verify the accuracy and stability of our proposed method. Our tests were carried out in the MATLAB R2014b software.

In the numerical examples, we selected the discrete interval to be $D=[-6-\gamma,6+\gamma]\times[-6-\gamma,6+\gamma] (\gamma=1\times10^{-3})$ and the measurement data $\phi\delta(x,y)$ was obtained as the following

$\phi^\delta(x,y)=\phi+\epsilon(2\mathrm{randn}(\mathrm{size}(\phi))-1)$,

where

$$\phi=(\phi(x_i,y_j))_{N\times N},\ x_i=-6+\frac{12(i-1)}{N-1},\ y_j=-6+\frac{12(j-1)}{N-1},\ i,j=1,2,\cdots,N.$$

The error level δ is given by

$$\delta=\|\phi-\phi^\delta\|_{l^2(D)}=\sqrt{\frac{1}{N\times N}\sum_{i=1}^{N}\sum_{j=1}^{N}(\phi(x_i,y_j)-\phi^\delta(x_i,y_j))^2}.$$

In the following numerical implementations, we need to take the two-dimensional discrete Fourier transform of the data vector $\phi^\delta(x,y)$ and the two-dimensional discrete inverse Fourier transform. We take $d=1$, $N=15$, and fix the reconstructed position z=0.3. The a priori mollification parameter α was determined by (**15**) and (**22**), where $\|\varphi(\cdot,\cdot,d)\|_{l2(D)}=E$ and $\|\varphi(\cdot,\cdot,d)\|_{Hp(D)}=E_p$. We define the relative error between the exact solution *u* and its approximate solution uα,δ as:

$$rel(u) = \frac{\|u - u^{\alpha,\delta}\|_{l^2(D)}}{\|u\|_{l^2(D)}}.$$

Example 1. *We chose the function* $g(x,y)=xye^{-x2-y2}$ *as the exact data for problem (3).*

Example 2. *We chose the function* $h(x,y)= e^{-\frac{36}{36+x^2}-\frac{36}{36+y^2}}$ *as the exact data for problem (4).*

We use different perturbation noise levels at the boundary z=1 with p=0.5 and p=3, respectively, in **Table 1** and **Table 2**. Note that the results for the relative error rel(u) at z=1 depended on the error level, δ, and *p*.

Table 1. Example 1: Relative Error at z=1, p=0.5 and p=3.

p	**Error Level** δ	**Regular Parameter** α	**Relative Error** $rel(u)$
0.5	1×10^{-8}	5.3392	0.2764
0.5	1×10^{-9}	5.9148	0.0955
3	1×10^{-7}	5.8034	0.1363
3	1×10^{-8}	6.3790	0.0777

Table 2. Example 2: Relative Error at z=1, p=0.5 and p=3.

p	**Error Level** δ	**Regular Parameter** α	**Relative Error** $rel(u)$
0.5	1×10^{-8}	5.4041	0.2404
0.5	1×10^{-9}	5.9798	0.0509
3	1×10^{-6}	5.2927	0.2684
3	1×10^{-7}	10.6332	0.0905

Figure 2 and Figure 3 show the re-constructed solution and exact solution of Example 1, corresponding to noise levels of $\delta=1\times10^{-10}$ and $\delta=2\times10^{-8}$, with z=0.3 and z=1, respectively. Figure 4 shows the corresponding error between (a) and (b) in Figure 2 and Figure 3.

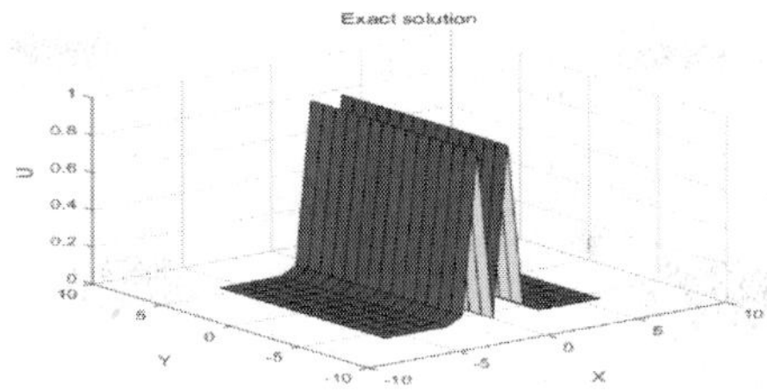

(a)

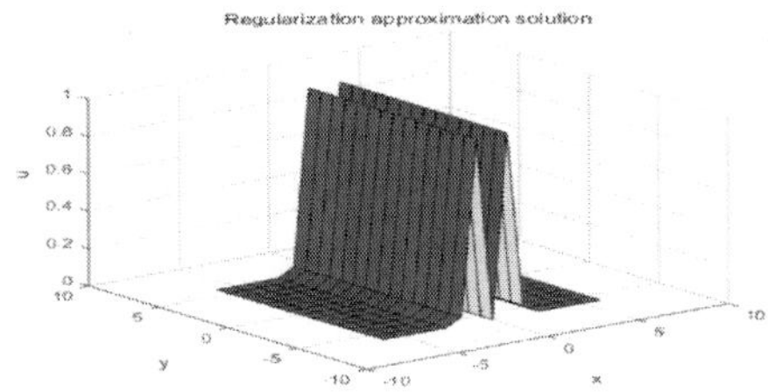

(b)

Figure 2. Example 1: δ=1×10^{-10}, z=0.3: (**a**) Exact solution, and (**b**) approximation solution.

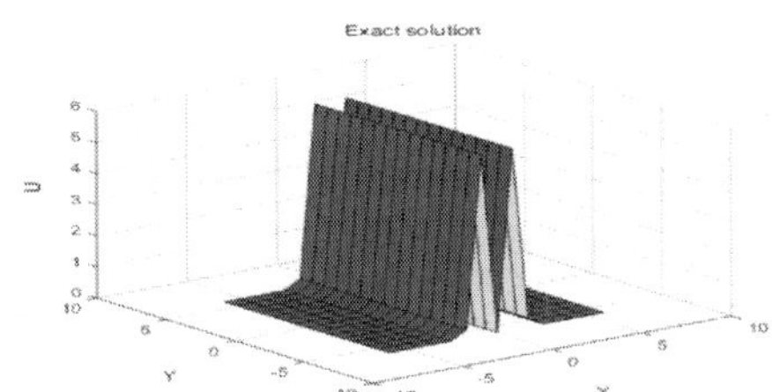

(a)

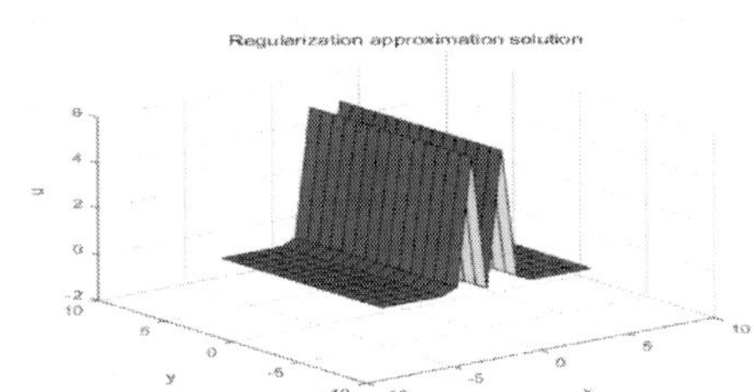

(b)

Figure 3. Example 1: For δ=2×10^{-8}, z=1: (**a**) Exact solution, and (**b**) approximation solution.

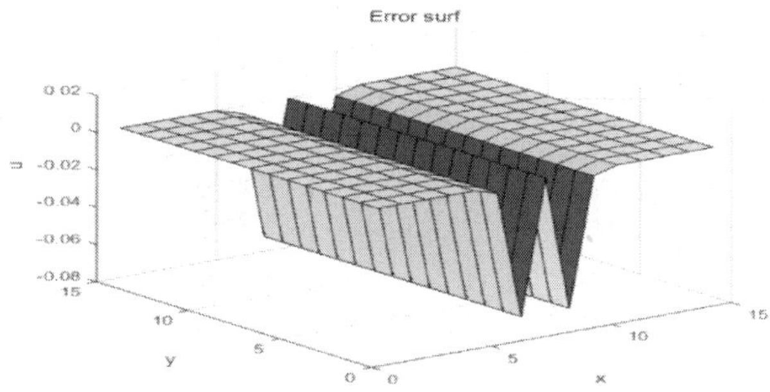

(a)

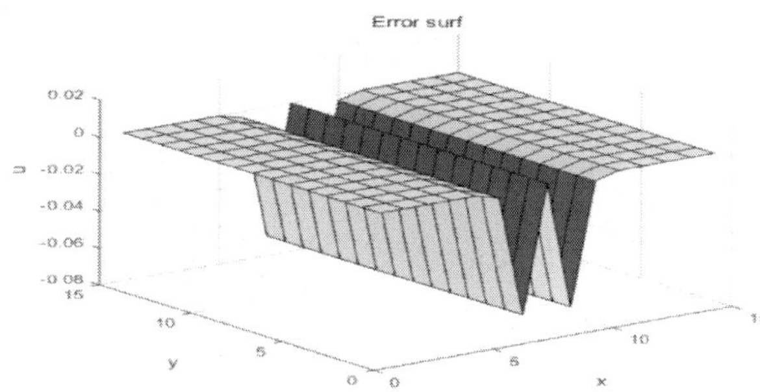

(b)

Figure 4. Example 1: (**a**) The error between (a) and (b) in **Figure 2**, and (**b**) the error between (a) and (b) in **Figure 3**.

Figure 5 and **Figure 6** show the regularization solution and exact solution of Example 2, corresponding to noise levels of $\delta=1\times10^{-10}$ and $\delta=2\times10^{-8}$ with z=0.3 and z=1, respectively. **Figure 7** shows the corresponding error between (a) and (b) in **Figure 5** and **Figure 6**.

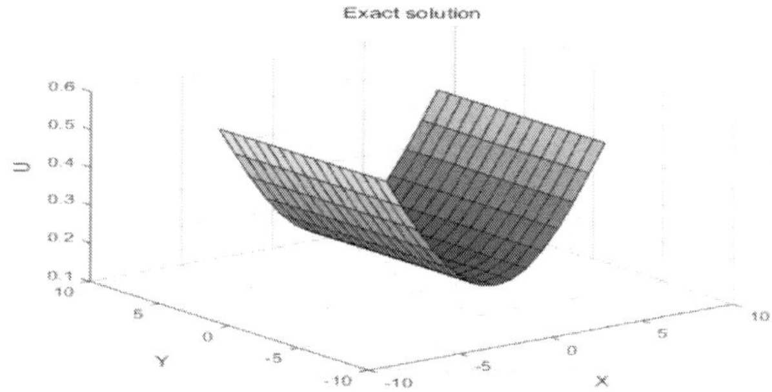

(a)

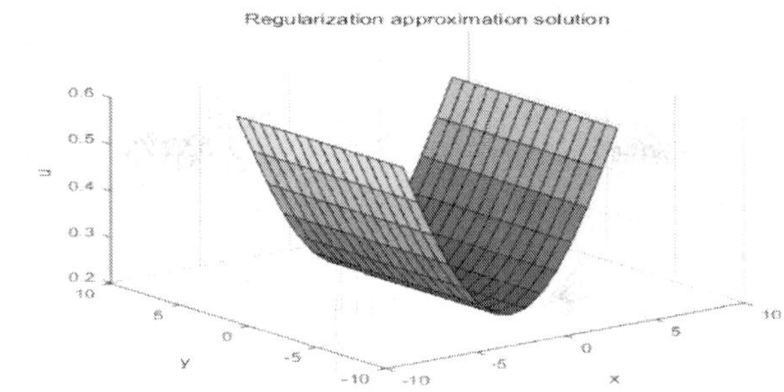

(b)

Figure 5. Example 2: δ=1×10^{-10}, z=0.3: (**a**) Exact solution, and (**b**) approximation solution.

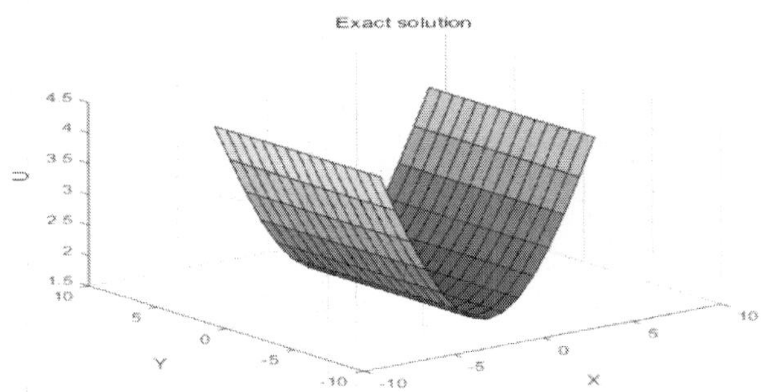

(a)

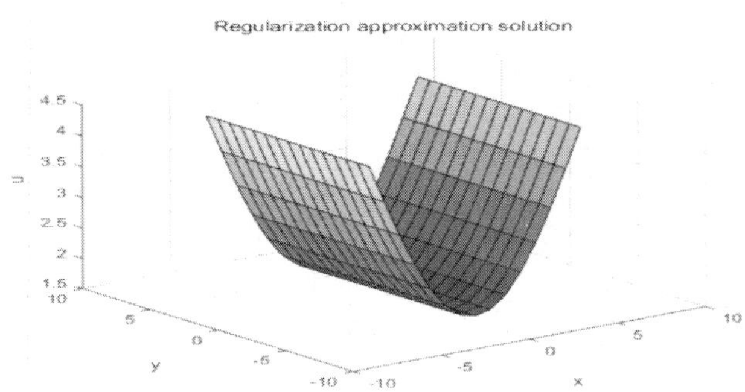

(b)

Figure 6. Example 2: For δ=2×10^{-8}, z=1: (**a**) Exact solution, and (**b**) approximation solution.

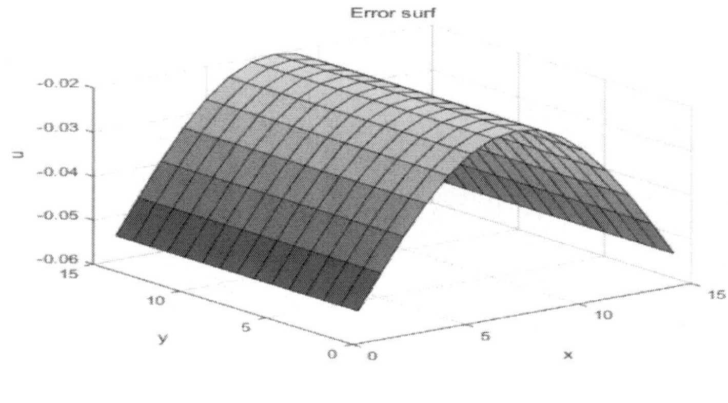

(a)

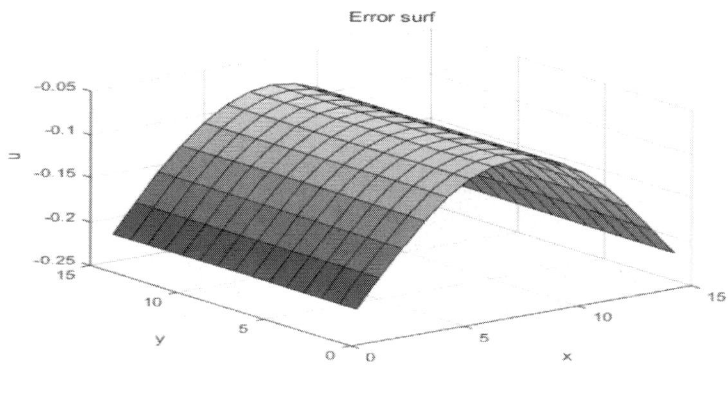

(b)

Figure 7. Example 2: (**a**) The error between (a) and (b) in **Figure 5**, and (**b**) the error between (a) and (b) in **Figure 6**.

In the two examples, we note that the methods which we adopted are stable and effective.

CONCLUSIONS

In this article, we use a regularization method to solve two Cauchy problems for the three-dimensional Laplace equationOe-. Stable approximate estimates are obtained under a priori bound assumptions and an appropriate choice of the regular parameter. Two numerical examples are investigated to verify the stability of our presented method.

We consider stability error estimates in the cases $0<z<d$ and $z=d$, respectively. In the interior, $0<z<d$, the convergence estimate is $O(\delta^{\lambda})(\lambda>0)$, which quickly converges to zero as $\delta \rightarrow 0+$. However, at the boundary, $z=d$, the error estimate is of logarithmic type with order $(\ln(E_p/\delta))^{-p}(p>0)$. In future work, we hope to find a new a priori assumption method, in order to obtain an error estimation which achieves better results.

AUTHOR CONTRIBUTIONS

All authors contributed equally and significantly in writing this article. All authors read and approved the final manuscript.

FUNDING

Thanks to the National Science Foundation of China (11161036). Thanks to the Natural Science Research Foundation of Ningxia Province, China (NR17260)(NR160117).

ACKNOWLEDGMENTS

The authors are deeply indebted to the anonymous referees for their very careful reading and valuable comments and suggestions which immensely improved the previous version of our manuscript.

CONFLICTS OF INTEREST

The authors declare that they have no competing interests.

REFERENCES

1. Tikhonov, A.N.; Arsenin, V.Y. *Solutions of Ill-Posed Problems*; Winson: Washington, DC, USA, 1977; pp. 10–80.
2. Colli-Franzone, P.; Magenes, E. On the inverse potential problem of electrocardiology. *Calcolo* **1980**, *10*, 459–538.
3. Alessandrini, G. Stable determination of a crack from boundary measurements. *Proc. R. Soc. Edinb.* **1993**, *123*, 497–516.
4. Hadamard, J. *Lectures on the Cauchy Problem in Linear Partial Differential Equations*; Yale University Press: New Haven, CT, USA, 1923.
5. Klibanov, M.V.; Santosa, F. A computational quasi-reversibility method for Cauchy problems for Laplace equation"\ ÿÿ. *SIAM J. Appl.* **1991**, *56*, 1653–1675.
6. Reinhardt, H.J.; Han, H.; Hào, D.N. Stability and regularization of a discrete approximation to the Cauchy problem for Laplace's equation. *SIAM J. Numer. Anal.* **1999**, *36*, 890–905.
7. Wang, W.F.; Wang, J.R. Wavelet solution to the three-Dimensional Cauchy problem for Laplace equation~\ ÿÿ. *J. Math.* **2012**, *36*, 239–248.
8. Hào, D.N.; Lesnic, D. The Cauchy problem for Laplace's equation via the conjugate gradient method. *IMA J. Appl. Math.* **2000**, *65*, 199–217.
9. Xiong, X.T.; Fu, C.L. Central difference regularization method for the Cauchy problem of the Laplace's equation. *Appl. Math. Comput.* **2006**, *181*, 675–684.
10. Fu, C.L.; Li, H.F.; Qian, Z.; Xiong, X.T. Fourier regularization method for solving a Cauchy problem for the Laplace equation□\ ÿÿ. *Inverse Probl. Sci. Eng.* **2008**, *16*, 159–169.
11. Li, Z.P.; Fu, C.L. A mollification method for a Cauchy problem for the Laplace equation□□\ÿÿ. *Appl. Math. Comput.* **2011**, *217*, 9209–9218.
12. Ang, D.D.; Nghia, N.H.; Tam, N.C. Regularized solutions of a Cauchy Problem for Laplace Equation in an irregular layer: A three dimensional model. *Acta Math. Vietnam.* **1998**, *23*, 65–74.
13. Cheng, J.; Hon, Y.C.; Wei, T.; Yamamoto, M. Numerical computation of a Cauchy Problem for Laplace's equation. *Zammz. Angew. Math. Mech.* **2001**, *81*, 665–674.

14. Manselli, P.; Miller, K. Calculation of the surface temperature and heat flux on one side of a wall from measurements on the opposite side. *Ann. Mat. Pura Appl.* **1980**, *123*, 161–183.
15. Murio, D.A. Numerical method for inverse transient heat conduction problems. *Rev. De La Unión Matemátic Argentina* **1981**, *30*, 25–46.
16. Murio, D.A. On the estimation of the boundary temperature on a sphere from measurements at its center. *J. Comp. Appl. Math.* **1982**, *8*, 111–119.
17. Hao, C.; Li, F.X.; Fu, C.L. A mollification regularization method for the Cauchy problem of an elliptic equation in a multi-dimensional case. *Inverse Probl. Sci. Eng.* **2010**, *18*, 971–982.
18. Xiong, X.T.; Shi, W.X.; Fan, X.Y. Two numerical methods for a Cauchy problem for modified Helmholtz equation. *Appl. Math. Model.* **2011**, *35*, 4915–4920.
19. Yang, F.; Fu, C.L.; Li, X.X. A mollification regularization method for identifying the time-dependent heat source problem. *J. Eng. Math.* **2016**, *100*, 67–80.
20. Xiong, X.T.; Mao D, L.; Cao, X.X. A mollification method for solving the Cauchy problem for the modified Helmholtz equation. *J. Northwest Univ. (Nat. Sci.)* **2018**, *53*, 1–7.
21. Ern, A.; Guermond, J.L. Mollification in strongly Lipschitz domains with application to continuous and discrete de Rham complexes. *Comput. Methods Appl. Math.* **2016**, *16*, 51–75.
22. Hào, D.N. A mollification method for ill-posed problems. *Numer. Math.* **1994**, *68*, 469–506.
23. Hào, D.N. A mollification method for a Noncharacteristic Cauchy Problem for a Parabolic Equation. *Math. Anal. Appl.* **1996**, *199*, 873–909.
24. Hào, D.N.; Hien, P.M.; Sahli, H. Stability results for a Cauchy problem for an elliptic equation. *Inverse Probl.* **2007**, *23*, 421–461.

CHAPTER

2

The Numerical Solution of Forward and Inverse Robin Problems for Laplace's Equation

Dan Qu and Yan-Bo Ma

College of Mathematics and Statistics, Hanshan Normal University, Chaozhou, P.R. China

ABSTRACT

The Robin boundary value problem for Laplace's equation in the elliptic region (which is a forward problem) and its related inverse problem can be used to reconstruct Robin coefficients from measurements on a partial boundary (inverse problem). We present a numerical solution of the forward problem that uses a boundary integral equation method, and we propose a fast solver based on one that reduces the computational complexity to O(Nlog(N)), where *N* is the size of the data. We compute the solution of the

Citation: Qu, D., & Ma, Y. B. (2019). The numerical solution of forward and inverse Robin problems for Laplace's equation. Boundary Value Problems, 2019(1), 114. (13 pages) DOI: https://doi.org/10.1186/s13661-019-1229-6

inverse problem using a preconditioned Krylov subspace method where the preconditioner is based on a block matrix decomposition. The structure of the matrix is then exploited to solve the direct problem. Numerical examples are presented to illustrate the effectiveness of the proposed approach.

Keywords: Robin inverse problem; Laplace equation; Boundary integral equations; Preconditioned Krylov subspace method

INTRODUCTION

Laplace's equation is of particular importance in applied mathematics because it is applicable to a wide range of different physical and mathematical phenomena, including electromagnetism, fluid and solid mechanics, conductivity. It has the special status of being the most straightforward elliptic partial differential equation for the description of all sorts of steady state phenomena. When attached to Robin boundary conditions it is able to handle the spherical geometry of the largest to the smallest structures in the universe [1]. If a domain Ω in R^2 is assumed to have a smooth boundary $\partial\Omega=\Gamma$, the Robin boundary value problem for Laplace's equation can be stated as follows:

$$\begin{cases} \Delta u = 0, & \text{in } \Omega, \\ \frac{\partial u}{\partial \nu} + pu = g, & \text{on } \partial\Omega = \Gamma, \end{cases} \tag{1.1}$$

where ν is an outward-facing unit normal vector on Γ; p=p(x) is the Robin coefficient, which is non-negative and not identical for $\Gamma_1 \subset \Gamma$; and g=g(x) is a given input function.

However, Laplace's equation is known to present a number of boundary value problems that can be complex to solve and, when it comes to computation, this can demand significant quantities of computational work. There is therefore an ongoing need to come up with more efficient algorithms to solve these kinds of boundary value problems. This paper addresses such a need.

If p in Eq. (1.1) is given, a unique solution, u, can be determined. This is known as the forward problem. The forward map from p to u has already been discussed in the literature, and a number of its properties has been identified, including uniqueness, continuity with respect to proper norms, differentiability and various forms of stability (see [2,3,4,5,6,7]).

However, a more challenging problem is the inverse problem, which aims to recover the Robin coefficient p from a partial boundary measurement of function *u*. In other words, how to obtain *p* by using $u=u_0$ on a part, Γ_0, of the boundary, where $\Gamma_0 \cap \Gamma_1 = \emptyset$. It is well-known that the inverse problem is currently ill-conditioned, which can cause significant computational difficulties. This issue is of concern because a number of applications seek a solution of the problem, notably various kinds of nondestructive evaluation (NDE) [8].

A lot of research over recent years has been devoted to developing a numerical solution to the inverse problem [9,10,11,12,13]. Lin and Fang transformed the Robin inverse problem into a linear integral equation by introducing a new variable *v*, then located a way of regularizing *v* [14]. From this, it was possible to derive a linear least-square-based methods for estimation of the Robin coefficient. Jin solved the Robin inverse problem by using conjugate gradient (CG) methods [15], then analyzing the convergence of different regular terms. Jin and Zou looked at ways of estimating piecewise constant Robin coefficients by using a concave-convex procedure to minimize the Modica–Mortola functional [16]. Chaabane et al. also considered the possibility of estimation of piecewise constant Robin coefficients, in their case by using the Kohn–Vogelius method [17]. Jiang et al. [18] discussed the convergence properties for elliptic and parabolic inverse robin problems.

The solution of an inverse problem usually depends upon a forward solver that is capable of expanding upon the current state to assess whether it can be proved given the current conditions. This makes the speed of forward solver a critical consideration. It is possible to discretize Eq. (1.1) directly by using a finite difference method or a finite element method (see, for instance, [19]) or wavelet method (see, for instance, [20, 21]). Equation (1.1) can also be reformulated as a boundary integral equation. The resulting integral equation can then be discretized by using a boundary element method or numerical quadratures [14, 15, 22, 23]. Out of these, using a boundary integral equation (BIE) appears to the best option because the resulting discrete system is much smaller than it is if one uses the original partial differential equation. This has led to a BIE approach being used by many researchers who take an interest in inverse problems, and we, too, will be adopting this method.

After Eq. (1.1) has been transformed into a BIE and the discretization scheme set up, it can be found that the structure of the coefficient matrix of the resulting discrete system is the sum of the circulant. This makes it possible for a fast algorithm for matrix-vector products to be developed by using a fast Fourier transform (FFT) approach where only O(Nlog(N)) operations are required instead of the standard $O(N^2)$. In this way, numerical methods for solving the inverse problem can be effectively sped up by using this faster forward solver. Aside from this, we shall be looking at a preconditioner that can exploit the structure of the forward problem matrix. A variant of the symmetric quasi-minimal residual (SQMR) method is employed to accomplish this. SQMR is known to offer significant economy in solving large-scale problems. As the preconditioner is not formed explicitly, we also present an algorithm that can obtain the product of the preconditoner with a given vector.

The paper is organized as follows. In Sect. 2, the Robin inverse problem is transformed into an equivalent BIE. We then derive a fast algorithm for discretization of its linear system by exploiting the special properties of certain relevant kernel functions. In Sect. 3, we look at how the preconditioner used in the Krylov subspace method for the Karush–Kuhn–Tucker (KKT) system might be used to further enhance our approach. In Sect. 4, we conclude with numerical examples to demonstrate the effectiveness of our proposed fast solver for the Robin inverse problem.

FAST SOLUTION OF THE FORWARD PROBLEM

Let Φ=Φ(x,y) be the fundamental solution for Laplace's equation in R^2. Thus:

$$\Phi(x, y) = \frac{1}{2\pi} \ln \frac{1}{|x - y|} \quad \text{for } x \neq y.$$

We will first introduce a formulation of the BIE for the boundary value problem presented in Eq. (1.1). Using Green's formula, the solution, u, to Eq. (1.1) in domain Ω can be represented in terms of its boundary value by the following:

$$u(x) = -\int_{\Gamma} \left(\frac{\partial \Phi(x,y)}{\partial \nu_y} + p(y)\Phi(x,y) \right) u(y)\, ds_y + \int_{\Gamma} \Phi(x,y) g(y)\, ds_y, \quad x \in \Gamma,$$

where ds_y denotes the arc length differential. After moving x in Ω to the boundary Γ, one can arrive at the following BIE on the boundary Γ:

$$\frac{1}{2} u(x) + \int_{\Gamma} \left(\frac{\partial \Phi(x,y)}{\partial \nu_y} + p(y)\Phi(x,y) \right) u(y)\, ds_y = \int_{\Gamma} \Phi(x,y) g(y)\, ds_y, \quad x \in \Gamma. \tag{2.1}$$

Hence, solving Eq. (1.1) in Ω can be reduced to solving the above BIE, (2.1), on Γ.

The double-layer and single-layer operators can be defined, respectively, by

$$(\mathcal{D}u)(x) = \int_{\Gamma} \frac{\partial \Phi(x,y)}{\partial \nu_y} u(y)\, ds_y, \qquad \text{for } x \in \Gamma,$$
$$(\mathcal{S}u)(x) = \int_{\Gamma} \Phi(x,y) u(y)\, ds_y,$$

and, by denoting A(p)(u)=(1/2 I+D)u+S(pu), we can re-write Eq. (2.1) in terms of its operators as follows:

A(p)(u)=f, (2.2)

where f=Sg.

Let us now suppose that Ω is an ellipse in R^2. So,

$$\Omega = \left\{ (x_1, x_2) : \left(\frac{x_1}{a}\right)^2 + \left(\frac{x_2}{b}\right)^2 < 1 \right\},$$

where a,b>0. The usual parametrization for Γ is

$$(x_1, x_2) = (\phi(t), \psi(t)) = (a\cos(2\pi t), b\sin(2\pi t)), \quad \text{for } 0 \le t \le 1.$$

The kernels in the integral operators D and S can be explicitly expressed as follows:

$$K_d(t,s) = -\frac{ab}{2(a^2\sin^2(\pi(t+s)) + b^2\cos^2(\pi(t+s)))},$$

$$K_s(t,s) = \ln(2|\sin(\pi(t-s))|)\sqrt{a^2\sin^2(\pi(t+s)) + b^2\cos^2(\pi(t+s))}$$
$$\times \sqrt{a^2\sin^2(2\pi s) + b^2\cos^2(2\pi s)}$$

for 0≤t, s≤1.

For discretization of the integral equation (2.1) with the given parametrization of Γ, a Nyström method can be applied by using the mid-point quadrature rule. If we partition the interval [0,1] into N uniform subintervals [(i−1)h,ih] (i=1,2,…,N) with h=1/N, then the quadrature points will be t_i=(i−1/2)h. With the mid-point quadrature rule, we can denote the matrix representation of the kernels K_d and K_s by D and S, respectively. As K_d(t,s) is smooth, the rectangular rule can be applied to obtain the following discrete representation:

$$D = h\left[-\frac{ab}{2(a^2\sin^2(\pi(t_i+t_j)) + b^2\cos^2(\pi(t_i+t_j)))}\right]_{i,j=1}^{n}.$$

However, because the kernel function SS is weakly singular at s=t and (t,s)=(0,1) and (1,0), discretizing $\int_0^1 K_s(t,s)u(s)ds$ requires special measures to avoid large errors. By denoting

$$K_{s1}(t,s) = \ln\left(2\left|\sin\left(\pi(t-s)\right)\right|\right),$$

$$K_{s2}(t,s) = \ln\sqrt{a^2\sin^2\left(\pi(t+s)\right) + b^2\cos^2\left(\pi(t+s)\right)},$$

$$K_{s3}(s) = -\sqrt{a^2\sin^2\left(2\pi(s)\right) + b^2\cos^2\left(2\pi(s)\right)},$$

it is possible to decompose $K_s(t,s)$ as follows:

$$K_s(t,s) = \left(K_{s1}(t,s) + K_{s2}(t,s)\right)K_{s3}(t,s).$$

If we now use a singularity subtraction technique, we get

$$S = h\left[K_s(t_i,t_j) - \delta_{i,j}\sum_{k=1}^{n} K_{s1}(t_i,t_k)K_{s3}(t_i)\right]_{i,j=1}^{n},$$

where δ_{ij} is the Kronecker delta.

Let us now denote Ks=hK_s(ti,tj), K_{s1}=hK_{s1}(ti,tj), K_{s2}=hK_{s2}(ti,tj), K_{s3}=K_{s3}(ti). For a given vector v, the product Sv can then be obtained as follows:

$$Sv = \left[hK_s(t_i,t_j) - \delta_{i,j}h\sum_{k=1}^{n} K_{s1}(t_i,t_k)K_{s3}(t_i)\right]_{i,j=1}^{n} v$$

$$:= K_s v - (K_{s1}1).*(K_{s3}.*v)$$

$$= (K_{s1} + K_{s2})(K_{s3}.*v) - (K_{s1}1).*(K_{s3}.*v)$$

$$= K_{s1}(K_{s3}.*v) + K_{s2}(K_{s3}.*v) - (K_{s1}1).*(K_{s3}.*v),$$

where 1 denotes the vector whose elements are all one and a.∗b denotes element-by-element multiplication of the vectors a and b. For more techniques to deal with the singular integral operators we refer to [24, 25].

In general, it is necessary to choose specific quadrature rules for the Nyström method to be able to discretize the BIEs. Nyström techniques can then be used to build faster algorithms [14]. In our case, we want to exploit the special structure of the matrices D and S to devise a fast algorithm that can solve Eq. (2.1) or (2.2).

From the above operator expression, it can be observed that the discretized matrix D is a Hankel matrix, where Di,j is an element of D and satisfies $D_{i,j}=D_{i-1,j+1}$. For a given vector v, we can use a fast algorithm to compute the product of D times vector v with the help of an FFT. Obviously, K_{s1} is a Toeplitz matrix, which has constant values along its negative-sloping diagonals, whilst K_{s2} is a Hankel matrix. This enables us to get the product of S with a given vector v by using the FFT.

Generally, the discretized form of the direct problem (2.2) is as follows:

$$A(\mathbf{p})\mathbf{u} := \left(\frac{1}{2}I + D + S\mathbf{p}\right)\mathbf{u} = \mathbf{f},$$

where p=diag(p), which can be solved by using a Gaussian elimination method (GE). However, in this paper we will be using an iterative method to solve the problem. For each iteration, we compute the product of the matrix ½ I+D+Sp with a given vector x by using the FFT algorithm. For the given vector x:

$$\begin{aligned} A(\mathbf{p})\mathbf{u} &= \left(\frac{1}{2}I + D + S\mathbf{p}\right)x \\ &= \frac{1}{2}x + Dx + S(\mathbf{p}x) \\ &= \frac{1}{2}x + Dx + K_{s1}(K_{s3}.*\mathbf{p}x) + K_{s2}(K_{s3}.*\mathbf{p}x) - (K_{s1}\mathbb{1}).*(K_{s3}.*\mathbf{p}x), \end{aligned} \tag{2.3}$$

where px=p.*x.

Note that all of the matrix-vector products involved can be rapidly obtained using FFT method, whereby the computational complexity can be reduced to O(Nlog(N)). This allows for a straightforward solution of the direct problem. As will be seen, the numerical results show that our fast algorithm is superior to the Gaussian elimination (GE) method.

THE ROBIN INVERSE PROBLEM AND PRECONDITIONER

The Robin inverse problem we specifically focus on was that, given $u=u_0$ on Γ_0, one recover the Robin coefficient p on Γ_1 with $\Gamma_0 \cap \Gamma_1 = \emptyset$. That is,

Find p such that:u,the solution of Eq. (2.1),also verifies $u|\Gamma_0=u_0$.

A solution to the inverse problem does not depend continuously on the data even if this does exist, because of the ill-conditioned nature of the Robin inverse problem for Laplace's equation. For this reason, the precess of regularization is required to obtain a relatively smooth solution to a nearby problem [10]. In view of this, the inverse problem can be transformed into the following constrained minimization problem:

$$\begin{cases} \min_{p,u} \frac{1}{2} \|\mathcal{R}_0 u - u_0\|^2_{L^2(\Gamma_0)} + \frac{\alpha}{2} J(p), \\ \text{s.t.} \quad A(p)(u) = f, \end{cases} \tag{3.1}$$

where R_0 is restriction operator from Γ to Γ_0. As for regularization function, we choose the H^1 semi-norm of p(x):

$$J(p) = \int_{\Gamma_1} |p'(x)|^2 \, ds_x,$$

and the regularization parameter α>0 to balance the data fidelity $\|\mathcal{R}_0 u - u_0\|^2_{L^2(\Gamma_0)}$ and regularization terms J(p).

Various nonlinear programming methods have been developed for constrained optimization. These methods seek a critical point for the Lagrangian function:

$$\mathcal{L}(u, p, \lambda) = \frac{1}{2} \|\mathcal{R}_0 u - u_0\|^2_{L^2(\Gamma_0)} + \frac{\alpha}{2} J(p) + \lambda^T [A(p)(u) - f], \tag{3.2}$$

where λ is a Lagrangian multiplier. The goal here is to get a solution to the large system of nonlinear equations:

$$\mathcal{L}_{\mathbf{u}} = \mathcal{R}_0^T (\mathcal{R}_0 u - u_0) + A^T \lambda = 0, \tag{3.3a}$$

$$\mathcal{L}_{\mathbf{p}} = \alpha J_p + G^T \lambda = 0, \tag{3.3b}$$

$$\mathcal{L}_{\lambda} = Au - f = 0, \tag{3.3c}$$

where

$$J_p = J_p(p) = \frac{\partial J}{\partial p}, \qquad G = G(u, p) = \frac{\partial (Au)}{\partial p}.$$

To solve these nonlinear equations, we can use a Newtonian method to obtain the following KKT system:

$$H_{kkt} \begin{pmatrix} \delta u \\ \delta p \\ \delta \lambda \end{pmatrix} = - \begin{pmatrix} \mathcal{L}_u \\ \mathcal{L}_p \\ \mathcal{L}_\lambda \end{pmatrix}, \tag{3.4}$$

where

$$H_{kkt} = \begin{pmatrix} R_0^T R_0 & K & A^T \\ K^T & \alpha J_{pp} + T & G^T \\ A & G & 0 \end{pmatrix}$$

with

$$K = K(p, \lambda) = \frac{\partial (A(p)^T \lambda)}{\partial p}, \qquad J_{pp} = J_{pp}(p) = \frac{\partial J_p}{\partial p},$$

Therefore, the main computational step in the solution process is the repeated solution of large linear systems in the KKT system. To solve the discretized KKT system expressed in Eq. (3.4), there are two basic approaches—a reduced sequential quadratic programming method (SQP) [26] and a full space method [27]. SQP seems preferable because it significantly reduces the dimensions of the problem. However, the major disadvantage here is that it is necessary to solve the variables for u and the Lagrange multipliers λ at each iteration.

The full space method, by contrast, is able to solves KKT systems in such a way that u, λ and p can be obtained simultaneously. The KKT system expressed in (3.4) is relatively straightforward to solve by using Krylov subspace methods. In this paper, we apply a preconditioned Krylov subspace method directly to the KKT system (3.4), and compute a solution to the inverse problem and to the direct problem simultaneously. The sky step here is choosing an appropriate preconditioner for the fast solver of the above KKT system. Recently, the idea of using indefinite preconditioners for KKT systems has come to the fore [28, 29]. However, although it is no longer necessary for the product of the preconditioner within the KKT system (3.4) to be symmetrical, a symmetrical algorithm still be applied to solve the system. In this paper, we assume the preconditioners are symmetrical but far from being positive definite. This is possible because the matrix H_{kkt} of the KKT system has similar properties. To this end, we shall be looking at how to use a fast symmetrical quasi-minimal residual algorithm (SQMR) to solve the system. We can find that it is possible to apply a symmetric QMR variant to the solution of the system, although the product of the preconditioner with the KKT system (3.3a)–(3.3c) is no longer symmetric.

A Preconditioned Reduced Method for the KKT System

In view of the similarity between a preconditioned Krylov subspace approach and a preconditioned reduced approach, we will first of all introduce a reduced Hessian system. In this way, we can eliminate δ_u, then δ_λ, and finally solve for δ_p in (3.4):

$$\delta_p = -H_{\text{red}}^{-1} p, \tag{3.5}$$

where

$$H_{\text{red}} = H_{\text{red}}(u,p,\lambda) = M^T M + \alpha J_{pp} + T - S - S^T, \tag{3.6a}$$

$$M = M(u,p) = -R_0 A(p)^{-1} G, \tag{3.6b}$$

$$S = S(u,p,\lambda) = K^T A(p)^{-1} G, \tag{3.6c}$$

and

$$p = p(u,p,\lambda) = \alpha J_p + M^T\left(\mathcal{R}_0 A^{-1} f - u_0\right) - K\left(u - A^{-1} f\right).$$

In the Gauss–Newton approximation, second-order information can be dropped by setting K=0 and T=0. This produces a symmetric positive-definite matrix:

$$H_{\text{red}} = M^T M + \alpha J_{pp}. \tag{3.7}$$

Thus, we are able to draw upon the classical Gauss–Newton method. In this paper, we will use a preconditioned conjugate gradient (PCG) method to solve Eq. (3.5).

A Preconditioned Krylov Subspace Method for the KKT System

As the invertible block in H_{kkt} is only on the cross-diagonal axis, the matrix H_{kkt} is singular. To find an appropriate preconditioner, we can first of all permute the block rows and columns in matrix H_{kkt} so that there is an invertible block on the diagonal, and obtain an invertible matrix:

$$H = \begin{pmatrix} A & 0 & G \\ R_0^T R_0 & A^T & K \\ K^T & G^T & \alpha J_{pp} + T \end{pmatrix}.$$

For the permuted matrix H we can obtain a block LU decomposition as follows:

H=LU, (3.8)

where

$$L = \begin{pmatrix} I & 0 & 0 \\ \mathcal{R}_0^T \mathcal{R}_0 A^{-1} & I & 0 \\ K^T A^{-1} & G^T A^{-T} & I \end{pmatrix}, \qquad U = \begin{pmatrix} A & 0 & G \\ 0 & A^T & K + \mathcal{R}_0^T M \\ 0 & 0 & H_{\text{red}} \end{pmatrix}. \tag{3.9}$$

Note that H is invertible, such that

$H^{-1} = U^{-1} L^{-1},$

with

$$L^{-1} = \begin{pmatrix} I & 0 & 0 \\ -\mathcal{R}_0^T \mathcal{R}_0 A^{-1} & I & 0 \\ -(K^T + M^T \mathcal{R}_0) A^{-1} & -G^T A^{-T} & I \end{pmatrix},$$

$$U^{-1} = \begin{pmatrix} A^{-1} & 0 & -A^{-1} G H_{\text{red}}^{-1} \\ 0 & A^{-T} & -A^{-T}(K + \mathcal{R}_0^T M) H_{\text{red}}^{-1} \\ 0 & 0 & H_{\text{red}}^{-1} \end{pmatrix}.$$

If we let B be an approximation of A^{-1} and M_{red} be an approximation of H^{-1}_{red}, we can get the following definitions:

$$\hat{L} = \begin{pmatrix} I & 0 & 0 \\ -\mathcal{R}_0^T \mathcal{R}_0 B & I & 0 \\ -(K^T + M^T \mathcal{R}_0) B & -G^T B^T & I \end{pmatrix},$$

$$\hat{U} = \begin{pmatrix} B & 0 & -BGM_{\text{red}} \\ 0 & B^T & -B^T(K + \mathcal{R}_0^T \bar{M}) M_{\text{red}} \\ 0 & 0 & M_{\text{red}} \end{pmatrix},$$

where $\bar{M} = -R_0 B$. These two matrices, $\hat{L}$ and $\hat{U}$, are approximations of the matrices L^{-1} and U^{-1}, respectively. So, if we let

$$M = \hat{U}\hat{L}, \tag{3.10}$$

we can obtain an approximate inverse, M, of the permuted KKT matrix H. Furthermore, if B and M_{red} are chosen to satisfy a product with a vector that can be rapidly calculated, we can also obtain the product with a vector of the approximate inverse, M, rapidly.

However, we do need not to specify the preconditioner, M, explicitly to able to calculate the product of the matrix with a given vector. In fact, given a vector $v=[v^T_{\lambda}, v^T_u, v^T_p]^T$, the product $x=[x^T_u, x^T_{\lambda}, x^T_p]^T=Mv$ can be obtained in six stages:

(1) $w_1 = Bv_\lambda$;

(2) $w_2 = B^T(v_u - \mathcal{R}_0^T \mathcal{R}_0 w_1)$;

(3) $w_3 = v_p - G^T w_2 - K^T w_1$;

(4) $x_p = M_{\text{red}} w_3$;

(5) $x_u = w_1 - BGx_p$;

(6) $x_\lambda = B^T(v_u - \mathcal{R}_0^T \mathcal{R}_0 x_u - Kx_p)$.

In this paper, we have chosen $B=A^{-1}$ because it enables a fast solution of the direct problem. On the basis of this, we can rapidly calculate the first, second and fifth stages. In our algorithm, we need to reorder the components in v and x to get the corresponding preconditioner for (3.4). This reorder ensures that the corresponding preconditioning matrix of (3.4) is symmetrical. Then we can use a preconditioned symmetric QMR algorithm to solve the KKT system.

NUMERICAL EXAMPLES

We will first of all consider a numerical solution of the direct problem.

Example 1

(Fast algorithm of the direct problem)

For this numerical experiment, we assumed that there was an ellipse with the following standard parametrization:

$$x = x(t) = \big(a\cos(2\pi t), b\sin(2\pi t)\big), \quad 0 \le t \le 1,$$

where a=1 and b=0.1. The two segments Γ_0 and Γ_1 were

$$\Gamma_0 = \big\{\big(a\cos(2\pi t), b\sin(2\pi t)\big) : t \in [0.55, 0.85]\big\},$$

$$\Gamma_1 = \big\{\big(a\cos(2\pi t), b\sin(2\pi t)\big) : t \in [0.15, 0.45]\big\}.$$

The function for g(t) was

$$g\big(a\cos(2\pi t)\big), b\sin(2\pi t)) = \begin{cases} 1 & \text{if } t \in [0.4, 0.6], \\ 0 & \text{elsewhere on } \Gamma. \end{cases}$$

The Robin coefficient p(t)p(t) was

$$p(t) = \begin{cases} -(\frac{20}{3})^6 (t-0.15)^3 (t-0.45)^3 & \text{if } t \in [0.15, 0.45], \\ 0 & \text{else other.} \end{cases}$$

For the interval [0,1], we set n equal-length intervals with $\{t_i\}_{i=0}^{n}$ nodes. The tests were carried out by using Matlab. In all of the tests, we first selected p(t), then obtained approximate values of the solution, u(t), at the various grid points by solving the linear system as a the direct problem. For comparison, the tests were also conducted using conventional GE.

Table 1 shows the results of the fast algorithm for the forward equation when using the discretized matrices. Uniform partitions were set on the parameter domain [0,1] at n=800. As can be seen from the table, when n exceeded 400, the proposed algorithm was faster than GE, it becoming proportionally faster as n increased.

Table 1. Temporal performance for the proposed (fast) algorithm and Gaussian elimination (GE) when solving the direct problem

n	**400**	**600**	**800**	**1000**	**1200**
Fast	0.27623	0.02356	0.02615	0.02743	0.03581
GE	0.02820	0.06229	0.13598	0.24962	0.65398

For the other examples, we will look at the numerical test results for our preconditioned symmetrical quasi-minimal residual method (PSQMR) for the recovery of the Robin coefficients from measurements on Γ_0. The boundary conditions were the same as those given for the direct problem.

For the approximate inverse M_{red}, an incomplete LU decomposition was recalled with a threshold (ILU(t)) of M^TM.

Example 2

(Comparing of the PCG and PSQMR)

In this example, we compare the PCG for the reduced Hessian system with the PSQMR for the KKT system. The related iterative processes were stopped when the residual was below 10^{-5}.

Table 2 presents iteration counts ('itns') for the various values of α because the regular parameters play an important role and need to be quite small. More computations are typically involved in a PCG iteration than in a PSQMR iteration, even though the same types of matrix-vector product are used in both algorithms. We also display flop counts ('work') to indicate the total computational work required by each algorithm. The results shown in this table indicate that the number of PSQMR iterations was larger than the number of PCG iteration. However, the overall work when using PSQMR was substantially lower than it was with PCG. This was especially the case when the ILU preconditioner was utilized, because more matrix-vector products were required to arrive at a solution to the forward problem.

Table 2. Iterations and flop counts for the reduced Hessian and KKT solvers

α	10^{-1}	10^{-2}	10^{-3}	10^{-4}
PSQMR itns	34	49	93	204
PCG itns	7	12	21	56
PSQMR work	1.2e9	3.1e9	5.4e9	1.8e10
PCG work	2.0e10	3.5e10	6.8e10	1.5e11

Example 3

(Recovery of the Robin coefficient)

The performance of the two methods was also compared in the elliptic domain. In Fig. 1, the exact p profile generated when using the PSQMR and PCG is plotted together with two reconstructed profiles. For both methods, the regularization parameter α was obtained by employing a discrepancy principle based on our knowledge of noise levels and statistics. The stop criterion was when the norm of the gradient gets below 10^{-5}. It can be seen that under low noise conditions, both method performed well. However, as the noise increased, PCG performed less well, while the PSQMR method remained robust.

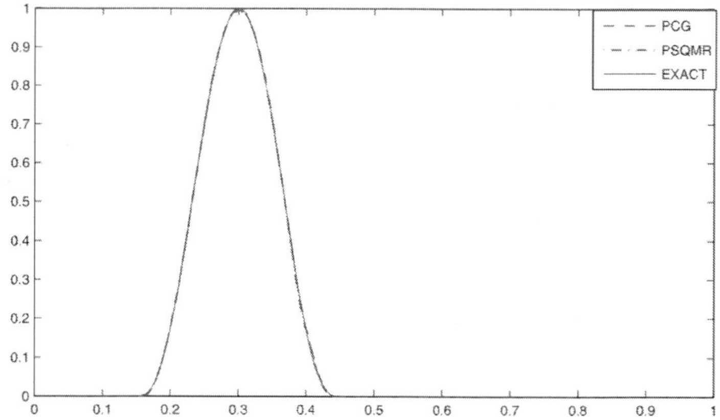

(a)

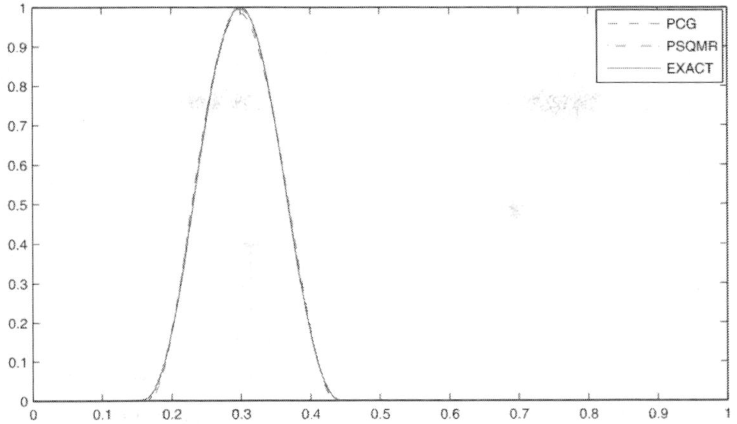

(b)

Figure 1. Comparison of effectiveness at recovery of the Robin coefficient for PCG and PSQMR

ACKNOWLEDGEMENTS

The authors thank the anonymous referees for their valuable comments and suggestions, which improved the quality of the paper.

FUNDING

This work was supported by Guangdong Provincial Department of education innovating strong school project, Grant No. 2016KTSCX085. Parts of this work were supported by the Doctoral Fund of Hanshan Normal University.

CONTRIBUTIONS

YM conceived of the study, designed the study and collected the data. All authors analyzed the data and were involved in writing the manuscript. All authors read and approved the final manuscript.

REFERENCES

1. Mottin, S.: An analytical solution of the Laplace equation with Robin conditions by applying Legendre transform. Integral Transforms Spec. Funct. **27**(4), 289–306 (2016)
2. Alessandrini, G., Piero, L.D., Rondi, L.: Stable determination of corrosion by a single electrostatic boundary measurement. Inverse Probl. **19**(4), 973–984 (2003)
3. Chaabane, S., Fellah, I., Jaoua, M., Leblond, J.: Logarithmic stability estimates for a Robin coefficient in two-dimensional Laplace inverse problems. Inverse Probl. **20**(1), 47–59 (2004)
4. Chaabane, S., Ferchichi, J., Kunisch, K.: Differentiability of the L1L1-tracking functional linked to the Robin inverse problem. C. R. Math. **337**(12), 771–776 (2003)
5. Sincich, E.: Stability for the determination of unknown boundary and impedance with a Robin boundary condition. SIAM J. Math. Anal. **42**(6), 2922–2943 (2010)
6. Papageorgiou, N.S., Radulescu, V.D., Repovs, D.D.: Asymmetric Robin problems with indefinite potential and concave terms. Adv. Nonlinear Stud. **19**, 69–87 (2019)
7. Papageorgiou, N.S., Scapellato, A.: Nonlinear Robin problems with general potential and crossing reaction. Rend. Lincei Mat. Appl. **30**, 1–29 (2019)
8. Cabib, E., Fasino, D., Sincich, E.: Linearization of a free boundary problem in corrosion detection. J. Math. Anal. Appl. **378**(2), 700–709 (2011)
9. Cakoni, F., Kress, R.: Integral equations for inverse problems in corrosion detection from partial Cauchy data. Inverse Probl. Imaging **1**(2), 229–245 (2007)
10. Chaabane, S., Elhechmi, C., Jaoua, M.: A stable recovery method for the Robin inverse problem. Math. Comput. Simul. **66**(4), 367–383 (2004)
11. Fang, W., Lu, M.: A fast collocation method for an inverse boundary value problem. Int. J. Numer. Methods Eng. **59**, 1563–1585 (2004)
12. Fang, W., Zeng, S.: A direct solution of the Robin inverse problem. J. Integral Equ. Appl. **21**(4), 545–557 (2009)
13. Haber, E., Ascher, U.M., Oldenburg, D.: On optimization techniques

for solving nonlinear inverse problems. Inverse Probl. **16**(5), 1263–1280 (2000)

14. Lin, F., Fang, W.: A linear integral equation approach to the Robin inverse problem. Inverse Probl. **21**(5), 1757–1772 (2005)
15. Jin, B.: Conjugate gradient method for the Robin inverse problem associated with the Laplace equation. Int. J. Numer. Methods Eng. **71**(4), 433–453 (2007)
16. Jin, B., Zou, J.: Numerical estimation of piecewise constant Robin coefficient. SIAM J. Control Optim. **48**(3), 1977–2002 (2009)
17. Chaabane, S., Feki, I., Mars, N.: Numerical reconstruction of a piecewise constant Robin parameter in the two- or three-dimensional case. Inverse Probl. **28**(6), 065016 (2012)
18. Daijun, J., Hui, F., Zou, J.: Quadratic convergence of Levenberg–Marquardt method for elliptic and parabolic inverse Robin problems. ESAIM: Math. Model. Numer. Anal. **52**, 07 (2015)
19. Jin, B., Zou, J.: Inversion of Robin coefficient by a spectral stochastic finite element approach. J. Comput. Phys. **227**(6), 3282–3306 (2008)
20. Lepik, D.: Solving PDEs with the aid of two-dimensional Haar wavelets. Comput. Math. Appl. **61**(7), 1873–1879 (2011)
21. Guariglia, E., Silvestrov, S.: Fractional-Wavelet Analysis of Positive Definite Distributions and Wavelets on DD'(CC), pp. 337–353. Springer, Berlin (2016)
22. Ma, Y., Lin, F.: Conjugate gradient method for estimation of Robin coefficients. East Asian J. Appl. Math. **4**(02), 189–204 (2014)
23. Ma, Y.: Newton method for estimation of the Robin coefficient. J. Nonlinear Sci. Appl. **08**(05), 660–669 (2015)
24. Ragusa, M.A.: On some trends on regularity results in Morrey spaces. AIP Conf. Proc. **1493**(1), 770–777 (2012)
25. Ragusa, M.A.: Regularity of solutions of divergence form elliptic equations. Proc. Am. Math. Soc. **128**(2), 533–540 (2000)
26. Benzi, M., Golub, G.H., Liesen, J.: Numerical solution of saddle point problems. Acta Numer. **14**(2), 1–137 (2005)
27. Prudencio, E., Byrd, R., Cai, X.: Parallel full space SQP Lagrange–Newton–Krylov–Schwarz algorithms for PDE-constrained optimization problems. SIAM J. Sci. Comput. **27**(4), 1305–1328 (2006)

28. Biros, G., Ghattas, O.: Parallel Lagrange–Newton–Krylov–Schur methods for PDE-constrained optimization. Part I: the Krylov–Schur solver. SIAM J. Sci. Comput. **27**, 687–713 (2005)
29. Stoll, M., Wathen, A.J.: Preconditioning for partial differential equation constrained optimization with control constraints. Numer. Linear Algebra Appl. **19**(1), 53–71 (2012)

CHAPTER

3

Moving Least Squares Method for a One-dimensional Parabolic Inverse Problem

Baiyu Wang

College of Mathematics and Econometrics, Hunan University, Changsha, Hunan 410082, China

ABSTRACT

This paper investigates the numerical solution of a class of one-dimensional inverse parabolic problems using the moving least squares approximation; the inverse problem is the determination of an unknown source term depending on time. The collocation method is used for solving the equation; some numerical experiments are presented and discussed to illustrate the stability and high efficiency of the method.

Citation: Wang, B. (2014). Moving least squares method for a one-dimensional parabolic inverse problem. In Abstract and Applied Analysis (Vol. 2014). Hindawi. (20 pages) DOI: https://doi.org/10.1155/2014/686020

INTRODUCTION

The inverse problem of parabolic equations appears in a wide variety of physical and engineering settings, such as elasticity, plasticity, hydrology, material sciences, heat transfer, combustion systems, medical imaging, transport problems, and control theory. Many researchers solved this problem using different methods [1–10].

For many decades, many numerical methods have been developed to solve partial differential equations, such as finite difference method, finite elements method, and boundary elements method.

However, these above methods are mesh dependent methods, and, for some problems such as large deformation, the mesh generation is very difficult, especially with the irregular geometries. As a result, the meshless method appeared.

Compared with these mesh dependent numerical methods, in the meshless methods, mesh generation on the spatial domain of the problem is not needed; this property is the main advantage of these techniques over the mesh dependent methods, and it is a technique on scattered data; this method uses the distributed nodal points to approximate the unknown function; the distribution of nodes could be selected regularly or randomly in the analyzed domain, and there is no need for the geometry of the domain.

Many meshless methods have been discussed, such as smooth particle hydrodynamics method [11], the finite point method [12], diffuse element method [13], element-free Galerkin method [14], hp-clouds method [15], and meshless local Petrov-Galerkin method [16].

Moving least squares is very important for constructing the shape functions in meshless method, and it had been studied by a number of investigators, Lancaster and Salkauskas [17], Levin [18], Mirzaei and Dehghan [19], Armentano [20, 21], and so forth.

In this paper, a collocation meshless method based on moving least squares approximation is used for numerical solution of a class of inverse problems in parabolic equation.

The rest of the paper is organized as follows. In Section 2, we will give an outline of the moving least squares. In Section 3, we solve the inverse problem using the meshless method based on the moving least squares. Numerical experiments will be given in Section 4 in order to illustrate the feasibility of the method.

AN OUTLINE OF THE MOVING LEAST SQUARES (MLS)

The MLS method as an approximation and interpolating method has been introduced by Lancaster and Salkauskas.

Given data values u = $\{u_j\}_{j=1}^{N}$ at nodes x_j, N is the number of scattered nodes, the MLS method produces a function uh that approximates data u in a weighted square sense.
It can be defined as

$$u^h(\mathbf{x}) = \sum_{i=1}^{m} p_i(\mathbf{x}) a_i(\mathbf{x}) = P^T(\mathbf{x})\,\mathbf{a}(\mathbf{x}), \quad \forall \mathbf{x} \in \overline{\Omega}, \tag{1}$$

where m is the number of terms in the basis, $p_i(\mathrm{x})$ are the monomial basis functions, $a_i(\mathrm{x})$ are the coefficients of the basis functions, and

$$P^T(\mathbf{x}) = [p_1(\mathbf{x}), p_2(\mathbf{x}), \ldots, p_m(\mathbf{x})],$$

$$a(\mathbf{x}) = [a_1(\mathbf{x}), a_2(\mathbf{x}), \ldots, a_m(\mathbf{x})]^T. \tag{2}$$

The unknown coefficient a(x) is determined by minimizing the functional J, which is defined as

$$\begin{aligned} J &= \sum_{j=1}^{n} \omega(\mathbf{x}-\mathbf{x}_j)\left(u^h(\mathbf{x}) - u_j\right)^2 \\ &= \sum_{j=1}^{n} \omega(\mathbf{x}-\mathbf{x}_j)\left(\sum_{i=1}^{m} p_i(\mathbf{x}_j) a_i(\mathbf{x}) - u_j\right)^2 \\ &= \sum_{j=1}^{n} \omega(\mathbf{x}-\mathbf{x}_j)\left(P^T(\mathbf{x}_j)\mathbf{a}(\mathbf{x}) - u_j\right)^2, \end{aligned} \tag{3}$$

where n is the number of nodes in the support domain of the point x, $\omega(\mathrm{x} - \mathrm{x}_j)$ is the weight function, and x_j are the nodes in the influence domain of x.

Equation (3) can be rewritten in the vector form

$$J = (P\mathbf{a} - \mathbf{u})^T W (P\mathbf{a} - \mathbf{u}), \tag{4}$$

where

$$P = \begin{bmatrix} p_1(\mathbf{x}_1) & p_2(\mathbf{x}_1) & \cdots & p_m(\mathbf{x}_1) \\ p_1(\mathbf{x}_2) & p_2(\mathbf{x}_2) & \cdots & p_m(\mathbf{x}_2) \\ \vdots & \vdots & & \vdots \\ p_1(\mathbf{x}_n) & p_2(\mathbf{x}_n) & \cdots & p_m(\mathbf{x}_n) \end{bmatrix},$$

$$\mathbf{u} = [u_1, u_2, \ldots, u_n]^T, \tag{5}$$

$$W = \begin{bmatrix} \omega(\mathbf{x}-\mathbf{x}_1) & 0 & \cdots & 0 \\ 0 & \omega(\mathbf{x}-\mathbf{x}_2) & \cdots & 0 \\ \vdots & \vdots & & \vdots \\ 0 & 0 & \cdots & \omega(\mathbf{x}-\mathbf{x}_n) \end{bmatrix}.$$

According to the above conditions (4)-(5), taking the derivative a(x) to zero, we have

$$\frac{\partial J}{\partial \mathbf{a}} = A(\mathbf{x})\,\mathbf{a}(\mathbf{x}) - B(\mathbf{x})\,\mathbf{u} = 0, \tag{6}$$

which is equivalent to

$$A(\mathbf{x})\,\mathbf{a}(\mathbf{x}) = B(\mathbf{x})\,\mathbf{u}, \tag{7}$$

where

$$A(\mathbf{x}) = P^T W P,$$

$$B(\mathbf{x}) = P^T W. \tag{8}$$

Then, we get

$$\mathbf{a}(\mathbf{x}) = A^{-1}(\mathbf{x})\,B(\mathbf{x})\,u, \tag{9}$$

and, by substituting (9) into (1), we have

$$u^h(\mathbf{x}) = P^T(\mathbf{x})\,\mathbf{a}(\mathbf{x}) = \Phi^T(\mathbf{x})\,\mathbf{u} = \sum_{j=1}^{N} \phi_j(\mathbf{x})\,u_j, \tag{10}$$

where

$$\Phi^T(\mathbf{x}) = P^T(\mathbf{x})\,A^{-1}(\mathbf{x})\,B(\mathbf{x}) \tag{11}$$

and ϕ_j(x) is called the shape function.

THE INVERSE PROBLEM AND ITS NUMERICAL SOLUTION

In this section, the collocation meshless method based on moving least squares approximation is implemented for the numerical solution of the inverse problem of parabolic equations, and the problem is as follows.

Find $u = (x, t)$ and $f(t)$ which satisfy

$$u_t(x,t) = u_{xx}(x,t) + f(t), \quad (x,t) \in Q_T = (0,l) \times (0,T),$$
$$u(x,0) = \varphi(x), \quad x \in (0,l),$$
$$u(0,t) = h_0(t), \quad u(l,t) = h_l(t), \quad t \in (0,T). \tag{12}$$

And the additional specification is

$$u(x^*,t) = E(t), \quad x^* \in (0,l), t \in (0,T), \tag{13}$$

where $\varphi(x)$, $h_0(t)$, $h_l(t)$, and $E(t)$ are known functions and x^* is a fixed prescribed point in the admissible set.

Assume that these known functions satisfy the compatibility conditions; the unique solvability of this inverse problem is presented in [5]. However, this inverse problem is ill-posed; that is to say, the solution is unstable, and any small change in the additional specification may result in a dramatic change in the numerical solution. To obtain a stable numerical solution, some regularization strategies should be applied.

According to (13), consider the following transformation [22]:

$$E'(t) = u_{xx}|_{x=x^*} + f(t), \tag{14}$$

and using (14), we get

$$f(t) = E'(t) - u_{xx}|_{x=x^*}. \tag{15}$$

By substituting (14) into (12), (12) is transformed to the following equation:

$$u_t = u_{xx} + \left(E'(t) - u_{xx}|_{x=x^*}\right), \quad (x,t) \in Q_T,$$
$$u(x,0) = \varphi(x), \quad x \in (0,l),$$
$$u(0,t) = h_0(t), \quad u(l,t) = h_l(t), \quad t \in (0,T). \tag{16}$$

So the inverse problem is transformed to a direct problem.

The approximate function $(x, t)^\wedge$ of $u(x, t)$ can be represented as

$$\hat{u}(x,t) = \sum_{j=1}^{n} \lambda_j(t)\,\phi_j(x), \tag{17}$$

where $\phi j(x)$ is the shape function described in Section 2.

By substituting (17) into (16), the system (16) can be rewritten as

$$\sum_{j=1}^{n} \lambda'_j(t)\,\phi_j(x) = \sum_{j=1}^{n} \lambda_j(t)\,\phi''_j(x) + \left(E'(t) - \sum_{j=1}^{n} \lambda_j(t)\,\phi''_j(x^*) \right), \quad (x,t) \in Q_T,$$

$$\sum_{j=1}^{n} \lambda_j(0)\,\phi_j(x) = \varphi(x), \quad x \in (0,l),$$

$$\sum_{j=1}^{n} \lambda_j(t)\,\phi_j(0) = h_0(t), \quad \sum_{j=1}^{n} \lambda_j(t)\,\phi_j(l) = h_l(t), \quad t \in (0,T). \qquad (18)$$

It is well known that the step lengths in the standard numerical method have regularization effect; hence, we can choose appropriate step lengths to compute the derivative of t. We apply a simple one-step forward difference formula, and we get

$$\sum_{j=1}^{n} \frac{\lambda_j(t_{m+1}) - \lambda_j(t_m)}{\Delta t}\phi_j(x) = \sum_{j=1}^{n} \lambda_j(t_m)\,\phi''_j(x) + \left(\frac{E(t_{m+1}) - E(t_m)}{\Delta t} - \sum_{j=1}^{n} \lambda_j(t_m)\,\phi''_j(x^*) \right), \quad (x,t) \in Q_T,$$

$$\sum_{j=1}^{n} \lambda_j(0)\,\phi_j(x) = \varphi(x), \quad x \in (0,l),$$

$$\sum_{j=1}^{n} \lambda_j(t_m)\,\phi_j(0) = h_0(t_m), \quad \sum_{j=1}^{n} \lambda_j(t_m)\,\phi_j(l) = h_l(t_m), \quad t \in (0,T). \qquad (19)$$

Then, from the first formula of (19), we obtain

$$\sum_{j=1}^{n}\lambda_j(t_{m+1})\phi_j(x)$$

$$=\sum_{j=1}^{n}\lambda_j(t_m)\phi_j(x)$$

$$+\Delta t\left[\sum_{j=1}^{n}\lambda_j(t_m)\phi_j''(x)\right.$$

$$\left.+\left(\frac{E(t_{m+1})-E(t_m)}{\Delta t}-\sum_{j=1}^{n}\lambda_j(t_m)\phi_j''(x^*)\right)\right];$$

$$(x,t)\in Q_T, \tag{20}$$

by substituting each x_k for x in (20), we get

$$\sum_{j=1}^{n}\lambda_j(t_{m+1})\phi_j(x_k)$$

$$=\sum_{j=1}^{n}\lambda_j(t_m)\phi_j(x_k)$$

$$+\Delta t\left[\sum_{j=1}^{n}\lambda_j(t_m)\phi_j''(x_k)\right.$$

$$\left.+\left(\frac{E(t_{m+1})-E(t_m)}{\Delta t}-\sum_{j=1}^{n}\lambda_j(t_m)\phi_j''(x^*)\right)\right].$$

$$(x,t)\in Q_T. \tag{21}$$

Let

$$\widehat{u}(x_k,t_m)=\sum_{j=1}^{n}\lambda_j(t_m)\phi_j(x_k); \tag{22}$$

then (21) can be rewritten as follows:

$$\hat{u}(x_k, t_{m+1}) = \sum_{j=1}^{n} \lambda_j(t_{m+1}) \phi_j(x_k)$$

$$= \hat{u}(x_k, t_m) + \Delta t \sum_{j=1}^{n} \lambda_j(t_m) \phi_j''(x_k)$$

$$+ E(t_{m+1}) - E(t_m) - \Delta t \sum_{j=1}^{n} \lambda_j(t_m) \phi_j''(x^*); \tag{23}$$

using the iterative method, by (19) and (22), we can obtain $\lambda_j(t_m)$, $j = 1, 2, \ldots, n$, $m = 1, 2, \ldots, M$, and the numerical solution $u(x\hat{}\ k, t_m)$; then we get the numerical solution

$$\hat{f}(t_{m+1}) = \frac{E(t_{m+1}) - E(t_m)}{\Delta t} - \sum_{j=1}^{n} \lambda_j(t_{m+1}) \phi_j''(x^*). \tag{24}$$

NUMERICAL EXPERIMENTS AND DISCUSSIONS

In this section, we give two examples to illustrate the correctness of the theoretical result and the feasibility of the method.

Example 1. Consider (12)-(13), with the conditions

$$u(x,0) = \varphi(x) = 2 + \sin x,$$

$$u(0,t) = h_0(t) = (2+t)e^{-t},$$

$$u(l,t) = h_l(t) = (2 + t + \sin l)e^{-t},$$

$$u(x^*,t) = E(t) = (2 + t + \sin x^*)e^{-t}, \tag{25}$$

and we let l=2, T=2, and $x^* = 1$.

The exact solutions are

$$u(x,t) = (2 + t + \sin x)e^{-t},$$

$$f(t) = -(1+t)e^{-t}. \tag{26}$$

At first, in order to get the approximation effect of $(t)\hat{}$ according to different nodes of x and t, we let $\Delta x = 0.5$, $\Delta t = 0.01$, 0.001 and $\Delta x = 0.1$, $\Delta t = 0.01$, 0.001, respectively; the exact solution $f(t)$, the numerical solution $f(t)\hat{}$, and the error $f(t) - f(t)\hat{}$ are plotted in Figures 1–4.

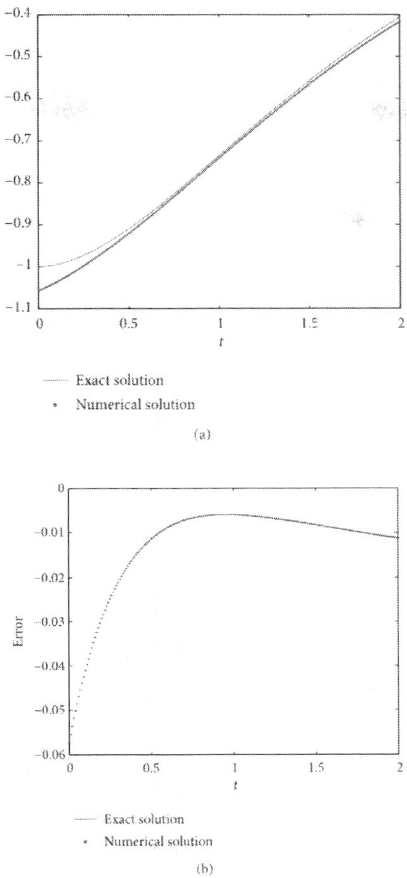

Figure 1. The exact solution (t) and the numerical solution of $f(t)\hat{}$ (a), the error $f(t) - f(t)\hat{}$ (b) with $\Delta x = 0.5$, $\Delta t = 0.01$.

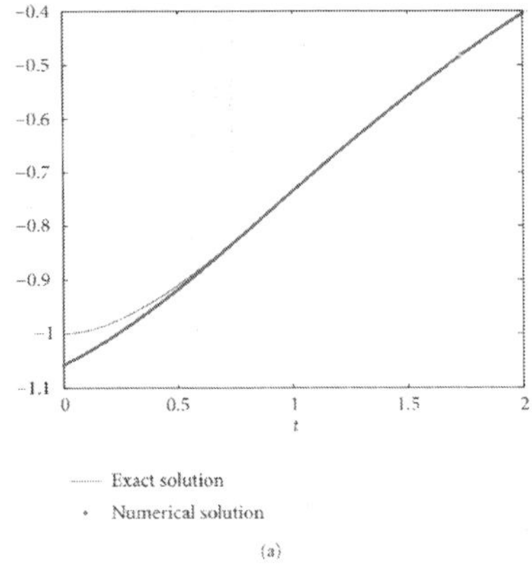

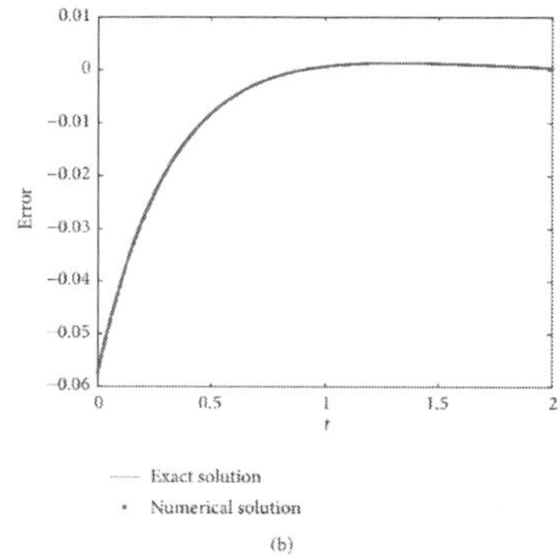

Figure 2. The exact solution (t) and the numerical solution of $f(t)\hat{}$ (a), the error $f(t) - f(t)\hat{}$ (b) with $\Delta x = 0.5$, $\Delta t = 0.001$.

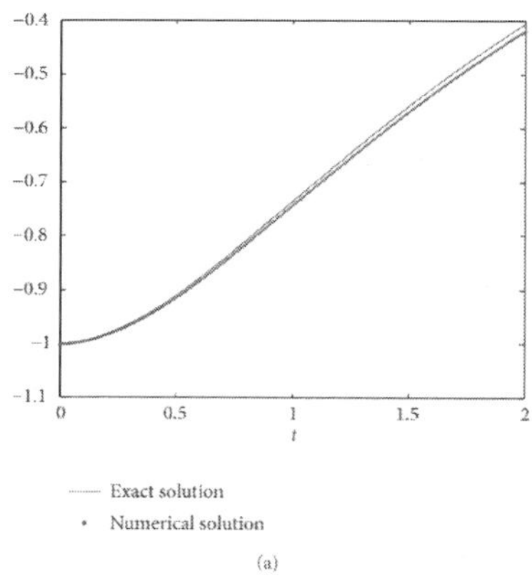

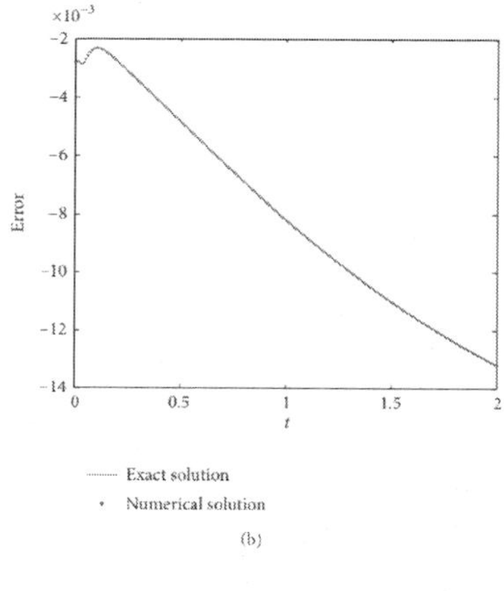

Figure 3. The exact solution (t) and the numerical solution of $f(t)\hat{}$ (a), the error $f(t) - f(t)\hat{}$ (b) with $\Delta x = 0.1$, $\Delta t = 0.01$

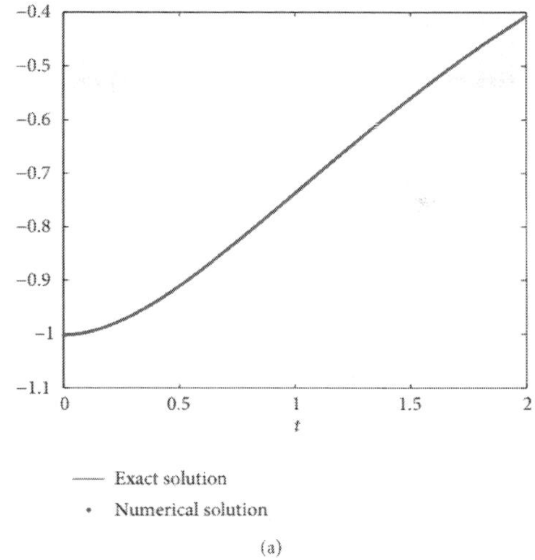

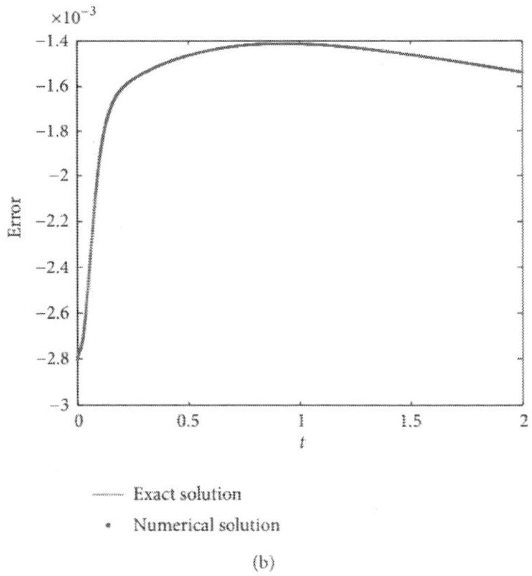

Figure 4 The exact solution (t) and the numerical solution of $f(t)\hat{}$ (a), the error $f(t) - f(t)\hat{}$ (b) with $\Delta x = 0.1, \Delta t = 0.001$.

As seen from Figures 1 and 2 if the Δx is fixed, the error decreases as Δt decreases, but around the initial point, the error changes a little. When we decrease Δx, from Figures 3 and 4, it is clear that the error around the initial point decreases obviously, and with the decrease of Δt, the total error decreases too.

In Figure 5, we plot the exact solution u and the numerical solutions $\hat{u}$ at $t = 0.1$, 1, 1.5, 2, respectively, with $\Delta x = 0.1$ and $\Delta t = 0.001$.

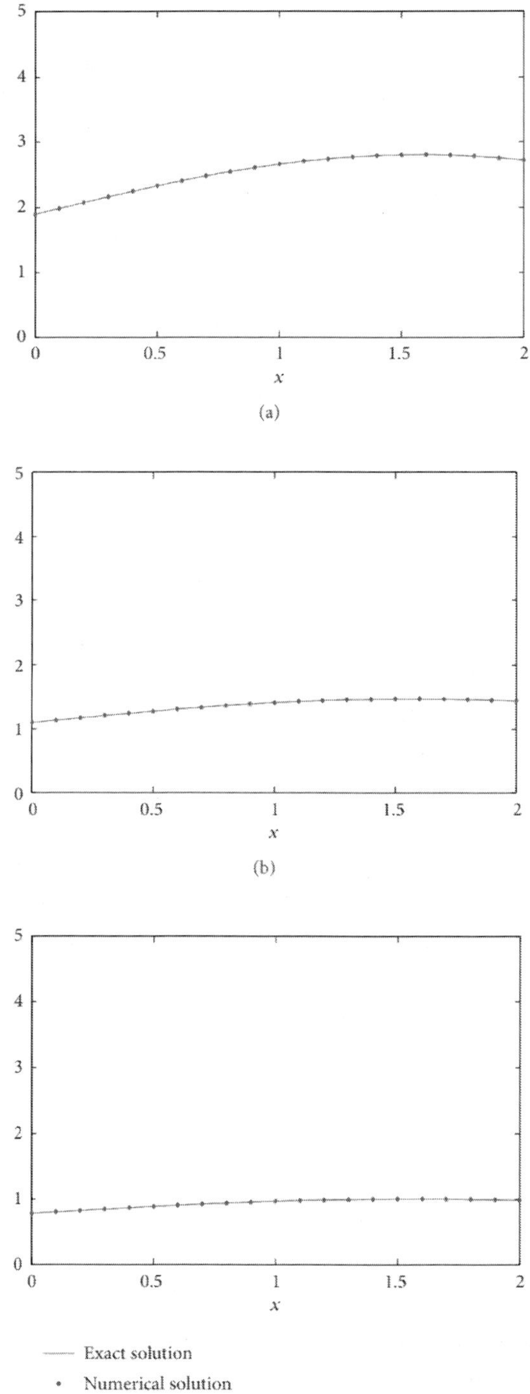
5
4
3
2
1
0
0
0.5
1
1.5
2
x
(a)
5
4
3
2
1
0
0
0.5
1
1.5
2
x
(b)
5
4
3
2
1
0
0
0.5
1
1.5
2
x
Exact solution
Numerical solution
(c)

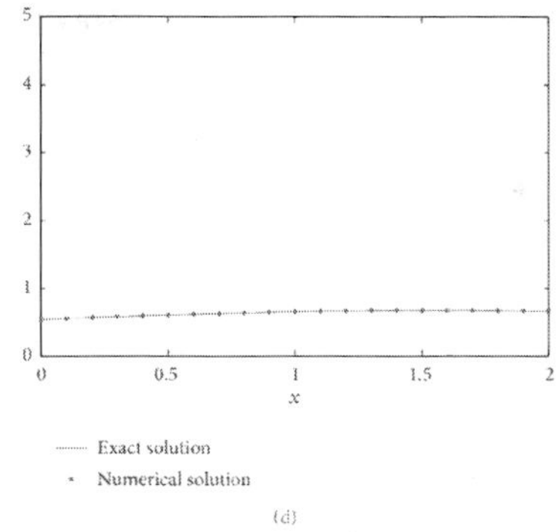

Figure 5. The exact solution (x, t) and numerical solution $u(x, t)$ˆ at $t = 0.1$ (a), 1 (b), 1.5 (c), and 2 (d).

Figure 5 shows clearly that the approximate solution $\hat{u}$ is well for different value of t.

Figures 6 and 7 give the numerical solution $\hat{u}$ and the error u−uˆ with $\Delta x = 0.1$, $\Delta t = 0.01$, 0.001, respectively.

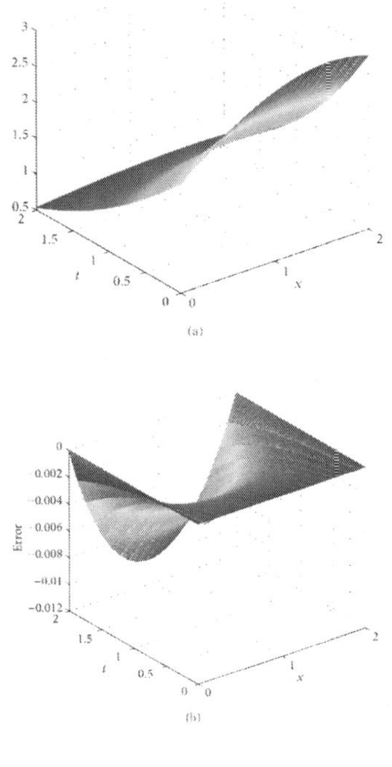

Figure 6. The numerical solution of (x, t)ˆ (a), the error $u(x, t) - u(x, t)$ˆ (b) with $\Delta x = 0.1, \Delta t = 0.01$.

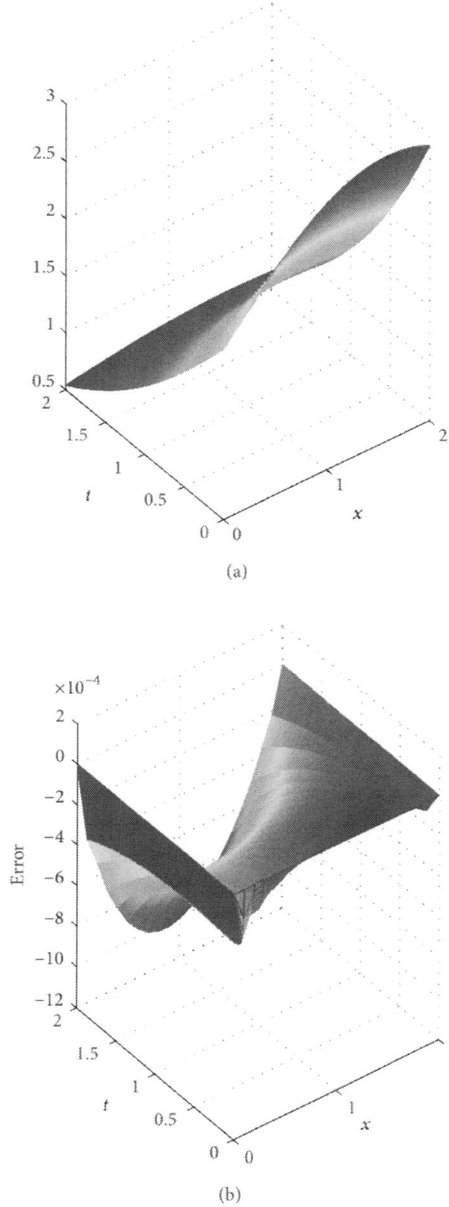

Figure 7. The numerical solution of $(x, t)\hat{}$ (a), the error $u(x, t) - u(x, t)\hat{}$ (b) with $\Delta x = 0.1, \Delta t = 0.001$.

As seen from Figures 6 and 7 the effect of different Δt to the approximate solution $\hat{u}$ is the same to f; that is, the error decreases according to decreasing Δt.

Next, in order to illustrate the effect of the noisy data on numerical solution, we define the artificial error function as follows:

$$E_\gamma(t) = E(t)(1+\gamma), \qquad (27)$$

where γ represents the noisy parameter, and we consider two cases in this paper as follows:

(i) γ is const;

(ii) γ is random uniformly distributed on $(-q, q)$.

Let $\gamma = 0.01, 0.001$, respectively; in case (i), we plot the exact solution (t), the numerical solution $f(t)^\wedge$, and the error $f(t) - f(t)^\wedge$ with $\Delta x = 0.1, \Delta t = 0.001$, in Figures 8 and 9.

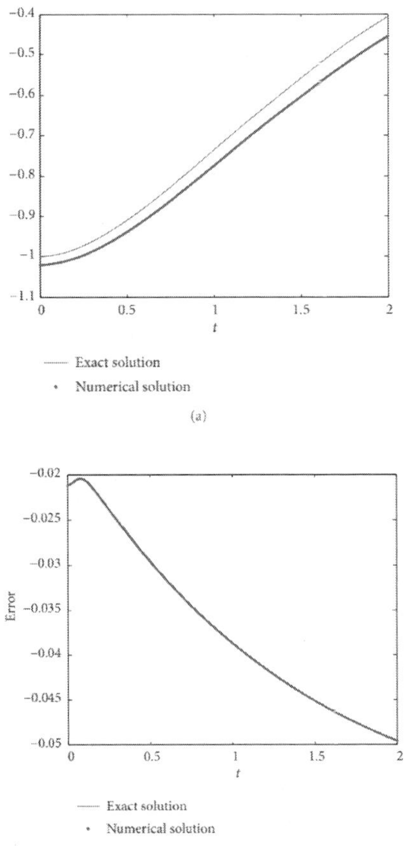

Figure 8. The exact solution (t) and the numerical solution of $f(t)^\wedge$ (a), the error $f(t) - f(t)^\wedge$ (b) with $\gamma = 0.01$.

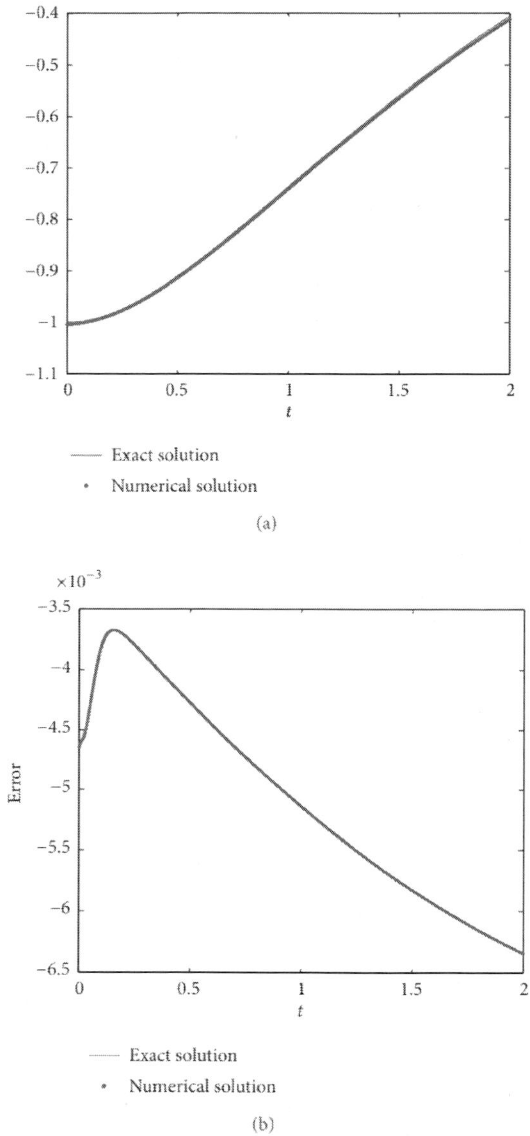

Figure 9. The exact solution (t) and the numerical solution of $f(t)\hat{}$ (a), the error $f(t) - f(t)\hat{}$ (b) with $\gamma = 0.001$.

From Figures 8 and 9, we can see that when the noisy data exist in additional specification, the effect of numerical solution is worse than when there is no noisy data. However, the effect of approximation becomes better with the noisy parameter γ decrease; this suggests that the method in this paper is stable.

In case (ii), we let the parameter $q = 0.01$ and 0.001, respectively, and plot the exact solution (t), the numerical solution $f(t)\hat{}$, and the error $f(t) - f(t)$ ^ with $\Delta x = 0.1$, $\Delta t = 0.001$, and the results are shown in Figures 10 and 11.

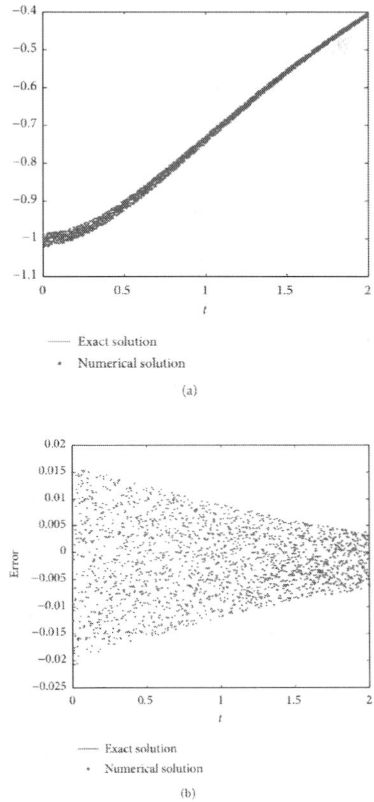

Figure 10. The exact solution (t) and the numerical solution of $f(t)\hat{}$ (a), the error $f(t) - f(t)\hat{}$ (b) with $q = 0.01$.

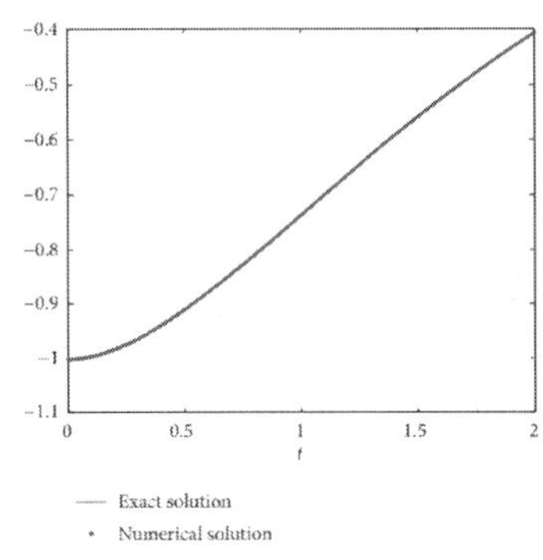

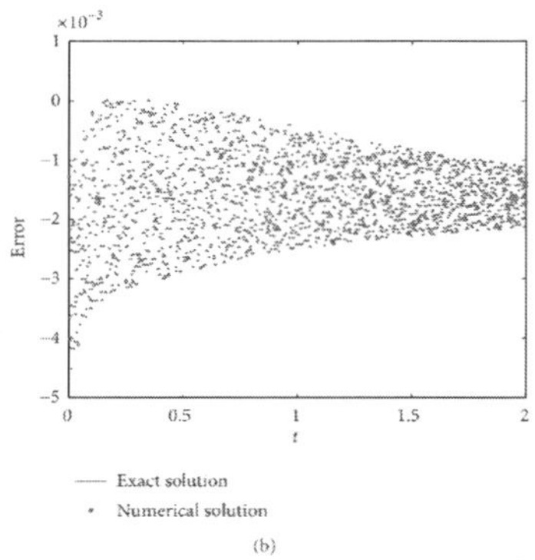

Figure 11. The exact solution (t) and the numerical solution of $f(t)$ˆ (a), the error $f(t) - f(t)$ˆ (b) with $q = 0.001$.

Figures 10 and 11 show that when the noisy parameter is random, the result of numerical solution is also bad, but it is the same to the case that γ is const.; with the parameter q decrease, the effect of approximation becomes better. Moreover, comparing with the fact that γ is const, when the noisy parameter is random, although the result fluctuation is very big, they can be controlled in a certain range.

At last, in order to illustrate the approximate effect, we define the root mean square error of the functions u and f as follows:

$$Eu = \sqrt{\frac{\sum_{i=1}^{M}\sum_{j=1}^{N}\left(u(x_i,t_j) - \hat{u}(x_i,t_j)\right)^2}{MN}},$$

$$Ef = \sqrt{\frac{\sum_{j=1}^{N}\left(f(t_j) - \hat{f}(t_j)\right)^2}{N}}, \tag{28}$$

where $u(xi, tj)$ and $f(tj)$ are the exact solution and $u(x\hat{}\ i, tj)$ and $f(t\hat{}\ j)$ are the approximate numerical solution. The error results for different Δx and Δt with no noisy data are listed in Table 1.

Table 1. The errors for different Δx and Δt

Δt	$\Delta x = 0.5$		$\Delta x = 0.1$	
	Eu	Ef	Eu	Ef
0.01	1.1428×10^{-2}	6.1902×10^{-3}	3.0506×10^{-3}	3.0599×10^{-3}
0.005	1.0569×10^{-2}	4.1804×10^{-3}	1.5394×10^{-3}	1.5832×10^{-3}
0.001	9.9744×10^{-3}	1.8376×10^{-3}	3.4687×10^{-4}	4.1021×10^{-4}
0.0005	9.9068×10^{-3}	1.2985×10^{-3}	2.1206×10^{-4}	2.5863×10^{-4}
0.0001	9.8539×10^{-3}	5.8058×10^{-4}	1.3264×10^{-4}	1.0545×10^{-4}

We will give the error results for different noisy parameter with $\Delta x = 0.1$, $\Delta t = 0.001$ in Tables 2 and 3 and for different Δt with the same noisy parameter in Table 4.

Table 2. The errors for different γ.

γ	Eu	Ef
0.01	1.0799×10^{-2}	2.0392×10^{-3}
0.005	5.5640×10^{-3}	1.4692×10^{-3}
0.001	1.3792×10^{-3}	7.4892×10^{-4}

Table 3. The errors for different q.

q	Eu	Ef
0.01	3.4147×10^{-4}	8.7343×10^{-4}
0.005	3.8703×10^{-4}	6.4143×10^{-4}
0.001	3.2703×10^{-4}	4.1964×10^{-4}

Table 4. The errors for different Δt.

	$\gamma = 0.01$	$\gamma = 0.001$	$q = 0.01$	$q = 0.001$
$\Delta t = 0.01$	7.0116×10^{-3}	3.6493×10^{-3}	3.6401×10^{-3}	3.0238×10^{-3}
$\Delta t = 0.001$	2.0392×10^{-3}	7.4892×10^{-4}	8.5168×10^{-4}	4.2884×10^{-4}
$\Delta t = 0.0001$	6.3895×10^{-4}	2.2068×10^{-4}	2.7201×10^{-4}	1.1466×10^{-4}

From Tables 2 and 3, we see that the errors vary with the noisy parameter. In Table 4, we find that the approximation effect varies with Δt for the same noisy parameter, and these show that the length of Δt has the effect of regularization, and our method in this paper is stable.

In the next example, we will compare the numerical solution using the finite difference method (FDM) and the meshless collocation method based on moving least squares (MLS), respectively.

Example 2. Consider the conditions

$$u(x,0) = \varphi(x) = x^2,$$

$$u(0,t) = h_0(t) = 2t + \sin(\pi t),$$

$$u(l,t) = h_l(t) = 4 + 2t + \sin(\pi t),$$

$$u(x^*,t) = E(t) = 1 + 2t + \sin(\pi t), \qquad (29)$$

and we let l=2, T=2, and $x* = 1$.

The exact solutions are

$$u(x,t) = x^2 + 2t + \sin(\pi t),$$

$$f(t) = \pi \cos(\pi t). \quad (30)$$

We plot the exact solution and numerical solution of (t) in three different cases using FDM and MLS, respectively.

Let $\Delta x = 0.1$, $\Delta t = 0.01$; first, we plot the result in Figure 12 with no noisy data.

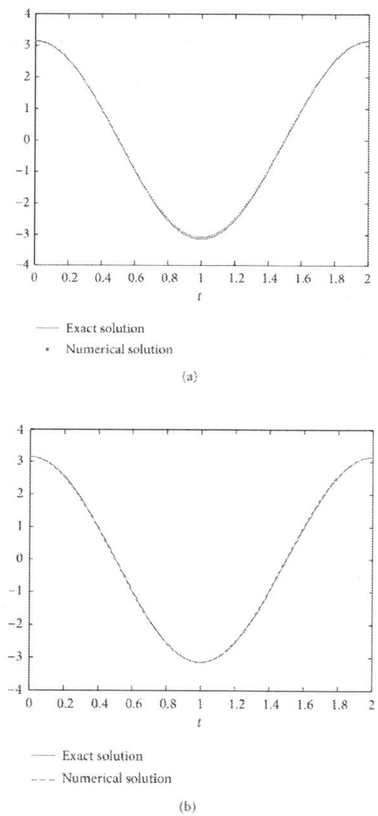

Figure 12. The exact solution (t) and the numerical solution of $f(t)$^ with no noisy data, (a) MLS and (b) FDM.

In Figures 13 and 14, we let $\gamma = 0.001$ and $q = 0.001$, respectively.

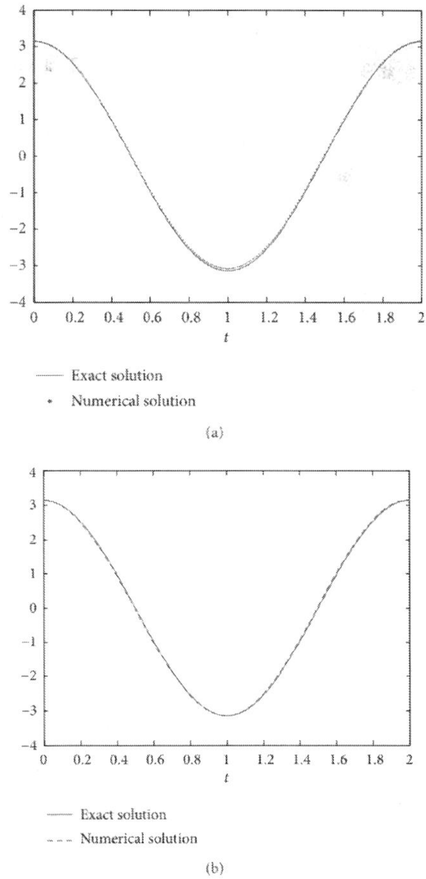

Figure 13. The exact solution (t) and the numerical solution of $f(t)\hat{}$ with γ = 0.001, (a) MLS and (b) FDM.

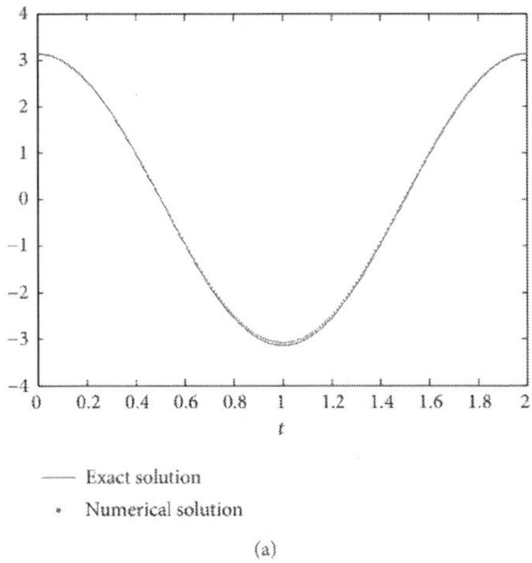

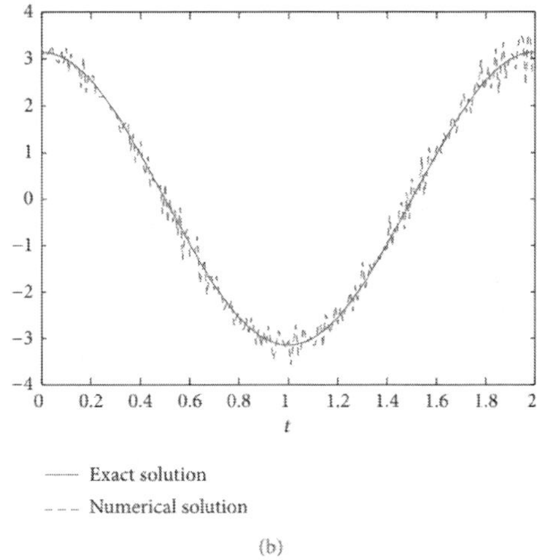

Figure 14. The exact solution (t) and the numerical solution of $f(t)^{\wedge}$ with q = 0.001, (a) MLS and (b) FDM.

From Figures 12, 13, and 14, we can see that when there is no noisy data, the two methods are stable relatively, and, when there exist noisy data, the method FDM is unstable, especially in the case that the noisy parameter is random; but the method MLS is always stable.

Finally, we give the errors of numerical solution with different cases using these two methods in Table 5.

Table 5. The errors for different Δx, Δt and noisy parameter

	FDM	MLS
$\Delta x = 0.5, \Delta t = 0.01, \gamma = 0$	3.4720×10^{-2}	9.0561×10^{-3}
$\Delta x = 0.5, \Delta t = 0.001, \gamma = 0$	3.4877×10^{-3}	9.0857×10^{-4}
$\Delta x = 0.1, \Delta t = 0.01, \gamma = 0$	3.4720×10^{-2}	6.4498×10^{-3}
$\Delta x = 0.1, \Delta t = 0.001, \gamma = 0$	3.4877×10^{-3}	6.4677×10^{-4}
$\Delta x = 0.5, \Delta t = 0.01, \gamma = 0.001$	3.5700×10^{-2}	9.5460×10^{-3}
$\Delta x = 0.5, \Delta t = 0.001, \gamma = 0.001$	7.3814×10^{-3}	1.4210×10^{-3}
$\Delta x = 0.1, \Delta t = 0.01, \gamma = 0.001$	3.5691×10^{-2}	6.7775×10^{-3}
$\Delta x = 0.1, \Delta t = 0.001, \gamma = 0.001$	7.2994×10^{-3}	9.9116×10^{-4}
$\Delta x = 0.5, \Delta t = 0.01, q = 0.001$	2.5285×10^{-1}	8.9877×10^{-3}
$\Delta x = 0.5, \Delta t = 0.001, q = 0.001$	2.4914×10^{0}	9.7315×10^{-4}
$\Delta x = 0.1, \Delta t = 0.01, q = 0.001$	2.5354×10^{-1}	6.4462×10^{-3}
$\Delta x = 0.1, \Delta t = 0.001, q = 0.001$	2.5664×10^{0}	6.7999×10^{-4}

From Table 5, we obtain that the method MLS is more stable and efficient than the method FDM in any case.

CONCLUSION

In this paper, the meshless method based on the moving least squares is used for solving the inverse problem of parabolic equation with the time-dependent source term. We use some transformations to change the problem, and, in order to overcome the ill-posedness, we use the appropriate length of time. From the experiments, we find that the numerical solution is stable. Moreover, compared with the method FDM, we see that our method is stable and highly efficient.

CONFLICT OF INTERESTS

The author declares that there is no conflict of interests regarding the publication of this paper.

ACKNOWLEDGMENTS

This work was supported by the Hunan Provincial Innovation Foundation for Postgraduate (CX2011B156) and NNSF of China (nos. 11271117 and 11201136).

REFERENCES

1. Z. H. Liu and B. Y. Wang, "Coefficient identification in parabolic equations," *Applied Mathematics and Computation*, vol. 209, no. 2, pp. 379–390, 2009.
2. J. B. Liu, B. Y. Wang, and Z. H. Liu, "Determination of a source term in a heat equation," *International Journal of Computer Mathematics*, vol. 87, no. 5, pp. 969–975, 2010.
3. A. Hasanov and Z. H. Liu, "An inverse coefficient problem for a nonlinear parabolic variational inequality," *Applied Mathematics Letters*, vol. 21, no. 6, pp. 563–570, 2008.
4. Z. H. Liu and S. Tatar, "Analytical solutions of a class of inverse coefficient problems," *Applied Mathematics Letters*, vol. 25, no. 12, pp. 2391–2395, 2012.
5. A. G. Fatullayev and E. Can, "Numerical procedures for determining unknown source parameter in parabolic equations," *Mathematics and Computers in Simulation*, vol. 54, no. 1–3, pp. 159–167, 2000.
6. A. G. Fatullayev, "Numerical solution of the inverse problem of determining an unknown source term in a heat equation," *Mathematics and Computers in Simulation*, vol. 58, no. 3, pp. 247–253, 2002.
7. A. Farcas and D. Lesnic, "The boundary-element method for the determination of a heat source dependent on one variable," *Journal of Engineering Mathematics*, vol. 54, no. 4, pp. 375–388, 2006.
8. A. E. Badia and T. Ha-Duong, "On an inverse source problem for the heat equation. Application to a pollution detection problem," *Journal of Inverse and Ill-Posed Problems*, vol. 10, no. 6, pp. 585–599, 2002.
9. J. R. Cannon, "Determination of an unknown heat source from overspecified boundary data," *SIAM Journal on Numerical Analysis*, vol. 5, pp. 275–286, 1968.
10. M. Choulli and M. Yamamoto, "Conditional stability in determining a heat source," *Journal of Inverse and Ill-Posed Problems*, vol. 12, no. 3, pp. 233–243, 2004.
11. J. J. Monaghan, "An introduction to SPH," *Computer Physics Communications*, vol. 48, no. 1, pp. 89–96, 1988.
12. E. Oate, F. Perazzo, and J. Miquel, "A finite point method for elasticity problems," *Computers and Structures*, vol. 79, no. 22–25, pp. 2151–2163, 2001.

13. B. Nayroles, G. Touzot, and P. Villon, "Generalizing the finite element method: diffuse approximation and diffuse elements," *Computational Mechanics*, vol. 10, no. 5, pp. 307–318, 1992.
14. T. Belytschko, Y. Y. Lu, and L. Gu, "Element-free Galerkin methods," *International Journal for Numerical Methods in Engineering*, vol. 37, no. 2, pp. 229–256, 1994.
15. C. A. Duarte and J. T. Oden, "Hp clouds—an meshless method to solve boundary-value problems," Texas Institute for Computational and Applied Mathematics, University of Texas at Austin, 1995.
16. S. N. Atluri and S. Shen, *The Meshless Local Petro-Galerkin (MLPG) Method*, Tech Science Press, Stuttgart, Germany, 2002.
17. P. Lancaster and K. Salkauskas, "Surfaces generated by moving least squares methods," *Mathematics of Computation*, vol. 37, no. 155, pp. 141–158, 1981.
18. D. Levin, "The approximation power of moving least-squares," *Mathematics of Computation*, vol. 67, no. 224, pp. 1517–1531, 1998.
19. D. Mirzaei and M. Dehghan, "A meshless based method for solution of integral equations," *Applied Numerical Mathematics*, vol. 60, no. 3, pp. 245–262, 2010.
20. M. G. Armentano, "Error estimates in Sobolev spaces for moving least square approximations," *SIAM Journal on Numerical Analysis*, vol. 39, no. 1, pp. 38–51, 2001.
21. M. G. Armentano and R. G. Durán, "Error estimates for moving least square approximations," *Applied Numerical Mathematics*, vol. 37, no. 3, pp. 397–416, 2001.
22. D. Colton, R. Ewing, and W. Rundell, *Inverse Problems in Partial Differential Equation*, SIAM, Philadelphia, Pa, USA, 1990.

CHAPTER

4

The Inverse Problem of the Heat Equation with Periodic Boundary And Integral Overdetermination Conditions

Fatma Kanca

Department of Information Technologies, Kadir Has University, Istanbul, 34083, Turkey

ABSTRACT

In this paper the inverse problem of finding the time-dependent coefficient of heat capacity together with the solution of a heat equation with periodic boundary and integral overdetermination conditions is considered. The conditions for the existence and uniqueness of a classical solution of the problem under consideration are established. A numerical example using the Crank-Nicolson finite-difference scheme combined with the iteration method is presented.

Citation: Kanca, F. (2013). The inverse problem of the heat equation with periodic boundary and integral overdetermination conditions. Journal of Inequalities and Applications, 2013(1), 108. (9 pages) DOI: https://doi.org/10.1186/1029-242X-2013-108

Keywords: Inverse Problem, Finite Difference Method, Continuous Dependence, Fourier Method, Parabolic Differential Equation

INTRODUCTION

In this paper we consider an inverse problem of simultaneously finding the unknown coefficient p(t) and the temperature distribution u(x,t) that satisfy the equation

$$u_t = u_{xx} - p(t)u + F(x,t), \quad 0 < x < 1, 0 < t \leq T, \tag{1}$$

with the initial condition

$$u(x,0) = \varphi(x), \quad 0 \leq x \leq 1, \tag{2}$$

the periodic boundary condition

$$u(0,t) = u(1,t), \qquad u_x(0,t) = u_x(1,t), \quad 0 \leq t \leq T, \tag{3}$$

and the overdetermination condition

$$\int_0^1 xu(x,t)\,dx = E(t), \quad 0 \leq t \leq T. \tag{4}$$

Denote the domain Q_T by

$$Q_T = \left\{(x,t) : 0 < x < 1, 0 < t \leq T\right\}.$$

Definition 1 The pair {p(t),u(x,t)} from the class $C[0,T] \times C^{2,1}(D_T) \cap C^{1,0}(\overline{D}_T)$, for which conditions (1)-(4) are satisfied and p(t)≥0 on the interval [0,T], is called a classical solution of the inverse problem (1)-(4).

The parameter identification in a parabolic differential equation from the data of integral overdetermination condition plays an important role in engineering and physics [1–6]. This integral condition in parabolic problems is also called heat moments are analyzed in [4].

Boundary value problems for parabolic equations in which one or two local classical conditions are replaced by heat moments [5–9]. In [9], a physical-mechanical interpretation of the integral conditions was also given. Various statements of inverse problems on determination of this coefficient in a one-dimensional heat equation were studied in [1–3, 5, 6, 10, 11]. In the papers [1, 3, 5], the coefficient is determined from heat moment. Boundary value problems and inverse problems for parabolic equations with periodic boundary conditions are investigated in [10, 12].

In the present work, one heat moment is used with a periodic boundary condition for the determination of a source coefficient. The existence and uniqueness of the classical solution of the problem (1)-(4) is reduced to fixed point principles by applying the Fourier method.

The paper organized as follows. In Section 2, the existence and uniqueness of the solution of the inverse problem (1)-(4) is proved by using the Fourier method. In Section 3, the continuous dependence upon the data of the inverse problem is shown. In Section 4, the numerical procedure for the solution of the inverse problem using the Crank-Nicolson scheme combined with the iteration method is given. Finally, in Section 5, numerical experiments are presented and discussed.

EXISTENCE AND UNIQUENESS OF THE SOLUTION OF THE INVERSE PROBLEM

We have the following assumptions on the data of the problem (1)-(4).

(A_1) $E(t) \in C^1[0,T], E(t) < 0, E'(t) \geq 0, \forall t \in [0,T]$;

(A_2) $\varphi(x) \in C^4[0,1]$;

(1) $\varphi(0) = \varphi(1), \varphi'(0) = \varphi'(1), \varphi''(0) = \varphi''(1), \int_0^1 x\varphi(x)\,dx = E(0)$;

(2) $\varphi_n \geq 0, n = 1,2,\ldots$;

(A_3) $F(x,t) \in C(\overline{D}_T)$; $F(x,t) \in C^4[0,1]$ for arbitrary fixed $t \in [0,T]$;

(1) $F(0,t) = F(1,t), F_x(0,t) = F_x(1,t), F_{xx}(0,t) = F_{xx}(1,t)$;

(2) $F_n(t) > 0, n = 1,2,\ldots$;

(3) $\sum_{n=1}^{\infty} 2\pi n(\varphi_n + \int_0^T F_n(\tau)\,d\tau) \leq E'(t), \forall t \in [0,T]$;

where $\varphi_n = \int_0^1 \varphi(x)\sin(2\pi nx)\,dx, F_n(t) = \int_0^1 F(x,t)\sin(2\pi nx)\,dx, n = 0,1,2,\ldots.$

Theorem 2 *Let the assumptions* (A_1)-(A_3) *be satisfied. Then the following statements are true*:

1. *The inverse problem* (1)-(4) *has a solution in* Q_T;
2. *The solution of the inverse problem* (1)-(4) *is unique in* QT_0, *where the number* T0 ($0<T_0<T$) *is determined by the data of the problem*.

Proof By applying the standard procedure of the Fourier method, we obtain the following representation for the solution of (1)-(3) for arbitrary $p(t) \in C[0,T]$:

$$x,t) = \sum_{n=1}^{\infty}\left[\varphi_n e^{-(2\pi n)^2 t - \int_0^t p(s)\,ds} + \int_0^t F_n(\tau)e^{-(2\pi n)^2(t-\tau)-\int_\tau^t p(s)\,ds}\,d\tau\right]\sin(2\pi nx). \tag{5}$$

Under the conditions (1) of (A_2) and (1) of (A_3), the series (5) and its x-partial derivative are uniformly convergent in $\bar{Q}_T$ since their majorizing sums are absolutely convergent. Therefore, their sums u(x,t) and ux(x,t) are continuous in $\bar{Q}_T$. In addition, the t-partial derivative and the xx-second order partial derivative series are uniformly convergent in Q_T. Thus, we have $u(x,t) \in C^{2,1}(QT) \cap C^{1,0}(\bar{Q}_T)$. In addition, ut(x,t) is continuous in $\bar{Q}_T$. Differentiating (4) under the assumption (A_1), we obtain

$$\int_0^1 x u_t(x,t)\, dx = E'(t), \tag{6}$$

and this yields

p(t)=K[p(t)], (7)

where

$$K[p(t)] = \left(-E'(t) + \sum_{n=1}^{\infty} 2\pi n\left(\varphi_n e^{-(2\pi n)^2 t - \int_0^t p(s)\,ds} + \int_0^t F_n(\tau) e^{-(2\pi n)^2 (t-\tau) - \int_\tau^t p(s)\,ds}\, d\tau\right)\right.$$
$$\left. - \sum_{n=1}^{\infty} \frac{1}{2\pi n} F_n(t)\right) / E(t). \tag{8}$$

Denote

$$C_1 = \frac{-\max_{t\in[0,T]} E'(t) - E(0) - \max_{t\in[0,T]}\left(\sum_{n=1}^{\infty} \frac{1}{2\pi n} F_n(t)\right)}{\max_{t\in[0,T]} E(t)},$$

$$C_2 = \frac{-\min_{t\in[0,T]} E'(t) + \sum_{k=1}^{\infty} 2\pi n\left(\varphi_n + \int_0^T F_n(\tau)\, d\tau\right) - \min_{t\in[0,T]}\left(\sum_{n=1}^{\infty} \frac{1}{2\pi n} F_n(t)\right)}{\min_{t\in[0,T]} E(t)}.$$

Using the representation (7), the following estimate is true:

$$0 < C_1 \le p(t) \le C_2.$$

Introduce the set M as follows:

$$M = \{p(t) \in C[0,T] : C_1 \le p(t) \le C_2\}.$$

It is easy to see that

K:M→M.

Compactness of K is verified by analogy to [6]. By virtue of the Schauder fixed point theorem, we have a solution $p(t) \in C[0,T]$ of equation (7).

Now let us show that there exists Q_{T0} ($0<T_0 \le T$) for which the solution (p,u) of the problem (1)-(4) is unique in Q_{T0}. Suppose that (q,v) is also a solution pair of the problem (1)-(4). Then, from the representations (5) and

(7) of the solution, we have

$$u(x,t) - v(x,t)$$
$$= \sum_{n=1}^{\infty} \varphi_n \left(e^{-(2\pi n)^2 t - \int_0^t p(s)\,ds} - e^{-(2\pi n)^2 t - \int_0^t q(s)\,ds}\right) \sin 2\pi n(x)$$
$$+ \sum_{n=1}^{\infty} \left(\int_0^t F_n(\tau)\left(e^{-(2\pi n)^2(t-\tau) - \int_\tau^t p(s)\,ds} - e^{-(2\pi n)^2(t-\tau) - \int_\tau^t q(s)\,ds}\right) d\tau \right) \sin 2\pi n(x), \tag{9}$$

$$p(t) - q(t) = K[p(t)] - K[q(t)], \tag{10}$$

where

$$K[p(t)] - K[q(t)]$$
$$= \frac{\sum_{n=1}^{\infty} 2\pi n \varphi_n (e^{-(2\pi n)^2 t - \int_0^t p(s)\,ds} - e^{-(2\pi n)^2 t - \int_0^t q(s)\,ds})}{E(t)}$$
$$- \frac{\sum_{n=1}^{\infty} 2\pi n \int_0^t F_n(\tau)(e^{-(2\pi n)^2(t-\tau) - \int_\tau^t p(s)\,ds} - e^{-(2\pi n)^2(t-\tau) - \int_\tau^t q(s)\,ds})\,d\tau}{E(t)}.$$

Using the estimates

$$\left|e^{-(2\pi n)^2 t - \int_0^t p(s)\,ds} - e^{-(2\pi n)^2 t - \int_0^t q(s)\,ds}\right| \le (2\pi n)^2 T \max_{0 \le t \le T} |p(t) - q(t)|,$$

$$\left|e^{-(2\pi n)^2(t-\tau) - \int_\tau^t p(s)\,ds} - e^{-(2\pi n)^2(t-\tau) - \int_\tau^t q(s)\,ds}\right| \le (2\pi n)^2 T \max_{0 \le t \le T} |p(t) - q(t)|,$$

we obtain

$$\max_{0 \le t \le T} \left|K[p(t)] - K[q(t)]\right| \le \alpha \max_{0 \le t \le T} |p(t) - q(t)|.$$

Let $\alpha \in (0,1)$ be an arbitrary fixed number. Fix a number T_0, $0 < T_0 \le T$, such that

$$\frac{C_3 + C_4}{C_0} T_0 \le \alpha,$$

where

$$C_0 = \min_{t \in [0,T]} E(t), \qquad C_3 = \sum_{n=1}^{\infty} 2\pi n \varphi_n, \qquad C_4 = T \max_{t \in [0,T]} \left(\sum_{n=1}^{\infty} 2\pi n F_n(t) \right).$$

Then, from equality (10), we obtain

$$\|p - q\|_{C[0,T_0]} \le \alpha \|p - q\|_{C[0,T_0]},$$

which implies that p=q. By substituting p=q in (9), we have u=v.

CONTINUOUS DEPENDENCE OF (P,U) UPON THE DATA

Theorem 3 *Under assumptions* (A_1)-(A_3), *the solution* (p,u) *of the problem* (1)-(4) *depends continuously upon the data for small T.*

The proof of the theorem is verified by analogy to [5].

NUMERICAL METHOD

We use the finite difference method with a predictor-corrector type approach that is suggested in [2]. Apply this method to the problem (1)-(4).

We subdivide the intervals [0,1] and [0,T] into N_x and N_t subintervals of equal lengths $h=1/N_x$ and $\tau=T/N_t$, respectively. Then we add two lines x=0 and $x=(N_x+1)h$ to generate the fictitious points needed for dealing with the boundary conditions. We choose the Crank-Nicolson scheme, which is absolutely stable and has a second-order accuracy in both *h* and *τ*. [13] The Crank-Nicolson scheme for (1)-(4) is as follows:

$$\frac{1}{\tau}\left(u_i^{j+1} - u_i^j\right) = \frac{1}{2h^2}\left[\left(u_{i-1}^j - 2u_i^j + u_{i+1}^j\right) + \left(u_{i-1}^{j+1} - 2u_i^{j+1} + u_{i+1}^{j+1}\right)\right] + \frac{1}{4}\left(p^{j+1} + p^j\right)\left(u_i^{j+1} + u_i^j\right) + \frac{1}{2}\left(F_i^{j+1} + F_i^j\right), \tag{11}$$

$$u_i^0 = \phi_i, \tag{12}$$

$$u_0^j = u_{N_x+1}^j, \tag{13}$$

$$\frac{u_1^j + u_{N_x}^j}{2} = u_{N_x+1}^j, \tag{14}$$

where $1 \le i \le N_x$ and $0 \le j \le N_t$ are the indices for the spatial and time steps respectively, $u_i^j = u(x_i, t_j)$, $\phi_i = \varphi(x_i)$, $F_i^j = F(x_i, t_j)$, $x_i = ih$, $t_j = j\tau$. At the t=0 level, adjustment should be made according to the initial condition and the compatibility requirements.

Equations (11)-(14) form an $N_x \times N_x$ linear system of equations

$$AU^{j+1} = b, \tag{15}$$

where

$$U^j = (u_1^j, u_2^j, \ldots, u_{N_x}^j)^{tr}, \quad 1 \le j \le N_t, b = (b_1, b_2, \ldots, b_{N_x})^{tr},$$

$$A = \begin{bmatrix} R + R1 + \frac{3}{2} & -1 & 0 & & 0 & -\frac{1}{2} \\ -1 & R + R1 + 2 & -1 & & & 0 \\ 0 & -1 & R + R1 + 2 & \cdots & & 0 \\ \vdots & & & & & \\ & & & & & \\ & & & -1 & R + R1 + 2 & -1 \\ -\frac{1}{2} & & & 0 & -1 & R + R1 + \frac{3}{2} \end{bmatrix},$$

$$R = \frac{2h^2}{\tau}, \qquad R1 = \frac{h^2}{2}(p^{j+1} + p^j), \quad j = 0, 1, \ldots, N_t,$$

$$b_1 = -\left(-R + R1 + \frac{3}{2}\right)u_1^j + u_2^j + \frac{1}{2}u_{N_x}^j + h^2(F_1^{j+1} + F_1^j), \quad j = 0, 1, \ldots, N_t,$$

$$b_{N_x} = u_{N_x-1}^j - \left(-R + R1 + \frac{3}{2}\right)u_{N_x}^j + \frac{1}{2}u_1^j + h^2(F_{N_x}^{j+1} + F_{N_x}^j), \quad j = 0, 1, \ldots, N_t,$$

$$b_i = u_{i-1}^j - (-R + R1 + 2)u_i^j + u_{i+1}^j + h^2(F_i^{j+1} + F_i^j), \quad i = 2, 3, \ldots, N_x - 1, j = 0, 1, \ldots, N_t.$$

Now, let us construct the predicting-correcting mechanism. First, integrating equation (1) with respect to x from 0 to 1 and using (3) and (4), we obtain

$$p(t) = \frac{-E'(t) + \int_0^1 xF(x,t)\,dx + u_x(1,t)}{E(t)}. \tag{16}$$

The finite difference approximation of (16) is

$$p^j = \frac{-((E^{j+1} - E^j)/\tau) + (\text{Fin})^j + (u_{N_x+1}^j - u_{N_x}^j)/h}{E^j},$$

where $E^j = E(t_j)$, $(\text{Fin})^j = \int_0^1 xF(x,t_j)\,dx$, $j = 0, 1, \ldots, N_t$.

For j=0,

$$p^0 = \frac{-((E^1 - E^0)/\tau) + (\text{Fin})^0 + (\phi_{N_x+1} - \phi_{N_x})/h}{E^j},$$

and the values of ϕ_i allow us to start our computation. We denote the values of p^j, u_i^j at the s th iteration step $p^{j(s)}$, $u_i^{j(s)}$, respectively. In numerical computation, since the time step is very small, we can take $p^{j+1(0)}=p^j$, $u_i^{j+1(0)}=u_i^j$, $j=0,1,2,\ldots,N_t$, $i=1,2,\ldots,N_x$. At each (s+1)th iteration step, we first determine $p^{j+1(s+1)}$ from the formula

$$p^{j+1(s+1)} = \frac{-((E^{j+2} - E^{j} + 1)/\tau) + (\text{Fin})^{j+1} + (u_{N_x+1}^{j+1(s)} - u_{N_x}^{j+1(s)})/h}{E^{j}}.$$

Then from (11)-(14) we obtain

$$\frac{1}{\tau}\left(u_i^{j+1(s+1)} - u_i^{j+1(s)}\right) = \frac{1}{2h^2}\left[\left(u_{i-1}^{j+1(s+1)} - 2u_i^{j+1(s+1)} + u_{i+1}^{j+1(s+1)}\right) + \left(u_{i-1}^{j+1(s)} - 2u_i^{j+1(s)} + u_{i+1}^{j+1(s)}\right)\right] + \frac{1}{4}\left(p^{j+1(s+1)} + p^{j+1(s)}\right)\left(u_i^{j+1(s+1)} + u_i^{j+1(s)}\right) + \frac{1}{2}\left(F_i^{j+1} + F_i^{j}\right), \quad (17)$$

$$u_0^{j+1(s)} = u_{N_x+1}^{j+1(s)}, \quad (18)$$

$$\frac{u_1^{j+1(s)} + u_{N_x}^{j+1(s)}}{2} = u_{N_x+1}^{j+1(s)}. \quad (19)$$

The system of equations (17)-(19) can be solved by the Gauss elimination method, and $u_i^{j+1(s+1)}$ is determined. If the difference of values between two iterations reaches the prescribed tolerance, the iteration is stopped, and we accept the corresponding values $p^{j+1(s+1)}$, $u^{j+1(s+1)}_i$ (i=1,2,…,N_x as p^{j+1}, u^{j+1}_i (i=1,2,…,N_x), on the ((j+1)th time step, respectively. By virtue of this iteration, we can move from level *j* to level j+1.

NUMERICAL EXAMPLE AND DISCUSSIONS

In this section, we present examples to illustrate the efficiency of the numerical method described in the previous section.

Example Consider the inverse problem (1)-(4) with

$$F(x,t) = \left(\left(1 + \sin(2\pi x)\right)\left(11 + \exp(13t)\right) + (2\pi)^2 \sin(2\pi x)\right)\exp(10t),$$

$$\varphi(x) = 1 + \sin(2\pi x), \qquad E(t) = \left(\frac{1}{2} - \frac{1}{2\pi}\right)\exp(10t), \quad x \in [0,1], t \in [0,T].$$

It is easy to check that the analytical solution of this problem is

$\{p(t),u(x,t)\}=\{1+\exp(13t),(1+\sin(2\pi x))\exp(10t)\}$. (20)

Let us apply the scheme which was explained in the previous section for the step sizes h=0.01, τ=h/8.

In the case when T=1/2, the comparisons between the analytical solution (19) and the numerical finite difference solution are shown in Figures 1 and 2.

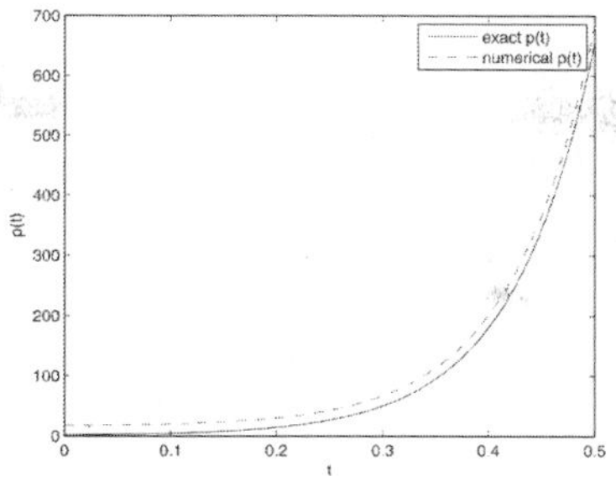

Figure 1. The analytical and numerical solutions of p(t) when T=1/2 . The numerical solution is shown by a dashed line.

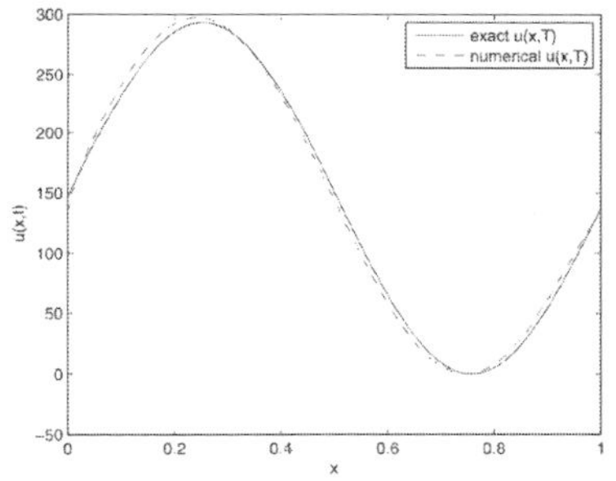

Figure 2. The analytical and numerical solutions of u(x,t) at the T=1/2 . The numerical solution is shown by a dashed line.

CONCLUSIONS

The inverse problem regarding the simultaneous identification of the time-dependent coefficient of heat capacity together with the temperature distribution in a one-dimensional heat equation with periodic boundary and integral overdetermination conditions has been considered. This inverse problem has been investigated from both theoretical and numerical points of view. In the theoretical part of the article, the conditions for the existence, uniqueness and continuous dependence upon the data of the problem have been established. In the numerical part, a numerical example using the Crank-Nicolson finite-difference scheme combined with the iteration method is presented.

ACKNOWLEDGEMENTS

Dedicated to Professor Hari M Srivastava.

REFERENCES

1. Cannon J, Lin Y, Wang S: Determination of a control parameter in a parabolic partial differential equation. *J. Aust. Math. Soc. Ser. B, Appl. Math* 1991, 33: 149–163. 10.1017/S0334270000006962
2. Cannon J, Lin Y, Wang S: Determination of source parameter in a parabolic equations. *Meccanica* 1992, 27: 85–94. 10.1007/BF00420586
3. Fatullayev A, Gasilov N, Yusubov I: Simultaneous determination of unknown coefficients in a parabolic equation. *Appl. Anal.* 2008, 87: 1167–1177. 10.1080/00036810802140616
4. Ivanchov M, Pabyrivska N: Simultaneous determination of two coefficients of a parabolic equation in the case of nonlocal and integral conditions. *Ukr. Math. J.* 2001, 53: 674–684. 10.1023/A:1012570031242
5. Ismailov M, Kanca F: An inverse coefficient problem for a parabolic equation in the case of nonlocal boundary and overdetermination conditions. *Math. Methods Appl. Sci.* 2011, 34: 692–702. 10.1002/mma.1396
6. Kanca F, Ismailov M: Inverse problem of finding the time-dependent coefficient of heat equation from integral overdetermination condition data. *Inverse Probl. Sci. Eng.* 2012, 20: 463–476. 10.1080/17415977.2011.629093
7. Ionkin N: Solution of a boundary-value problem in heat conduction with a nonclassical boundary condition. *Differ. Equ.* 1977, 13: 204–211.
8. Vigak V: Construction of a solution of the heat-conduction problem with integral conditions. *Dokl. Akad. Nauk Ukr. SSR* 1994, 8: 57–60.
9. Ivanchov N: Boundary value problems for a parabolic equation with integral conditions. *Differ. Equ.* 2004, 40: 591–609.
10. Choi J: Inverse problem for a parabolic equation with space-periodic boundary conditions by a Carleman estimate. *J. Inverse Ill-Posed Probl.* 2003, 11: 111–135. 10.1515/156939403766493519
11. Ashyralyev A, Erdogan A: On the numerical solution of a parabolic inverse problem with the Dirichlet condition. *Int. J. Math. Comput.* 2011, 11: 73–81.
12. Sakinc I: Numerical solution of the quasilinear parabolic problem with periodic boundary condition. *Hacet. J. Math. Stat.* 2010, 39: 183–189.
13. Samarskii AA: *The Theory of Difference Schemes*. Dekker, New York; 2001.

CHAPTER 5

An Inverse Problem for a Two-Dimensional Time-Fractional Sideways Heat Equation

Songshu Liu[1] and Lixin Feng[2]

[1]School of Mathematics and Statistics, Northeastern University at Qinhuangdao, Qinhuangdao 066004, China

[2]School of Mathematical Sciences, Heilongjiang University, Harbin 150080, China

ABSTRACT

In this paper, we consider a two-dimensional (2D) time-fractional inverse diffusion problem which is severely ill-posed; i.e., the solution (if it exists) does not depend continuously on the data. A modified kernel method is presented for approximating the solution of this problem, and the convergence estimates are obtained based on both a priori choice and a posteriori choice

Citation: Liu, S., & Feng, L. (2020). An Inverse Problem for a Two-Dimensional Time-Fractional Sideways Heat Equation. Mathematical Problems in Engineering, 2020.. (13 pages) DOI: https://doi.org/10.1155/2020/5865971

of regularization parameters. The numerical examples illustrate the behavior of the proposed method.

INTRODUCTION

In the past decades, fractional differential equations have attracted wide attention. Various models using fractional partial differential equations have been successfully applied to describe a range of problems in mechanical engineering [1], viscoelasticity [2], electron transport [3], dissipation [4], heat conduction [5, 6], and high-frequency financial data [7].

The time-fractional diffusion equationore is deduced by replacing the standard time derivative with a time-fractional derivative and can be used to describe the superdiffusion and subdiffusion phenomena [8–13]. The direct problems, i.e., initial value problems and initial boundary value problems, for the time-fractional diffusion equation have been studied extensively in recent years. For instance, a well-posedness analysis [14–18] and numerical methods and simulations [19–24] are typical examples of this kind of analysis.

However, in some practical situations, part of boundary data, or initial data, or diffusion coefficient, or source term, or the order of fractional derivative may not be given and we want to find them by additional measured data which will yield some inverse problems of the fractional diffusion equation. The early papers on inverse problems of time-fractional diffusion equationT were provided by Murio in [25, 26] for solving the time-fractional diffusion equation by mollification methods. Cheng et al. [27] gave the uniqueness in determining the parameter α and p(x) by means of observation data at one end point. Liu and Yamamoto [28] considered a backward problem in time for a time-fractional partial diffusion equation in one-dimensional case. Zhang and Xu [29] studied an inverse source problem in the time-fractional diffusion equation and proved uniqueness for identifying a space-dependent source term by using analytic continuation and Laplace transform. Zheng and Wei [30] applied a new regularization method to solve an inverse problem for a time-fractional diffusion equation in a one-dimensional semi-infinite domain. Jin and Rundell [31] studied an inverse problem of recovering a spatially varying potential term in the 1D time-fractional diffusion equation from the flux measurements and proposed a quasi-Newton-type reconstruction algorithm. Recently, Wei and Wang [32] proposed a modified quasi-boundary value method for an inverse source problem of the time-fractional diffusion equation. Wei et al. [33] identified a time-dependent source term in a multidimensional time-fractional diffusion equation from boundary Cauchy data.

As for time-fractional inverse diffusion problem for one-dimensional (1D) models, theoretical concepts and computational implementation have been discussed by many authors, and a number of solution methods have been proposed, i.e., spectral regularization method [34], iteration regularization method [35], optimal regularization method [36], and a new dynamic method [37]. However, for the two-dimensional time-fractional inverse diffusion problem, few results are available. Xiong et al. [38] gave a conditional stability estimate for the inverse heat conduction problem in the 2D time-fractional heat equation and studied a dynamic spectral regularization method with numerical testification. However, error estimates were given in [38] by only choosing the regularization parameter by a priori choice rule.

In the present paper, we will consider the 2D time-fractional inverse diffusion problem by using a modified kernel method and a posteriori parameter choice rule is given. For a priori choice and a posteriori choice of the regularization parameter, we obtain the convergence estimates, respectively.

The rest of this paper is organized as follows. In Section 2, we describe the 2D time-fractional inverse diffusion problem and give an analysis on the ill-posedness of this problem. In Section 3, we propose a modified kernel method and prove the convergence estimates under a priori and a posteriori parameter choice rule, respectively. Section 4 is the numerical aspect of the proposed method. Finally, we give a brief conclusion in Section 5.

DESCRIPTION OF THE 2D TIME-FRACTIONAL INVERSE DIFFUSION PROBLEM

We consider the following 2D time-fractional inverse diffusion problem:

$$\frac{\partial^\alpha u}{\partial t^\alpha} - u_{xx} - u_{yy} = 0, \quad x>0,\ y>0,\ t>0, \tag{1}$$

$$u(1,y,t) = g(y,t), \quad y \ge 0,\ t \ge 0, \tag{2}$$

with the corresponding measured data function $g^\delta(y, t)$ and initial and boundary conditions

$$u(x,y,0) = 0, \quad x>0,\ y\ge 0, \tag{3}$$

$$u(x,0,t) = 0, \quad x > 0,\ t \geq 0, \tag{4}$$

$$u_y(x,0,t) = 0, \quad x > 0,\ t \geq 0, \tag{5}$$

$$u(x,y,t)|_{x \longrightarrow \infty} \text{ bounded}, \quad y > 0,\ t > 0, \tag{6}$$

where the time-fractional derivative ($\partial^{\alpha} u / \partial t^{\alpha}$) is the Caputo fractional derivative of order $\alpha (0 < \alpha \leq 1)$ defined by (see [39])

$$\frac{\partial^\alpha u}{\partial t^\alpha}(x,y,t) = \frac{1}{\Gamma(1-\alpha)} \int_0^t \frac{\partial u(x,y,s)}{\partial s} \frac{ds}{(t-s)^\alpha}, \quad 0 < \alpha < 1,$$

$$\frac{\partial^\alpha u}{\partial t^\alpha}(x,y,t) = \frac{\partial u(x,y,t)}{\partial t}, \quad \alpha = 1. \tag{7}$$

Here, we wish to determine the temperature u(x, y, t) for $0 \leq x < 1$ from the temperature measurements $g^\delta(y, t)$.

In order to apply the Fourier transform techniques, we extend the functions u(x, ·, ·), g(·, ·), and $g^\delta(\cdot, \cdot)$ to the whole plane $-\infty < y < +\infty$ and $-\infty < t < +\infty$ by setting the functions to be zero for $y < 0$ or $t < 0$. We assume that these functions are in $L^2(R^2)$ and use the corresponding L^2-norm, defined as follows:

$$\|g\| = \left(\int_{\mathbb{R}^2} |g(y,t)|^2 dy\, dt \right)^{1/2}. \tag{8}$$

We also assume that the measured data function $g^\delta(y, t)$ satisfies

$$\left\| g^\delta - g \right\| \leq \delta, \tag{9}$$

where the constant $\delta > 0$ represents a bound on the measurement error.

Let

$$\hat{g}(\xi,\eta) = \frac{1}{2\pi} \int_{\mathbb{R}^2} g(y,t) e^{-i(\xi y + \eta t)} dy\, dt, \quad \xi, \eta \in \mathbb{R}, \tag{10}$$

be the Fourier transform of a function g(y, t). Taking the transform to equations (1)–(6) with respect to y and t, we can get the solution of equations (1)–(6) in the frequency domain (see [38]):

$$\hat{u}(x,\xi,\eta) = e^{(1-x)\sqrt{(i\eta)^\alpha + \xi^2}} \hat{g}(\xi,\eta), \tag{11}$$

where

$$(i\eta)^{\alpha} = |\eta|^{\alpha}\left(\cos\left(\frac{\alpha\pi}{2}\right) + i\,\mathrm{sign}(\eta)\sin\left(\frac{\alpha\pi}{2}\right)\right). \tag{12}$$

Denote γ as follows:

$$\gamma := \sqrt{(i\eta)^{\alpha} + \xi^2} = A + iB, \tag{13}$$

where

$$A := \sqrt{\frac{\sqrt{|\eta|^{2\alpha}\sin^2(\alpha\pi/2) + \left(\xi^2 + |\eta|^{\alpha}\cos(\alpha\pi/2)\right)^2} + \left(\xi^2 + |\eta|^{\alpha}\cos(\alpha\pi/2)\right)}{2}},$$

$$B := \mathrm{sign}\left(\mathrm{sign}(\eta)|\eta|^{\alpha}\sin\left(\frac{\alpha\pi}{2}\right)\right)$$

$$\times\sqrt{\frac{\sqrt{|\eta|^{2\alpha}\sin^2(\alpha\pi/2) + \left(\xi^2 + |\eta|^{\alpha}\cos(\alpha\pi/2)\right)^2} - \left(\xi^2 + |\eta|^{\alpha}\cos(\alpha\pi/2)\right)}{2}}. \tag{14}$$

For the above problem, since $|e^{(1-x)\sqrt{(i\eta)^{\alpha}+\xi^2}}|$ is unbounded with respect to variables ξ and η for fixed $0 \leq x < 1$, the small error in the high-frequency components will be amplified. +erefore, the 2D time-fractional inverse diffusion problem is severely ill-posed. To solve the 2D time-fractional inverse diffusion problem, a natural way to stabilize the problem is eliminate the high frequencies or to replace the "kernel" $e^{(1-x)\sqrt{(i\eta)^{\alpha}+\xi^2}}$ by a bounded approximation.

A MODIFIED METHOD AND CONVERGENCE ESTIMATES

In this section, we will give a modified "kernel" method and obtain the convergence estimates. The regularization solution is given by

$$\widehat{u}_{\beta}^{\delta}(x,\xi,\eta) = k_{\beta}(x,\xi,\eta)\widehat{g}^{\delta}(\xi,\eta), \tag{15}$$

Where

$$k_{\beta}(x,\xi,\eta) = \begin{cases} e^{(1-x)(A+iB)}, & e^{(1-x)A} \leq \beta(x), \\ \beta(x)e^{i(1-x)B}, & e^{(1-x)A} > \beta(x), \end{cases} \tag{16}$$

where $\beta(x) > 1$ is the regularization parameter.

To obtain the convergence estimates between the regularization solution and the exact solution, we need to assume a priori bound:

$$\|u(0,\cdot,\cdot)\|_p \leq E, \quad p > 0, \tag{17}$$

where E > 0 is a constant and $\|\cdot\|_p$ denotes the norm in Sobolov space $H^p(R^2)$ defined by

$$\|u(x,\cdot,\cdot)\|_p := \left(\int_{\mathbb{R}^2} \left(1 + \xi^2 + \eta^2\right)^p |\widehat{u}(0,\cdot,\cdot)|^2 d\xi \, d\eta \right)^{1/2}. \tag{18}$$

Remark 1. Obviously, when p = 0, we can know that $H^p(R^2) = H^0(R^2) = L^2(R^2)$ and formula (17) is bounded in the $L^2(R^2)$-norm.

A Priori Parameter Choice

In the following, we give the convergence estimate for $\|u_\beta^\delta(x,\cdot,\cdot) - u(x,\cdot,\cdot)\|$ by using a priori choice rule for the regularization parameter.

Theorem 1. *Suppose that $u^\delta_\beta(x, y, t)$ is the regularization solution with noisy data $g^\delta(y, t)$ and that $u(x, y, t)$ is the exact solution with the exact data $g(y, t)$. Let assumption (9) be satisfied and let $\|u(0, \cdot, \cdot)\| \leq E$. If we choose*

$$\beta(x) = x\left(\frac{\delta}{E}\right)^{x-1}, \tag{19}$$

then for every $x \in (0, 1)$, we obtain the error estimate

$$\left\|u_\beta^\delta(x,\cdot,\cdot) - u(x,\cdot,\cdot)\right\| \leq \delta^x E^{1-x}. \tag{20}$$

Proof. Due to Parseval's identity and triangle inequality, we can obtain

$$\begin{aligned}
&\left\|u_\beta^\delta(x,\cdot,\cdot) - u(x,\cdot,\cdot)\right\| \\
&= \left\|\widehat{u}_\beta^\delta(x,\cdot,\cdot) - \widehat{u}(x,\cdot,\cdot)\right\| \\
&= \left\|k_\beta(x,\xi,\eta)\widehat{g}^\delta - e^{(1-x)(A+iB)}\widehat{g}\right\| \\
&= \left\|k_\beta(x,\xi,\eta)\widehat{g}^\delta - k_\beta(x,\xi,\eta)\widehat{g} + k_\beta(x,\xi,\eta)\widehat{g} \right. \\
&\quad \left. - e^{(1-x)(A+iB)}\widehat{g}\right\| \\
&\leq \left\|k_\beta(x,\xi,\eta)\widehat{g}^\delta - k_\beta(x,\xi,\eta)\widehat{g}\right\| + \left\|k_\beta(x,\xi,\eta)\widehat{g} \right. \\
&\quad \left. - e^{(1-x)(A+iB)}\widehat{g}\right\| \\
&= I_1 + I_2.
\end{aligned} \tag{21}$$

For I_1, we have

$$I_1 = \left\|k_\beta(x,\xi,\eta)\widehat{g}^\delta - k_\beta(x,\xi,\eta)\widehat{g}\right\| \leq \delta \sup_{\xi,\eta \in \mathbb{R}} \left|k_\beta(x,\xi,\eta)\right| \leq \delta\beta(x).$$

(22)

From (11), we know

$$\widehat{g}(\xi,\eta) = e^{-(A+iB)}\widehat{u}(0,\xi,\eta). \tag{23}$$

We now estimate I_2; note that from (23),

$$I_2 = \left\|k_\beta(x,\xi,\eta)\widehat{g} - e^{(1-x)(A+iB)}\widehat{g}\right\|$$

$$= \left\|\frac{e^{(1-x)(A+iB)} - k_\beta(x,\xi,\eta)}{e^{(A+iB)}}\widehat{u}(0,\xi,\eta)\right\|$$

$$= \left\|\frac{e^{(1-x)(A+iB)} - \min\{e^{(1-x)A},\beta(x)\}e^{(1-x)iB}}{e^{(A+iB)}}\widehat{u}(0,\xi,\eta)\right\|$$

$$= \left\|\frac{(e^{(1-x)A} - \min\{e^{(1-x)A},\beta(x)\})e^{(1-x)iB}}{e^{(A+iB)}}\widehat{u}(0,\xi,\eta)\right\|$$

$$\leq \sup_{\xi,\eta\in\mathbb{R}}\left|\frac{(e^{(1-x)A} - \min\{e^{(1-x)A},\beta(x)\})e^{(1-x)iB}}{e^{(A+iB)}}\right|E$$

$$\leq E \sup_{\xi,\eta\in\mathbb{R},e^{(1-x)A}>\beta(x)} \frac{e^{(1-x)A} - \beta(x)}{e^{A}}. \tag{24}$$

Let

$$\varphi(A) = \frac{e^{(1-x)A} - \beta(x)}{e^{A}}. \tag{25}$$

Differentiating φ(A) and setting the derivative equal to zero, we find that

$$A^* = \frac{1}{1-x}\ln\left(\frac{1}{x}\beta(x)\right), \tag{26}$$

is the maximum value point of the function φ(A). Therefore, we know

$$I_2 \leq \frac{((1/x)-1)\beta(x)}{((1/x)\beta(x))^{1/(1-x)}}E. \tag{27}$$

Combining (19), (21), (22), and (27), we have the Hölder-type error estimate (20).

Remark 2. The error estimate in theorem 1 does not give any useful information on the continuous dependence of the solution at x = 0. To retain the continuous dependence of the solution at x = 0, one has to introduce a stronger a priori assumption.

Theorem 2. *Suppose that $u^{\delta}{}_{\beta}(0, y, t)$ is the regularization solution with noisy data $g^{\delta}(y, t)$ and that $u(0, y, t)$ is the exact solution with the exact data $g(y, t)$ at $x = 0$. Let assumption (9) be satisfied and let $\|u(0, \cdot, \cdot)\|p \leq E(p > 0)$. If we choose*

$$\beta = \left(\frac{\delta}{E}\right)^{-(1/2)}, \tag{28}$$

then for $x = 0$, we obtain the error estimate

$$\left\|u_{\beta}^{\delta}(0, y, t) - u(0, y, t)\right\| \leq \delta^{1/2} E^{1/2} + E\left(\frac{1}{4}\ln\frac{E}{\delta}\right)^{-p}. \tag{29}$$

Proof. Due to Parseval's identity and the triangle inequality, we have

$$\begin{aligned}
&\left\|u_{\beta}^{\delta}(0, y, t) - u(0, y, t)\right\| \\
&= \left\|\widehat{u}_{\beta}^{\delta}(0, \cdot, \cdot) - \widehat{u}(0, \cdot, \cdot)\right\| \\
&= \left\|k_{\beta}(0, \xi, \eta)\widehat{g}^{\delta} - e^{(A+iB)}\widehat{g}\right\| \\
&= \left\|k_{\beta}(0, \xi, \eta)\widehat{g}^{\delta} - k_{\beta}(0, \xi, \eta)\widehat{g} + k_{\beta}(0, \xi, \eta)\widehat{g} - e^{(A+iB)}\widehat{g}\right\| \\
&\leq \left\|k_{\beta}(0, \xi, \eta)\widehat{g}^{\delta} - k_{\beta}(0, \xi, \eta)\widehat{g}\right\| + \left\|k_{\beta}(0, \xi, \eta)\widehat{g} - e^{(A+iB)}\widehat{g}\right\| \\
&= I_3 + I_4,
\end{aligned} \tag{30}$$

where

$$\begin{aligned}
I_3 &= \left\|k_{\beta}(0, \xi, \eta)\widehat{g}^{\delta} - k_{\beta}(0, \xi, \eta)\widehat{g}\right\| \\
&= \left\|k_{\beta}(0, \xi, \eta)\left(g^{\delta} - g\right)\right\| \\
&\leq \delta \sup_{\xi, \eta \in \mathbb{R}} \left|k_{\beta}(0, \xi, \eta)\right| \\
&\leq \delta\beta,
\end{aligned} \tag{31}$$

$$\begin{aligned} I_4 &= \left\| k_\beta(0,\xi,\eta)\hat{g} - e^{(A+iB)}\hat{g} \right\| \\ &= \left\| \frac{e^{((A+iB)} - \min\{e^A, \beta\} e^{iB}}{e^{(A+iB)}} \hat{u}(0,\xi,\eta) \right\| \\ &= \left\| \frac{(e^A - \min\{e^A,\beta\})e^{iB}}{\left(1+\xi^2+\eta^2\right)^{p/2} e^{(A+iB)}} \left(1+\xi^2+\eta^2\right)^{p/2} \hat{u}(0,\xi,\eta) \right\| \\ &\le \sup_{\xi,\eta\in\mathbb{R}} \left| \frac{(e^A - \min\{e^A,\beta\})e^{iB}}{\left(1+\xi^2+\eta^2\right)^{p/2} e^{(A+iB)}} \right| E \\ &\le E \sup_{\xi,\eta\in\mathbb{R}, e^A>\beta} \frac{e^A-\beta}{\left(1+\xi^2+\eta^2\right)^{p/2} e^A}. \end{aligned} \tag{32}$$

Using the fact $|\gamma| = |\sqrt{(i\eta)^\alpha + \xi^2}| \le |\eta|^{\alpha/2} + |\xi|$, we can get

$$A \le \sqrt{A^2+B^2} = |\gamma| \le |\eta|^{\alpha/2} + |\xi|, \text{ i.e., } A \le |\eta|^{\alpha/2} + |\xi|. \tag{33}$$

Note that also $\beta < e^A$, and we have $A > \ln\beta$; therefore, we can derive

$$\ln\beta \le |\eta|^{\alpha/2} + |\xi|. \tag{34}$$

If $|\xi| > |\eta|^{\alpha/2} \ge 0$, we know that

$$\ln\beta \le |\eta|^{\alpha/2} + |\xi| \le 2|\xi|, \tag{35}$$

i.e., $|\xi| > (1/2)\ln\beta$. Therefore,

$$\left(1+\xi^2+\eta^2\right)^{-(p/2)} \le \left(\xi^2+\eta^2\right)^{-(p/2)} < |\xi|^{-p} < \left(\frac{1}{2}\ln\beta\right)^{-p}. \tag{36}$$

If $|\eta|^{\alpha/2} \ge |\xi| \ge 0$, we have

$$\ln\beta \le |\eta|^{\alpha/2} + |\xi| \le 2|\eta|^{\alpha/2}, \tag{37}$$

i.e., $|\eta| \ge ((1/2)\ln\beta)^{2/\alpha}$. Therefore

$$\left(1+\xi^2+\eta^2\right)^{-(p/2)} \le \left(\xi^2+\eta^2\right)^{-(p/2)} \le |\eta|^{-p} < \left(\frac{1}{2}\ln\beta\right)^{-((2p)/\alpha)}. \tag{38}$$

Combining (32), (36), and (38), we can obtain

$$I_4 \leq E \sup_{\xi,\eta \in \mathbb{R}, e^A > \beta} \frac{1}{\left(1+\xi^2+\eta^2\right)^{p/2}} \leq E \max$$

$$\cdot \left\{ \left(\frac{1}{2}\ln\beta\right)^{-p}, \left(\frac{1}{2}\ln\beta\right)^{-((2p)/\alpha)} \right\}. \tag{39}$$

Combining (28), (30), (31), and (39), we can obtain

$$\left\| u_\beta^\delta(0,y,t) - u(0,y,t) \right\| \leq I_3 + I_4$$

$$\leq \delta^{1/2}E^{1/2} + E\max\left\{ \left(\frac{1}{4}\ln\frac{E}{\delta}\right)^{-p}, \right.$$

$$\left. \left(\frac{1}{4}\ln\frac{E}{\delta}\right)^{-((2p)/\alpha)} \right\}$$

$$= \delta^{1/2}E^{1/2} + E\left(\frac{1}{4}\ln\frac{E}{\delta}\right)^{-p}. \tag{40}$$

A Posteriori Parameter Choice

In this section, we first give the following lemma.

Lemma 1 (see [40]). . *Let the function f(λ): (0, a] ⟶ R be given*

$$f(\lambda) = \lambda^b \left[d \ln\frac{1}{\lambda} \right]^{-c}, \tag{41}$$

with a constant c ∈ R and positive constants a < 1, b, and d; then, for the inverse function, we have

$$f^{-1}(\lambda) = \lambda^{1/b}\left[\frac{d}{b}\ln\frac{1}{\lambda}\right]^{c/b}(1+o(1)), \quad \text{for } \lambda \longrightarrow 0. \tag{42}$$

In this following, we give the convergence estimate for $\|u^\delta{}_\beta(x,\cdot,\cdot) - u(x,\cdot,\cdot)\|$ *by using a posteriori choice rule for the regularization parameter, i.e., Morozov's discrepancy principle.*

According to Morozov's discrepancy principle [41], we adopt the regularization parameter β as the solution of the following equation:

$$\left\| e^{-(A+iB)(1-x)} k_\beta(x,\xi,\eta)\hat{g}^\delta - \hat{g}^\delta \right\| = \tau\delta, \tag{43}$$

where $\tau > 1$ is a constant. Denote

$$T_\beta = e^{-(A+iB)(1-x)} k_\beta(x,\xi,\eta) - 1. \tag{44}$$

Here and in the following, β is a function of the variable x and sometimes the function β(x) is also used.

Lemma 2. *Let* $\rho(\beta) := \|e^{-(A+iB)(1-x)} k_\beta(x,\xi,\eta)\, \hat{g}^\delta - \hat{g}^\delta\|$, *and then, the following results hold:*

(a) $\rho(\beta)$ *is a continuous function*

(b) $\lim_{\beta \longrightarrow 1} \rho(\beta) = \|(e^{-A(1-x)} - 1)\hat{g}^\delta(\xi,\eta)\|$

(c) $\lim_{\beta \longrightarrow +\infty} \rho(\beta) = 0$

(d) $\rho(\beta)$ *is a strictly decreasing function for* $\beta \in (1,+\infty)$

The proof is obvious, and we omit it here.

Remark 3. According to Lemma 2, we find that, if $0 < \tau\delta < (e^{-A(1-x)} - 1)\, \hat{g}^\delta(\xi,\eta)$, equation (43) has a unique solution.

Lemma 3. *If* $\beta(x)$ *is the solution of equation (43), then the following inequality holds*

$$\beta(x) \le \left(\frac{(\tau-1)\delta}{E}\right)^{x-1} \left(\frac{1}{2}\ln\frac{E}{(\tau-1)\delta}\right)^{-p(1-x)} (1+o(1)). \tag{45}$$

Proof. Due to the triangle inequality and (43), note that $|T_\beta| \le 1$, and we obtain

$$\tau\delta = \left\|T_\beta \hat{g}^\delta\right\| \le \left\|T_\beta(\hat{g}^\delta - \hat{g})\right\| + \left\|T_\beta \hat{g} \le \delta\right\| + \left\|T_\beta \hat{g}\right\|. \tag{46}$$

We now estimate the second term on the right hand side of (46):

$$T_\beta \hat{g} = \left\|\left(e^{-(A+iB)(1-x)} k_\beta(x,\xi,\eta) - 1\right) e^{-(A+iB)} \hat{u}(0,\cdot,\cdot)\right\|$$

$$= \left\| \frac{1 - e^{-(A+iB)(1-x)} k_\beta(x,\xi,\eta)}{e^{A+iB}\left(1+\xi^2+\eta^2\right)^{p/2}} \left(1+\xi^2+\eta^2\right)^{p/2} \hat{u}(0,\cdot,\cdot)\right\|$$

$$\le E \sup_{\xi,\eta\in\mathbb{R}, e^{A(1-x)} > \beta(x)} \left| \frac{1 - e^{-A(1-x)}\beta(x)}{e^{A}\left(1+\xi^2+\eta^2\right)^{p/2}} \right|$$

$$= E \sup_{\xi,\eta\in\mathbb{R},e^{A(1-x)}>\beta(x)} \left| \frac{e^{A(1-x)} - \beta(x)}{e^{A(2-x)}\left(1+\xi^2+\eta^2\right)^{p/2}} \right|$$

$$= E \sup_{\xi,\eta\in\mathbb{R},e^{A(1-x)}>\beta(x)} \frac{e^{A(1-x)} - \beta(x)}{e^{A(1-x)}} e^{-A}\left(1+\xi^2+\eta^2\right)^{-(p/2)}$$

$$\le E \sup_{\xi,\eta\in\mathbb{R},e^{A(1-x)}>\beta(x)} e^{-A}\left(1+\xi^2+\eta^2\right)^{-(p/2)}. \tag{47}$$

Similar to the proof of Theorem 2, we can know

$$\left\|T_\beta \hat{g}\right\| \le E \sup_{\xi,\eta\in\mathbb{R},e^{A(1-x)}>\beta(x)} e^{-A}\left(1+\xi^2+\eta^2\right)^{-(p/2)}$$

$$\le E\beta(x)^{-(1/(1-x))}\max\left\{\left(\frac{1}{2(1-x)}\ln\beta(x)\right)^{-p},\right.$$

$$\left.\left(\frac{1}{2(1-x)}\ln\beta(x)\right)^{-((2p)/\alpha)}\right\}$$

$$= E\beta(x)^{-(1/(1-x))}\left(\frac{1}{2(1-x)}\ln\beta(x)\right)^{-p}. \tag{48}$$

Combining (46) and (48), we obtain

$$(\tau-1)\delta \le E\beta(x)^{-(1/(1-x))}\left(\frac{1}{2(1-x)}\ln\beta(x)\right)^{-p}. \tag{49}$$

Let $(\tau - 1)\delta = E\beta_*(x)^{-(1/(1-x))}((1/(2(1-x)))\ln \beta_*(x))^{-p}$. From Lemma 1, we have

$$\beta_*(x) = \left(\frac{(\tau-1)\delta}{E}\right)^{x-1}\left(\frac{1}{2}\ln\frac{E}{(\tau-1)\delta}\right)^{-p(1-x)}(1+o(1)). \tag{50}$$

Let $G(\beta) := E\beta(x)^{-(1/(1-x))}((1/(2(1-x)))\ln \beta(x))^{-p}$. For $\beta > 1$ and $0 \le x < 1$, the function $G(\beta)$ is strictly decreasing monotonically with β, so we have

$$\beta(x) \le \beta_*(x). \tag{51}$$

Lemma 4. *If $\beta(x)$ is the solution of equation (43), then the following inequality holds:*

$$\left\|T_\beta \hat{g}\right\| \le (\tau+1)\delta. \tag{52}$$

Proof. Using the triangle inequality and (43), we have

$$\|T_\beta \widehat{g}\| = \|T_\beta(\widehat{g} - \widehat{g}^\delta) + T_\beta \widehat{g}^\delta\| \le \|T_\beta(\widehat{g} - \widehat{g}^\delta)\| + \|T_\beta \widehat{g}^\delta\|$$
$$\le (\tau + 1)\delta. \tag{53}$$

Theorem 3. *Assume that conditions (9) and (17) hold and taking the solution of equation (43) as the regularization parameter, then there holds the following error estimate:*

$$\|u_\beta^\delta(x,\cdot,\cdot) - u(x,\cdot,\cdot)\| \le \delta^x E^{1-x}\left(\frac{1}{2}\ln\frac{E}{(\tau-1)\delta}\right)^{-p(1-x)}(C + o(1)), \tag{54}$$

where $C = (\tau - 1)^{x-1}(2\tau + 1),\ 0 \le x < 1.$

Proof. Due to the Parseval formula and the triangle inequality, we obtain

$$\|u_\beta^\delta(x,\cdot,\cdot) - u(x,\cdot,\cdot)\| = \|\widehat{u}_\beta^\delta(x,\cdot,\cdot) - \widehat{u}(x,\cdot,\cdot)\|$$
$$= \|k_\beta \widehat{g}^\delta - e^{(A+iB)(1-x)}\widehat{g}\|$$
$$\le \|k_\beta(\widehat{g}^\delta - \widehat{g})\| + \|(k_\beta - e^{(A+iB)(1-x)})\widehat{g}\|$$
$$\le \beta(x)\delta + \|(k_\beta e^{-(A+iB)(1-x)} - 1)\widehat{u}(x,\cdot,\cdot)\|$$
$$= \beta(x)\delta + \|T_\beta \widehat{u}(x,\cdot,\cdot)\|. \tag{55}$$

From Lemma 3, we can know that

$$\beta(x)\delta \le \left(\frac{E}{\tau-1}\right)^{1-x}\delta^x\left(\frac{1}{2}\ln\frac{E}{(\tau-1)\delta}\right)^{-p(1-x)}(1 + o(1)). \tag{56}$$

In the following, we estimate the second term of (55). Denote $\varphi = T_\beta \widehat{u}(x,\cdot,\cdot)$ and take $\beta(x) = (E/((\tau-1)\delta))^{1-x}((1/2)\ln(E/((\tau-1)\delta)))^{-p(1-x)}$; we have

$$\|\varphi\| = \|\varphi - k_\beta e^{-(A+iB)(1-x)}\varphi + k_\beta e^{-(A+iB)(1-x)}\varphi\|$$
$$\le \|\varphi - k_\beta e^{-(A+iB)(1-x)}\varphi\| + \|k_\beta e^{-(A+iB)(1-x)}\varphi\|$$
$$= M_1 + M_2. \tag{57}$$

Note that $\widehat{u}(x,\cdot,\cdot) = e^{(A+iB)(1-x)}\widehat{g}(\cdot,\cdot)$, we get

$$M_1 = \left\| T_\beta e^{(A+iB)(1-x)} \hat{g} \left(1 - k_\beta e^{-(A+iB)(1-x)}\right) \right\|$$

$$= \left\| T_\beta \left(e^{(A+iB)(1-x)} - k_\beta\right) \hat{g} \right\|$$

$$\le \left\| \left(e^{(A+iB)(1-x)} - k_\beta\right) \hat{g} \right\|. \tag{58}$$

Denote $M := \|(e^{(A+iB)(1-x)} - k_\beta)\hat{g}\|$, we have

$$M = \left\| \left(e^{(A+iB)(1-x)} - k_\beta\right) \hat{g} \right\| \le E \sup_{\xi,\eta \in \mathbb{R}} \left| \frac{e^{(A+iB)(1-x)} - k_\beta}{e^{(A+iB)} \left(1 + \xi^2 + \eta^2\right)^{p/2}} \right|$$

$$\le E \sup_{\xi,\eta \in \mathbb{R}} \left| \frac{e^{(A+iB)(1-x)} - \min\{e^{A(1-x)}, \beta(x)\} e^{iB(1-x)}}{e^{(A+iB)} \left(1 + \xi^2 + \eta^2\right)^{p/2}} \right|$$

$$\le E \sup_{\xi,\eta \in \mathbb{R}, e^{A(1-x)} > \beta(x)} \frac{e^{A(1-x)} - \beta(x)}{e^{A} \left(1 + \xi^2 + \eta^2\right)^{p/2}}$$

$$= E \sup_{\xi,\eta \in \mathbb{R}, e^{A(1-x)} > \beta(x)} \frac{e^{A(1-x)} - \beta(x)}{e^{A(1-x)}} \left(1 + \xi^2 + \eta^2\right)^{-(p/2)} e^{-Ax}$$

$$\le E \sup_{\xi,\eta \in \mathbb{R}, e^{A(1-x)} > \beta(x)} \left(1 + \xi^2 + \eta^2\right)^{-(p/2)} e^{-Ax}. \tag{59}$$

Similar to the proof of Theorem 2, we can know that

$$
\begin{aligned}
M &\le E \sup_{\xi,\eta\in\mathbb{R}, e^{A(1-x)} > \beta(x)} \left(1+\xi^2+\eta^2\right)^{-(p/2)} e^{-Ax} \\
&\le E\beta(x)^{-(x/(1-x))} \max\left\{ \left(\frac{1}{2(1-x)}\ln\beta(x)\right)^{-p}, \left(\frac{1}{2(1-x)}\ln\beta(x)\right)^{-((2p)/\alpha)} \right\} \\
&= E\beta(x)^{-(x/(1-x))} \left(\frac{1}{2(1-x)}\ln\beta(x)\right)^{-p} \\
&= E\left[\frac{1}{2(1-x)}\ln\left(\left(\frac{E}{(\tau-1)\delta}\right)^{1-x}\left(\frac{1}{2}\ln\frac{E}{(\tau-1)\delta}\right)^{-p(1-x)}\right)\right]^{-p} \\
&\quad\times\left[\left(\frac{E}{(\tau-1)\delta}\right)^{1-x}\left(\frac{1}{2}\ln\frac{E}{(\tau-1)\delta}\right)^{-p(1-x)}\right]^{-(x/(1-x))} \\
&= E\left[\frac{1}{2(1-x)}\ln\left(\left(\frac{E}{(\tau-1)\delta}\right)^{1-x}\left(\frac{1}{2}\ln\frac{E}{(\tau-1)\delta}\right)^{-p(1-x)}\right)\right]^{-p} \\
&\quad\times\left[\left(\frac{E}{(\tau-1)\delta}\right)^{1-x}\left(\frac{1}{2}\ln\frac{E}{(\tau-1)\delta}\right)^{-p(1-x)}\right]^{-(x/(1-x))} \\
&= E\left[\frac{1}{2}\ln\frac{E}{(\tau-1)\delta}+\ln\left(\frac{1}{2}\ln\frac{E}{(\tau-1)\delta}\right)^{-(p/2)}\right]^{-p}\left(\frac{E}{(\tau-1)\delta}\right)^{-x}\left(\frac{1}{2}\ln\frac{E}{(\tau-1)\delta}\right)^{px} \\
&= E\left(\frac{1}{2}\ln\frac{E}{(\tau-1)\delta}\right)^{-p}\left[1+\frac{\ln(1/2\ln E/(\tau-1)\delta)^{-(p/2)}}{1/2\ln E/(\tau-1)\delta}\right]^{-p}\left(\frac{(\tau-1)\delta}{E}\right)^{x}\left(\frac{1}{2}\ln\frac{E}{(\tau-1)\delta}\right)^{px} \\
&= (\tau-1)^x\delta^x E^{1-x}\left(\frac{1}{2}\ln\frac{E}{(\tau-1)\delta}\right)^{-p(1-x)}(1+o(1)).
\end{aligned}
\tag{60}
$$

For M_2, from Lemma 4 and assuming that $|k_\xi| \le \beta(x)$, we have

$$
\begin{aligned}
M_2 &\le \beta(x)\left\| e^{-(A+iB)(1-x)}\varphi\right\| \\
&= \beta(x)\left\| T_\beta \hat{g}\right\| \\
&\le (\tau+1)(\tau-1)^{x-1}\delta^x E^{1-x}\left(\frac{1}{2}\ln\frac{E}{(\tau-1)\delta}\right)^{-p(1-x)}.
\end{aligned}
\tag{61}
$$

Combining (56)–(61) with (55), the proof of Theorem 3 is completed.

NUMERICAL ASPECT

In this section, we present two numerical examples intended to illustrate the behavior of the proposed method.

The numerical examples are constructed in the following way. First, we present the initial data u(0, y, t) = f(y, t) of 2D time-fractional diffusion problem at x = 0 and computed the function u(1, y, t) = g(y, t) by solving a direct problem, which is a well-posed problem. Then, we added a random distributed perturbation to the data function obtaining vector gδ(y, t), i.e.,

$$g^{\delta}(y,t) = g(y,t) + \varepsilon\,\mathrm{randn}(\mathrm{size}(g(y,t))), \tag{62}$$

where

$$g(y,t) = (g(y_1,t_1), g(y_2,t_2), \ldots, g(y_n,t_n))^T,$$
$$y_i = (i-1)\Delta y, \quad \Delta y = \frac{1}{n-1}, \quad i = 1,2,\ldots,n,$$
$$t_j = (j-1)\Delta t, \quad \Delta t = \frac{1}{n-1}, \quad j = 1,2,\ldots,n, \tag{63}$$

where δ indicates the error level of , i.e.,

$$\delta := \sqrt{\frac{1}{n^2}\sum_{i=1}^{n}\sum_{j=1}^{n}\left(g^{\delta}(y_i,t_j) - g(y_i,t_j)\right)^2}. \tag{64}$$

In numerical implementations, we give the data f(y, t) and sample at an equidistant grid in the domain [0, 1] × [0, 1] with 64 × 64 grid points. +e function "randn(·)" generates arrays of random numbers whose elements are normally distributed with mean 0, variance $\sigma^2 = 1$, and standard deviation σ = 1; "randn size(g(y, t))" returns an array of random entries which has the same size as g(y, t). Let RMS denotes the mean square for a sampled function φ(·, ·) which is defined by

$$\mathrm{RMS}(\varphi) := \sqrt{\frac{1}{n^2}\sum_{i=1}^{n}\sum_{j=1}^{n}\left(\varphi(y_i,t_j)\right)^2}, \tag{65}$$

where *n* is the total number of test points. Similarly, we can define the mean square error (RMSE) between the computed data and the exact data.

Finally, we solved the 2D time-fractional inverse diffusion problem by the modified method.

Example 1. We consider a smooth function

$$u(0,y,t) = f(y,t) = 16\sin(2\pi y)\sin(2\pi t). \tag{66}$$

Example 2. Let Ω = {(y, t) | 0.1 ≤ y ≤ 0.3, 0.1 ≤ t ≤ 0.3} ∪ {(y, t) | 0.6 ≤ y ≤ 0.8, 0.6 ≤ t ≤ 0.8}. Consider a nonsmooth function

$$f(y,t) = \begin{cases} 1, & \text{if } (y,t) \in \Omega, \\ 0, & \text{if else.} \end{cases} \qquad (67)$$

Figures 1–3 are the comparison of the exact solution and the approximation solution for α = 0.1 with noise level ε = 0.001 at different points x = 0.9, 0.5, and 0.1 under a priori regularization parameter choice rule with Example 1. Figures 4–6 are the comparison of the exact solution and the approximation solution for α = 0.1 with noise level ε = 0.0001 at different points x = 0.9, 0.5, and0.1 under a posteriori regularization parameter choice rule with Example 1.

Table 1 shows the comparison of RMSE of Example 1 for different x with ε = 0.0001 and α = 0.1. We find that the smaller the x is, the worse the computed approximation will be

Table 2 shows the comparison of RMSE of Example 1 for different ε with α = 0.1 at x = 0.5. +e numerical error is decreasing as the level of noise becomes smaller.

Table 3 shows the comparison of RMSE of Example 1 for different α with ε = 0.0001 at x = 0.5. +e numerical accuracy is stable to the fractional order α.

Figure 7 is the comparison of the exact solution and the approximation solution for α = 0.1 with noise level ε = 0.0001 at x = 0.4 under both a priori and a posteriori regularization parameter choice rule with Example 2.

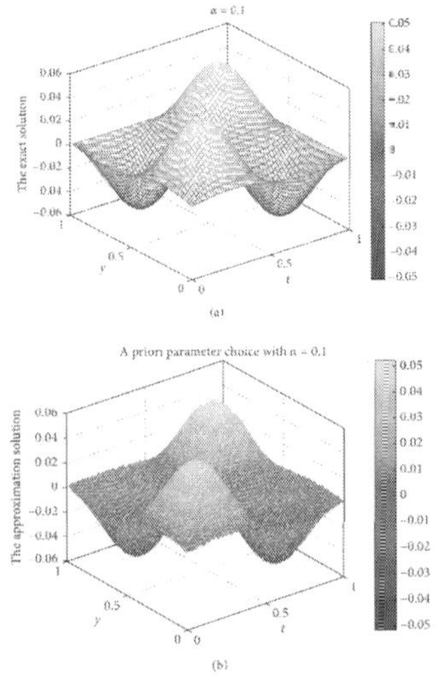

Figure 1 . The exact solution and its approximation solution at ε = 0.001 and α = 0.1: (a) the exact solution at x = 0.9; (b) its approximation solution at x = 0.9.

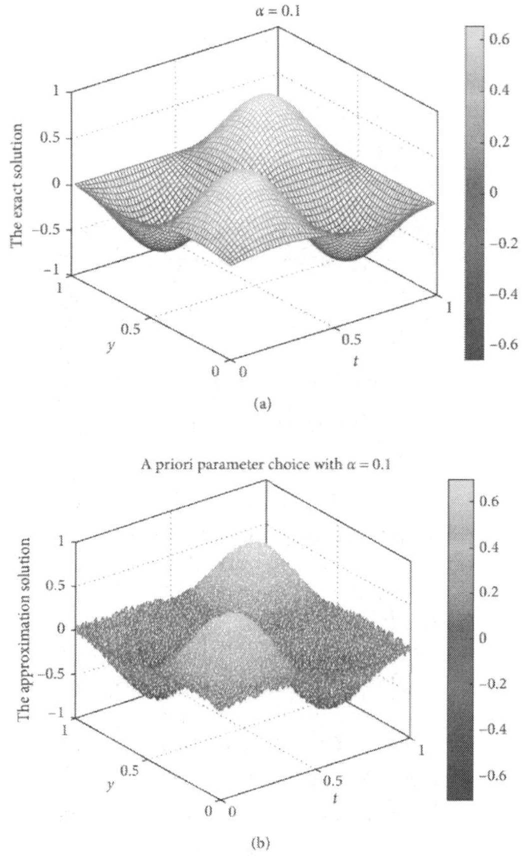

Figure 2. : The exact solution and its approximation solution at ε = 0.001 and α = 0.1: (a) the exact solution at x = 0.5; (b) its approximation solution at x = 0.5.

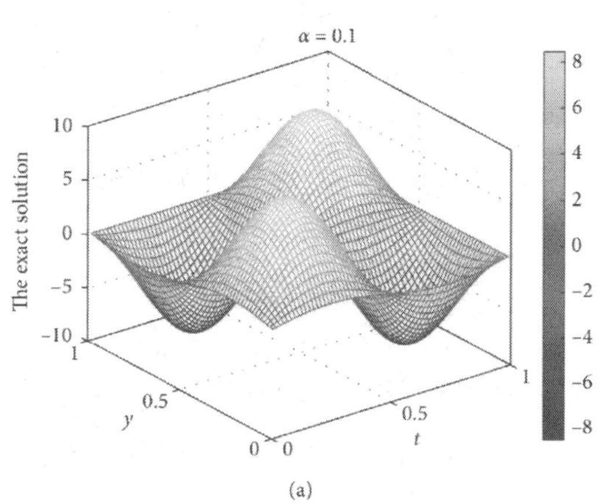

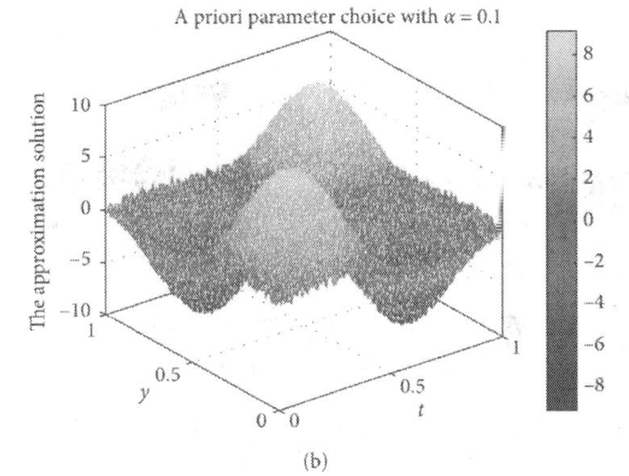

Figure 3. The exact solution and its approximation solution at ε = 0.001 and α = 0.1: (a) the exact solution at x = 0.1; (b) its approximation solution at x = 0.1.

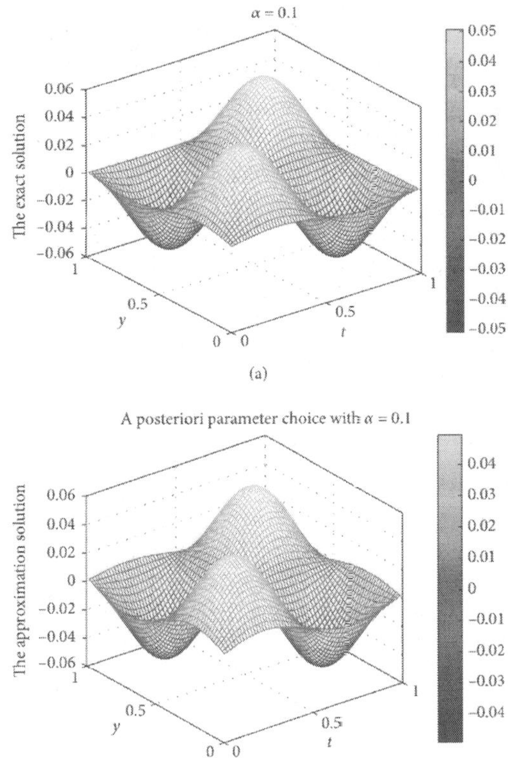

Figure 4. The exact solution and its approximation solution atε = 0.0001 and α = 0.1: (a) the exact solution at x = 0.9; (b) its approximation solution at x = 0.9.

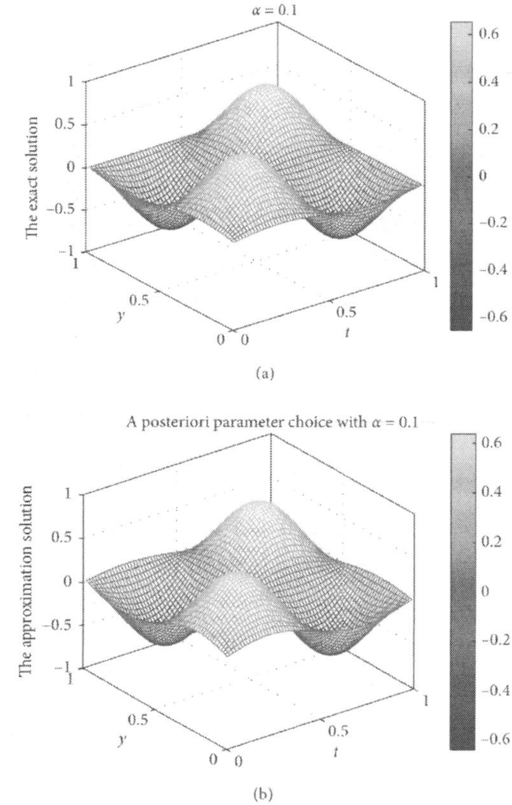

Figure 5. The exact solution and its approximation solution at ε = 0.0001 and α = 0.1: (a) the exact solution at x = 0.5; (b) its approximation solution at x = 0.5.

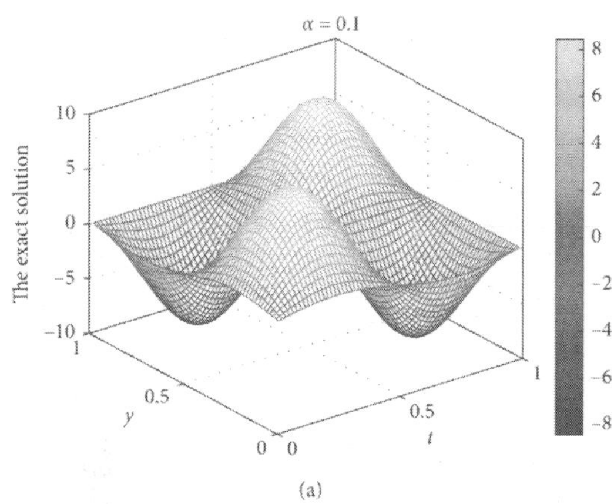

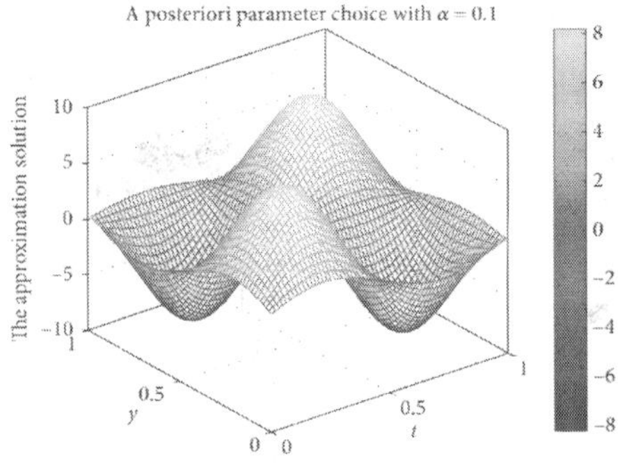

(b)

Figure 6. The exact solution and its approximation solution at ε = 0.0001 and α = 0.1: (a) the exact solution at x = 0.1; (b) its approximation solution at x = 0.1.

Table 1. Error behavior of Example 1 for different x with α = 0.1 and ε = 0.0001.

x	0.5	0.3	0.1
A priori choice rule: RMSE	4.59×10^{-2}	1.73×10^{-1}	6.37×10^{-1}
A posteriori choice rule: RMSE	4.55×10^{-2}	1.63×10^{-1}	5.88×10^{-1}

Table 2. Error behavior of Example 1 for different ε with α = 0.1 at x = 0.5

ε	0.01	0.001	0.0001
A priori choice rule: RMSE	9.22×10^{-2}	5.16×10^{-2}	4.59×10^{-2}
A posteriori choice rule: RMSE	8.94×10^{-2}	4.90×10^{-2}	4.55×10^{-2}

Table 3. Error behavior of Example 1 for different α with ε = 0.0001 at x = 0.5.

α	0.1	0.3	0.8
A priori choice rule: RMSE	4.59×10^{-2}	4.38×10^{-2}	3.89×10^{-2}
A posteriori choice rule: RMSE	4.55×10^{-2}	4.35×10^{-2}	3.88×10^{-2}

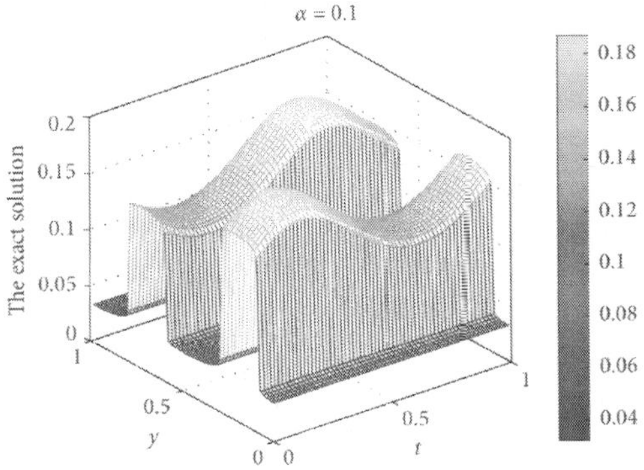

(a)

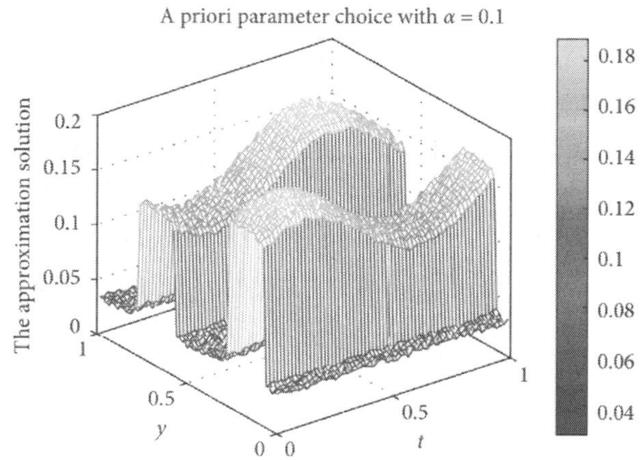

(b)

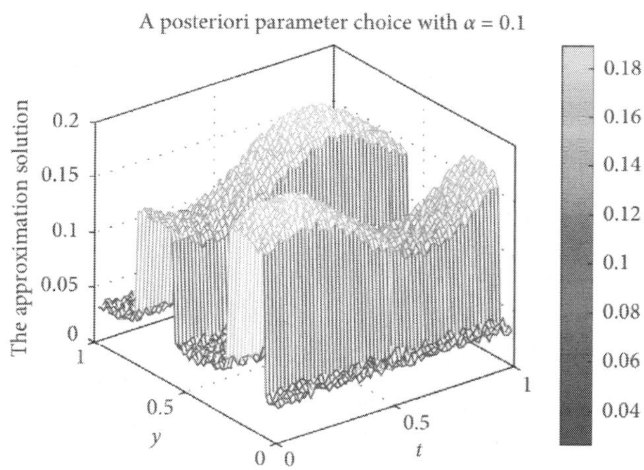

(c)

Figure 7. The exact solution and its approximation solution at $\varepsilon = 0.0001$ and $\alpha = 0.1$: (a) the exact solution at $x = 0.4$; (b) its approximation solution under a priori parameter choice rule at $x = 0.4$; (c) its approximation solution under a posteriori parameter choice rule at $x = 0.4$.

CONCLUSION

In this paper, we propose a modified method to solve the time-fractional inverse diffusion problem in the two-dimensional setting. For the choice of regularization parameter, we give not only a priori but also a posteriori rules. Moreover, under both a priori rule and a posteriori rule, we prove the error estimates from the viewpoint of theoretical analysis.

The numerical examples are presented to illustrate the validity and effectiveness of the proposed method. In general, $E > 0$ such that a priori condition (17) holds will not be known. But such a E has to be known if one wants to construct a priori parameter choice rule. We propose a posteriori parameter choice rule which is independent of E. So it is more implementable in practical application. Moreover, the numerical experiments show that a posteriori parameter choice rule also works well.

DATA AVAILABILITY

The data used to support the findings of this study are available from the corresponding author upon request.

CONFLICTS OF INTEREST

The authors declare that they have no conflicts of interest.

AUTHORS' CONTRIBUTIONS

All authors contributed equally to the writing of this paper. All authors read and approved the final manuscript.

ACKNOWLEDGMENTS

This work was partially supported by the National Natural Science Foundation of China (11871198), the Fundamental Research Funds for the Universities of Heilongjiang Province Heilongjiang University Special Project (RCYJTD201804), the National Science Foundation of Hebei Province (A2017501021), and the Fundamental Research Funds of the Central Universities (N182304024).

REFERENCES

1. W. Chen, L. Ye, and H. Sun, "Fractional diffusion equations by the Kansa method," *Computers & Mathematics with Applications*, vol. 59, no. 5, pp. 1614–1620, 2010.
2. Z. Yu and J. Lin, "Numerical research on the coherent structure in the viscoelastic second-order mixing layers," *Applied Mathematics and Mechanics*, vol. 19, no. 8, pp. 717–723, 1998.
3. H. Scher and E. W. Montroll, "Anomalous transit-time dispersion in amorphous solids," *Physical Review B*, vol. 12, no. 6, pp. 2455–2477, 1975.
4. T. L. Szabo and J. Wu, "A model for longitudinal and shear wave propagation in viscoelastic media," *The Journal of the Acoustical Society of America*, vol. 107, no. 5, pp. 2437–2446, 2000.
5. R. Gorenflo, F. Mainardi, D. Moretti, G. Pagnini, and P. Paradisi, "Discrete random walk models for space-time fractional diffusion," *Chemical Physics*, vol. 284, no. 1-2, pp. 521–541, 2002.
6. C. Wang, L. Ling, X. Xiong, and M. Li, "Regularization for 2-D fractional sideways heat equations," *Numerical Heat Transfer, Part B: Fundamentals*, vol. 68, no. 5, pp. 418–433, 2015.
7. R. V. Mendes, "A fractional calculus interpretation of the fractional volatility model," *Nonlinear Dynamics*, vol. 55, no. 4, pp. 395–399, 2009.
8. B. Berkowitz, H. Scher, and S. E. Silliman, "Anomalous transport in laboratory-scale, heterogeneous porous media," *Water Resources Research*, vol. 36, no. 1, pp. 149–158, 2000.
9. R. Metzler and J. Klafter, "The random walk's guide to anomalous diffusion: a fractional dynamics approach," *Physics Reports*, vol. 339, no. 1, pp. 1–77, 2000.
10. E. Scalas, R. Gorenflo, and F. Mainardi, "Fractional calculus and continuous-time finance," *Physica A: Statistical Mechanics and Its Applications*, vol. 284, no. 1–4, pp. 376–384, 2000.
11. I. M. Sokolov and J. Klafter, "From diffusion to anomalous diffusion: a century after Einsteins Brownian motion," *Chaos: An Interdisciplinary Journal of Nonlinear Science*, vol. 15, no. 2, Article ID 026103, 2005.
12. L. R. Evangelista and E. K. Lenzi, *Fractional Diffusion Equations and Anomalous Diffusion*, Cambridge University Press, Cambridge, UK, 2018.

13. A. A. Tateishi, H. V. Ribeiro, and E. K. Lenzi, "The role of fractional time-derivatives operators on anomalous diffusion," *Frontiers in Physics*, vol. 5, p. 52, 2017.
14. S. D. Eidelman and A. N. Kochubei, "Cauchy problem for fractional diffusion equations," *Journal of Differential Equations*, vol. 199, no. 2, pp. 211–255, 2004.
15. R. Gorenflo, Y. F. Luchko, and P. P. Zabrejko, "On solvability of linear fractional differential equations in Banach spaces," *Fractional Calculus and Applied Analysis*, vol. 2, pp. 163–176, 1999.View at: Google Scholar
16. A. Hanyga, "Multidimensional solutions of time-fractional diffusion-wave equations," *Proceedings of the Royal Society of London. Series A: Mathematical, Physical and Engineering Sciences*, vol. 458, no. 2018, pp. 933–957, 2002.
17. Y. Luchko, "Maximum principle for the generalized time-fractional diffusion equation," *Journal of Mathematical Analysis and Applications*, vol. 351, no. 1, pp. 218–223, 2009.
18. Y. Luchko, "Some uniqueness and existence results for the initial-boundary-value problems for the generalized time-fractional diffusion equation," *Computers & Mathematics with Applications*, vol. 59, no. 5, pp. 1766–1772, 2010.
19. F. Liu, V. Anh, and I. Turner, "Numerical solution of the space fractional Fokker-Planck equation," *Journal of Computational and Applied Mathematics*, vol. 166, no. 1, pp. 209–219, 2004.
20. F. Liu, P. Zhuang, V. Anh, I. Turner, and K. Burrage, "Stability and convergence of the difference methods for the space-time fractional advection-diffusion equation," *Applied Mathematics and Computation*, vol. 191, no. 1, pp. 12–20, 2007.
21. M. M. Meerschaert and C. Tadjeran, "Finite difference approximations for fractional advection-dispersion flow equations," *Journal of Computational and Applied Mathematics*, vol. 172, no. 1, pp. 65–77, 2004.
22. M. M. Meerschaert, Y. Zhang, and B. Baeumer, "Particle tracking for fractional diffusion with two time scales," *Computers & Mathematics with Applications*, vol. 59, no. 3, pp. 1078–1086, 2010.
23. Q. Yang, F. Liu, and I. Turner, "Numerical methods for fractional partial differential equations with Riesz space fractional derivatives,"

Applied Mathematical Modelling, vol. 34, no. 1, pp. 200–218, 2010.
24. Z. Yang, Y. Nie, Z. Yuan, and J. Wang, "Finite element methods for fractional PDEs in three dimensions," *Applied Mathematics Letters*, vol. 100, Article ID 106041, 2020.
25. D. A. Murio, "Stable numerical solution of a fractional-diffusion inverse heat conduction problem," *Computers & Mathematics with Applications*, vol. 53, no. 10, pp. 1492–1501, 2007.
26. D. A. Murio, "Time fractional IHCP with Caputo fractional derivatives," *Computers & Mathematics with Applications*, vol. 56, no. 9, pp. 2371–2381, 2008.
27. J. Cheng, J. Nakagawa, M. Yamamoto, and T. Yamazaki, "Uniqueness in an inverse problem for a one-dimensional fractional diffusion equation," *Inverse Problems*, vol. 25, no. 11, Article ID 115002, 2009.
28. J. J. Liu and M. Yamamoto, "A backward problem for the time-fractional diffusion equation," *Applicable Analysis*, vol. 89, no. 11, pp. 1769–1788, 2010.
29. Y. Zhang and X. Xu, "Inverse source problem for a fractional diffusion equation," *Inverse Problems*, vol. 27, no. 3, Article ID 035010, 2011.
30. G. H. Zheng and T. Wei, "A new regularization method for solving a time-fractional inverse diffusion problem," *Journal of Mathematical Analysis and Applications*, vol. 378, no. 2, pp. 418–431, 2011.
31. B. T. Jin and W. Rundell, "An inverse problem for a one-dimensional time-fractional diffusion problem," *Inverse Problems*, vol. 28, no. 7, Article ID 075010, 2012.
32. T. Wei and J. Wang, "A modified quasi-boundary value method for an inverse source problem of the time-fractional diffusion equation," *Applied Numerical Mathematics*, vol. 78, pp. 95–111, 2014.
33. T. Wei, X. L. Li, and Y. S. Li, "An inverse time-dependent source problem for a time-fractional diffusion equation," *Inverse Problems*, vol. 32, no. 8, Article ID 085003, 2016.
34. G. H. Zheng and T. Wei, "Spectral regularization method for solving a time-fractional inverse diffusion problem," *Applied Mathematics and Computation*, vol. 218, no. 2, pp. 396–405, 2011.
35. H. Cheng and C.-L. Fu, "An iteration regularization for a time-fractional inverse diffusion problem," *Applied Mathematical Modelling*, vol. 36, no. 11, pp. 5642–5649, 2012.
36. X. Xiong, H. Guo, and X. Liu, "An inverse problem for a fractional

diffusion equation," *Journal of Computational and Applied Mathematics*, vol. 236, no. 17, pp. 4474–4484, 2012.

37. M. Li, X.-X. Xi, and X.-T. Xiong, "Regularization for a fractional sideways heat equation," *Journal of Computational and Applied Mathematics*, vol. 255, pp. 28–43, 2014.
38. X. Xiong, Q. Zhou, and Y. C. Hon, "An inverse problem for fractional diffusion equation in 2-dimensional case: stability analysis and regularization," *Journal of Mathematical Analysis and Applications*, vol. 393, no. 1, pp. 185–199, 2012.
39. I. Podlubny, *Fractional Differential Equations*, Academic Press, San Diego, CA, USA, 1999.
40. U. Tautenhahn, "Optimality for ill-posed problems under general source conditions," *Numerical Functional Analysis and Optimization*, vol. 19, no. 3-4, pp. 377–398, 1998.
41. A. Kirsch, *An Introduction to the Mathematical Theory of Inverse Problem*, Springer, New York, NY, USA, 1999.

CHAPTER

6

Numerical Inversion for the Multiple Fractional Orders in the Multiterm TFDE

Chunlong Sun,[1,2] Gongsheng Li,[1] and Xianzheng Jia[1]

[1]School of Science, Shandong University of Technology, Zibo 255049, China

[2]Department of Mathematics, Southeast University, Nanjing 210096, China

ABSTRACT

The fractional order in a fractional diffusion model is a key parameter which characterizes the anomalous diffusion behaviors. This paper deals with an inverse problem of determining the multiple fractional orders in the multiterm time-fractional diffusion equation (TFDE for short) from numerics. The homotopy regularization algorithm is applied to solve the inversion problem using the finite data at one interior point in the space

Citation: Sun, C., Li, G., & Jia, X. (2017). Numerical inversion for the multiple fractional orders in the multiterm TFDE. Advances in Mathematical Physics, 2017. (7 pages) DOI: https://doi.org/10.1155/2017/3204959

domain. The inversion fractional orders with random noisy data give good approximations to the exact order demonstrating the efficiency of the inversion algorithm and numerical stability of the inversion problem.

INTRODUCTION

The partial differential equations of fractional order have played an important role in modeling of the anomalous phenomena and in the theory of the complex systems during the last two decades; see, for example, [1–8]. The so-called time-fractional diffusion equation (TFDE) that is obtained from the classical diffusion equation by replacing the firstorder time derivative by a fractional derivative of order $\alpha$with $0<\alpha< 1$ has to be especially mentioned. On the other hand, by the attempts to describe some real processes with the equations of the fractional order, several researches were confronted with the situation that the order α of the timefractional derivative from the corresponding models did not remain constant and changed, say, in the interval from 0 to 1, from 1 to 2, or even from 0 to 2. To manage these phenomena, several approaches were suggested. One of them introduces the fractional derivatives of the variable order, that is, the derivatives with the order that can change with the time or/and depending on the spatial coordinates [9–11], and the other way is to employ the distributed order TFDE, or the multiterm TFDE in discretization. Let Ω be a bounded domain in R^d ($d\geq 1$) with smooth boundary $\partial\Omega$, and let $T>0$; the multiterm homogeneous TFDE with variable coefficient in Ω is given as

$$\frac{\partial^{\alpha} u}{\partial t^{\alpha}} + \sum_{s=1}^{S} r_s \frac{\partial^{\beta_s} u}{\partial t^{\beta_s}} = \nabla \cdot (D \nabla u), \quad x \in \Omega,\ 0 < t \leq T, \tag{1}$$

where $u = u(x, t)$ denotes the state variable at space point x and time t, α denotes the principal fractional order, and $\beta_1, \beta_2, \ldots, \beta_S$ are the multiterm fractional orders of the time derivatives, which satisfy the condition

$$0 < \beta_S < \beta_{S-1} < \cdots < \beta_1 < \alpha < 1, \tag{2}$$

and $r_1, r_2, \ldots, r_S$ are positive constants, and $D(x) > 0$ is the smooth diffusion coefficient tensor. All of the above timefractional derivatives are defined in the sense of Caputo; for example, the fractional derivative of the order $\beta \in (0, 1)$ is given by

$$\frac{\partial^{\beta} u}{\partial t^{\beta}} = \frac{1}{\Gamma(1-\beta)} \int_0^t \frac{\partial u(x,s)}{\partial s} \frac{ds}{(t-s)^{\beta}}. \tag{3}$$

See, for example, Podlubny [12] and Kilbas et al. [13] for the definition and properties of Caputo's derivative.

There are still a few research works reported on the multiterm TFDE like (1). On theoretical analysis and analytical methods for the forward problem, we refer to Daftardar-Gejji and Bhalekar [14], Luchko [15, 16], Jiang et al. [17], Ding et al. [18, 19], and Li et al. [20], and for numerical methods and simulations we refer to [21–23], and so on.

However, for real problems, the fractional orders, the initial distribution, the diffusion coefficient, or the source term cannot be obtained directly and we have to determine them by some additional measurements, which contributes to inverse problems arising in the fractional diffusion models. There are still some researches on inverse problems for the one-term TFDE; see, for example, Murio [24], Liu et al. [25, 26], Sakamoto and Yamamoto [27], Tuan [28], Chi et al. [29], Yamamoto and Zhang [30], Luchko et al. [31], Wei et al. [32, 33], and Liu et al. [34]; also see Jin and Rundell [35] for a tutorial review on inverse problems for anomalous diffusion processes.

It is noted that the research works stated above are almost related to coefficient identification problems in the one-term time/space fractional diffusion equations. However, it is also important to deal with inverse problems of determining the fractional orders in the fractional differential equations since the fractional order is an essential index characterizing the anomalous diffusion.

As for inverse problems of determining fractional orders in the single-term time/space fractional diffusion models, we refer to [36–42], and so on. On the other hand, there are few literatures concerned with the inverse problems in the multiterm TFDEs to our knowledge. Li and Yamamoto [39] studied an inverse problem of identifying the multiple fractional orders in the multiterm TFDE, and they gave the uniqueness result using Laplace transform and analytical method, and later they considered the similar model [42], and also the uniqueness of determining the fractional orders, the number of the fractional terms, and the spatially varying coefficient simultaneously is proved. Recently, Sun et al. [43] considered a simultaneous inversion problem for determining the space-dependent diffusion and source coefficients in the multiterm TFDE using the optimal perturbation regularization algorithm, and quite a few numerical inversions are presented.

Based on the above analysis, we are to deal with the inverse problem of determining the multiple fractional orders in the multiterm TFDE with the additional measurements at the interior point from numerics. The uniqueness results for such kind of inverse problems have been obtained (c.f. [39, 42], e.g.), but numerical inversions are still open to be implemented. Based on the difference solution to the forward problem, we perform numerical inversions by utilizing the homotopy regularization algorithm not only with the accurate data but also with random noisy data. The inversion fractional orders approximate to the exact orders as the noise level gets smaller demonstrating a numerical stability of the inverse problem here.

The rest of the paper is organized as follows. In Section 2, an implicit finite difference solution to the forward problem is given and the inverse problem of determining the fractional orders is formulated. In Section 3, the homotopy regularization algorithm is introduced to solve the inversion problem and numerical inversions are presented, and concluding remarks are given in Section 4.

THE FORWARD PROBLEM AND THE INVERSE PROBLEM

Consider the forward problem given by (1) with the initial condition

$$u(x,0) = u_0(x), \quad x \in \Omega, \tag{4}$$

and the homogeneous Dirichlet boundary condition

$$u|_{\partial\Omega} = 0, \quad 0 < t \le T, \tag{5}$$

where the initial function is smooth enough and satisfies the consistency condition with the boundary condition. Let the diffusion coefficient $D(x)$ be C^1 -class and take positive values on $\overline{\Omega}$; then the forward problem (1), (4)-(5) has a unique solution for suitable initial functions (c.f. [20], e.g.).

Here we focus our attention on the finite difference solution to the forward problem. For completeness of the paper, we give an implicit finite difference scheme in 1D case for solving the forward problem. For further details, see [22, 23], and so on.

The Difference Scheme to the Forward Problem

Let $\Omega = (0, l)$ for $l>0$. For given integer numbers M and N, discretizing the space domain by $x_i = ih$ ($i = 0, 1, \ldots, M$) and the time domain by $t_n = n\tau$ ($n = 0, 1, \ldots, N$), we have by definition (3)

$$\frac{\partial^{\alpha} u}{\partial t^{\alpha}}\left(x_i, t_{n+1}\right) = \frac{\tau^{-\alpha}}{\Gamma(2-\alpha)}$$

$$\cdot \sum_{k=0}^{n} \left[u\left(x_i, t_{n+1-k}\right) - u\left(x_i, t_{n-k}\right)\right]\left[(k+1)^{1-\alpha} - k^{1-\alpha}\right]$$

$$+ O(\tau),$$

$$\frac{\partial^{\beta_s} u}{\partial t^{\beta_s}}\left(x_i, t_{n+1}\right) = \frac{\tau^{-\beta_s}}{\Gamma(2-\beta_s)}$$

$$\cdot \sum_{k=0}^{n} \left[u\left(x_i, t_{n+1-k}\right) - u\left(x_i, t_{n-k}\right)\right]\left[(k+1)^{1-\beta_s}\right.$$

$$\left. - k^{1-\beta_s}\right] + O(\tau), \tag{6}$$

for $s = 1, 2, \ldots, S$, respectively; here $h = l/M$ is the space mesh step and $\tau = T/N$ is the time mesh step.

Let $(x) \in C^1(\bar{\Omega})$ and $D(x) > 0$ for $x \in \bar{\Omega}$. By discretizing the term $(\partial/\partial x)(D(x)u_x)$ using the ordinary integer-order difference method and denoting $u^n{}_i \approx u(x_i, t_n)$ and $D_i = D(x_i)$, we get

$$\left(u_i^{n+1} - u_i^{n}\right)\left[1 + \sum_{s=1}^{S} r_i \tau^{\alpha-\beta_s} \frac{\Gamma(2-\alpha)}{\Gamma(2-\beta_s)}\right]$$

$$+ \sum_{k=1}^{n} \left(u_i^{n+1-k} - u_i^{n-k}\right)\left\{\left[(k+1)^{1-\alpha} - k^{1-\alpha}\right]\right.$$

$$\left. + \sum_{s=1}^{S} r_s \left[(k+1)^{1-\beta_s} - k^{1-\beta_s}\right] \tau^{\alpha-\beta_s} \frac{\Gamma(2-\alpha)}{\Gamma(2-\beta_s)}\right\}$$

$$= \frac{D_i \tau^{\alpha} \Gamma(2-\alpha)}{h^2}\left(u_{i-1}^{n+1} - 2u_i^{n+1} + u_{i+1}^{n+1}\right)$$

$$+ \frac{\left(D_i - D_{i-1}\right)\tau^{\alpha}\Gamma(2-\alpha)}{h^2}\left(u_i^{n+1} - u_{i-1}^{n+1}\right) + R^{n+1}, \tag{7}$$

where $R^n = O(\tau^{\alpha} h^2 + \tau^{\alpha+1})$ is the truncated term. Denoting $V = 1 + \sum_{s=1}^{S} r_s \tau^{\alpha-\beta_s} (\Gamma(2-\alpha)/\Gamma(2-\beta_s))$, dividing by V on two sides of (7), and ignoring the truncated term, we get the following:

$$(-p_i + q_i)\,u_{i-1}^{n+1} + (1 + 2p_i - q_i)\,u_i^{n+1} - p_i u_{i+1}^{n+1}$$
$$= u_i^n - \sum_{k=1}^{n} d_k \left(u_i^{n+1-k} - u_i^{n-k}\right), \tag{8}$$

Where

$$p_i = \frac{D_i \tau^{\alpha} \Gamma(2-\alpha)}{h^2 v},$$
$$q_i = \frac{(D_i - D_{i-1})\,\tau^{\alpha} \Gamma(2-\alpha)}{h^2 v},$$
$$d_k = \frac{(k+1)^{1-\alpha} - k^{1-\alpha}}{v}$$
$$+ \sum_{s=1}^{S} \frac{(k+1)^{1-\beta_s} - k^{1-\beta_s}}{v} r_s \tau^{\alpha-\beta_s} \frac{\Gamma(2-\alpha)}{\Gamma(2-\beta_s)} \tag{9}$$

for k = 1, 2, . . . , n. The initial boundary value conditions are discretized as

$$u_i^0 = u_0(x_i);$$
$$u_0^n = 0,$$
$$u_M^n = 0. \tag{10}$$

Let

$$U^n = (u_1^n, u_2^n, \ldots, u_{M-1}^n);$$
$$U^0 = (u_1^0, u_2^0, \ldots, u_{M-1}^0),$$
$$c_k = \begin{cases} 1 - d_1, & k = 1, \\ d_{k-1} - d_k, & k = 2, \ldots, n; \end{cases} \tag{11}$$

and $B = (b_{ij})_{(M-1)\times(M-1)}$, where b_{ij} is defined by

$$b_{ij} = \begin{cases} 0, & j > i+1, \\ -p_i, & j = i+1, \\ 1 + 2p_i - q_i, & j = i, \\ -(p_i - q_i), & j = i-1, \\ 0, & j < i-1, \end{cases} \tag{12}$$

for $i = 1, 2, \ldots, M-1$, $j = 1, 2, \ldots, M-1$.

Thus, we have the implicit finite difference scheme in the matrix form given as

$$BU^1 = U^0;$$

$$BU^{n+1} = c_1U^n + c_2U^{n-1} + \cdots + c_nU^1 + d_nU^0. \quad (13)$$

Theorem 1. The implicit difference scheme (13) has only one solution, and it is of unconditional stability and convergence for any finite time $T>0$.

Proof. By the assumptions for the diffusion coefficient (x), we have $p_i > 0$ and $p_i - q_i > 0$ for $i = 1, \ldots, M - 1$. So the coefficient matrix B of (13) is strictly diagonally dominant; then the difference equation (13) has only one solution. Moreover, thanks to the equality $\sum^n{}_{k=1} c_k + d_n = 1$ and with a similar method as used in [22], we get the unconditional stability and convergence of the difference scheme for any given finite time $T>0$.

The Inverse Problem

For the forward problem (1), (4)- (5), if the fractional orders α and β_s ($s = 1, \ldots, S$) in (1) are unknown, we encounter the inverse problem of determining these multiple fractional orders. Suppose that there are some measured points in the space domain; for example, let $x_0 \in \Omega$ be the measured point, and we have the additional information given as

$$u(x_0, t) = \theta(t), \quad 0 < t \leq T, \quad (14)$$

and the inverse problem is to determine $\alpha \in (0, 1)$ and $\beta_s \in (0, 1)$ ($s = 1, \ldots, S$) using the overposed condition (14) based on (1) and the initial boundary conditions (4) and (5).

As stated in Section 1, the general uniqueness results have been proved in [39, 42] by Dr. Li and Professor Yamamoto. We give the following lemma for the completeness of the paper.

Lemma 2 (see [39]). Suppose that the fractional orders α and β_s ($S = 1, \ldots, S$), the diffusion coefficient tensor $D(x)$, and the coefficients r_s ($s = 1, \ldots, S$) in (1) satisfy the conditions given in Section 1, and the initial function $u0(x)$ is smooth enough, and $u_0(x) \geq 0$ and $u0(x) \equiv 0$ for $x \in \Omega$. Then all the fractional orders can be determined uniquely by the additional data $\{(x0, t) = \theta(t)\}$ for $t \in (0, T)$.

The uniqueness result is very important for inverse problems in theory. By Lemma 2 we need to utilize the additional data (t_i) measured at x_0 for $i = 1, 2, \ldots$; however, in concrete computations, we find that the numerical inversions can also be performed only employing a few of the additional data. So it is still meaningful to study inverse problems from numerics.

THE INVERSION ALGORITHM AND NUMERICAL INVERSIONS

In this section, we present numerical inversions for the inverse problem of (1), (4)-(5) with (14). The inversion algorithm we utilize is the homotopy regularization algorithm (see [43, 44], e.g.), which is a combination of the homotopy method with the optimal perturbation algorithm. We give a sketch for the inversion algorithm in the following.

The Homotopy Regularization Algorithm

For the fractional orders α and β_s ($s = 1, 2, \ldots, S$) satisfying the order condition (2), let $I = (0, 1) \times (0, 1) \times \cdots \times (0, 1) \subset R^{S+1}$. We denote a vector

$$\mathbf{a} = (\alpha, \beta_1, \beta_2, \ldots, \beta_S) \in I \tag{15}$$

and equip the Euclidean norm

$$\|\mathbf{a}\| = \sqrt{\alpha^2 + \beta_1^2 + \cdots + \beta_S^2}. \tag{16}$$

For any given a $\in I$, denote (a)(x, t) as the unique solution to the forward problem. Combined with the additional information (14), solving the inverse problem is equivalent to a minimization problem

$$\min_{\mathbf{a}\in I} \|u(\mathbf{a})(x_0, t) - \theta(t)\|^2_{L^2(0,T)}, \tag{17}$$

where $\|u(\mathbf{a})(x_0,t) - \theta(t)\|^2_{L^2(0,T)} = \int_0^T [u(\mathbf{a})(x_0,t) - \theta(t)]^2 dt$.

Following the homotopy regularization idea, it turns out that the following minimization problem is solved instead of (17):

$$\min_{\mathbf{a}\in I} \left\{(1-\mu)\|u(\mathbf{a})(x_0, t) - \theta(t)\|^2_{L^2(0,T)} + \mu\|\mathbf{a}\|^2\right\}, \tag{18}$$

where $\mu \in (0, 1)$ is the homotopy parameter which takes values from near 1 decreasingly approximating to 0. By linearization and numerical differentiation approximations, solving (18) is transformed to solve a normal equation combined with an iteration process via

$$\left(\mu E + (1-\mu)\mathbf{G}^T\mathbf{G}\right)\delta\mathbf{a}_j = (1-\mu)\mathbf{G}^T(\eta - \xi),$$

$$\mathbf{a}_j + \delta\mathbf{a}_j \longrightarrow \mathbf{a}_{j+1}, \quad j = 0, 1, \ldots, \tag{19}$$

where δa_j is a perturbation vector for any given $a_j \in I$, j denotes the iterative number, and a_0 is the initial iteration, and

$$\mathbf{G} = \left(g_{qi}\right)_{Q*(S+1)},$$

$$g_{qi} = \frac{u\left(\mathbf{a}_j + \tau \mathbf{e}_i\right)\left(x_0, t_q\right) - u\left(\mathbf{a}_j\right)\left(x_0, t_q\right)}{\tau}; \tag{20}$$

here τ is the numerical differential step and ei ($i = 1, 2, \ldots, S+1$) is the unit basis vector in R^{S+1}, and

$$\xi = \left(u\left(\mathbf{a}_j\right)\left(x_0, t_1\right), \ldots, u\left(\mathbf{a}_j\right)\left(x_0, t_Q\right)\right)^T;$$

$$\eta = \left(\theta\left(t_1\right), \ldots, \theta\left(t_Q\right)\right)^T. \tag{21}$$

The algorithm can be terminated as long as an optimal perturbation satisfies the condition $\|a_j\|2 \leq$ eps; here eps is the given convergent precision. Like that done in [43, 44], we employed the homotopy parameter by

$$\mu = \mu(j) = \frac{1}{1 + \exp\left(\sigma\left(j - j_0\right)\right)}, \tag{22}$$

where j is the number of iterations, j_0 is a preestimated number, and $\sigma>0$ is an adjusted parameter

In the following, we will take two-term case (S=1) and three-term case (S=2) as example to perform the inversion algorithm in 1D case. We set the model parameters l=π, T = 1, and D=1 and $u_0(x)$ = sin(x) if there is no specification. In addition, we choose j_0 = 5 and σ = 0.5 in (22) to determine the homotopy parameter and utilize the grid steps h = l/100 and τ = T/100 in the computation of the forward problem by the difference scheme (13).

Inversion for S = 1

In the case of S=1, let $r1$ = 0.5, and the fractional orders we have to determine are α and β_1, and a = (α, β_1)is the exact solution of the fractional orders. By the exact fractional orders, we compute the forward problem to get the additional data at x_0 = l/2 with which we reconstruct the fractional orders. It is noticeable that the dimension of the additional data is related to the discretization of the time domain; that is, there are $u(l/2, t_i)$ for i = 0, 1, . . . , N; here t_i is the time at which the measurement is made, and we have a

N+1-dimensional vector representing the additional data given as

$$\left(u\left(\frac{l}{2},t_0\right),u\left(\frac{l}{2},t_1\right),\ldots,u\left(\frac{l}{2},t_N\right)\right). \tag{23}$$

However, the inversion could be performed using a few of the additional data, and it is meaningful to investigate the influence of the number (dimension) of the additional data on the inversion algorithm.

Inversion with Accurate Data

Taking two additional datasets at the time of $t = 0.3$ and $t = 0.4$ and choosing the initial iteration as $a_0 = 0$, the inversion results are listed in Table 1, where ainv is the inversion solution, Err = $\|a - a^{inv}\|/\|a\|$ denotes the solutions error, and j is the number of iterations.

Table 1. The inversion results with different fractional orders for S=1.

a	a^{inv}	Err	j
(0.9, 0.1)	(0.900000, 0.100000)	$1.7583e-9$	30
(0.8, 0.2)	(0.800000, 0.200000)	$1.2048e-9$	30
(0.7, 0.3)	(0.700000, 0.300000)	$1.5710e-8$	30
(0.6, 0.4)	(0.600000, 0.400000)	$2.0909e-8$	32
(0.4, 0.3)	(0.400000, 0.300000)	$5.9200e-8$	34
(0.3, 0.2)	(0.300000, 0.200000)	$1.9477e-8$	34
(0.2, 0.1)	(0.200000, 0.100000)	$1.5804e-7$	33

Without loss of generality, choosing the exact fractional orders as $\alpha = 0.8$ and $\beta_1 = 0.4$, we implement the inversion algorithm using different number of the additional data. The inversion results are listed in Table 2, where L denotes the number (dimension) of the additional data and $[t]$ denotes the measured time to get the additional data.

Table 2. The inversion results with number of the additional data for S=1.

L	$[t]$	a^{inv}	Err	j
2	[0.3, 0.4]	(0.800000, 0.400000)	$2.569e-8$	30
3	[0.3, 0.4, 0.5]	(0.800000, 0.400000)	$3.288e-9$	28
4	[0.3, 0.4, 0.5, 0.6]	(0.800000, 0.400000)	$1.042e-8$	26
5	[0.3, 0.4, 0.5, 0.6, 0.7]	(0.800000, 0.400000)	$3.712e-9$	25
6	[0.3, 0.4, 0.5, 0.6, 0.7, 0.8]	(0.800000, 0.400000)	$3.448e-9$	24
7	[0.2, 0.3, 0.4, 0.5, 0.6, 0.7, 0.8]	(0.800000, 0.400000)	$6.181e-9$	23
8	[0.2, 0.3, 0.4, 0.5, 0.6, 0.7, 0.8, 0.9]	(0.800000, 0.400000)	$1.885e-8$	22
9	[0.1, 0.2, 0.3, 0.4, 0.5, 0.6, 0.7, 0.8, 0.9]	(0.800000, 0.400000)	$9.076e-9$	22

From Table 1, it can be seen that the inversion can be realized only utilizing two additional datasets, and the inversion solutions are good approximations to the exact solutions. From Table 2, we can see that there are some influences for choosing the number of the additional data on the inversion algorithm, and the inversion results become a little better as the number of the additional data increases and the number of the iterations decreases. Moreover, if using much more additional data in the above, for example, taking all of the measured data given by (23), the inversion solution is still $a^{inv} = (0.800000, 0.400000)$, and the solutions error is Err = $2.4636e-8$, but the number of iterations decreases to $j = 17$.

Inversion with Noisy Data

It is difficult to perform an inversion algorithm in the case of using random noisy data, especially for inverse problems arising from the fractional diffusion. Noting computational errors and data noises, the additional information utilized for real inversions is often given as

$$\theta^{\varepsilon}(t) = \theta(t) + \varepsilon\zeta, \quad (24($$

where $\varepsilon>0$ is the noise level and ζ is a random vector ranging within $[-1, 1]$.

Also we take the exact fractional orders as $\alpha = 0.8$ and $\beta = 0.4$; that is, the exact solution of the inverse problem here is given as a = (0.8, 0.4). It is noticeable that the inversion with data noises could always fail if still using the above inversion parameters, and we perform the inversion using the completely additional data at ti for $i = 0, 1, . . . , N$ and choosing the initial iteration as nonzero vector of $a_0 = (0.1, 0.1)$. The average inversion results with continuous tentime inversions are listed in Table 3, where $\overline{a}^{inv}$ is the average inversion solution of the ten-time inversions and $Err = \| \overline{a}^{inv} - a\|_2/\|a\|_2$ is the relative average error in the solutions.

Table 3. The inversion results using noisy data for S=1.

ε	$\overline{a}^{inv}$	$\overline{Err}$	$\overline{j}$
5%	(0.789840, 0.422128)	2.7223e − 2	14.6
1%	(0.799757, 0.401140)	1.3032e − 3	14.3
0.1%	(0.799990, 0.400094)	1.0588e − 4	13.9
0.01%	(0.799999, 0.400009)	1.0086e − 5	15.0
0.001%	(0.800000, 0.400001)	9.6878e − 7	16.0

From Table 3, we find that the inversion is satisfactory in the case of using random noisy data, and the inversion errors become small when reducing the noise level.

Inversion for S=2

In the case of S=2, the fractional orders we are to determine are α, β_1 and β_2, and $a = (\alpha, \beta_1, \beta_2)$ is the exact solution in this case. Let $r_1 = 1$, $r_2 = 0.2$, and also choose $x_0 = l/2$ as the measured point. We utilize nine additional datasets at the time of $t_i = i/10$ for $i = 1, 2, . . . , 9$ and choose the initial iteration as a0 = (0.1, 0.05, 0.01) here. The inversion results with accurate data are listed in Table 4.

Table 4. The inversion results with different fractional orders for S=2.

a	a^{inv}	Err	j
(0.9, 0.7, 0.5)	(0.900000, 0.700000, 0.500000)	$1.0216e-9$	41
(0.8, 0.6, 0.4)	(0.800000, 0.600000, 0.400000)	$4.5193e-9$	40
(0.7, 0.5, 0.3)	(0.700000, 0.500000, 0.300000)	$4.3115e-8$	39
(0.6, 0.5, 0.4)	(0.600000, 0.500000, 0.400000)	$2.9329e-9$	45
(0.5, 0.4, 0.3)	(0.500000, 0.400000, 0.300000)	$2.4011e-9$	45
(0.4, 0.3, 0.2)	(0.400000, 0.300000, 0.200000)	$8.8878e-9$	45
(0.3, 0.2, 0.1)	(0.300000, 0.200000, 0.100000)	$9.9418e-8$	45

As done for *S*=1, we can also realize the inversion in the case of using noisy data for *S*=2 although the illposedness of the inversion becomes more serious than that of *S*=1. Taking a = (0.9, 0.7, 0.5) as an example and using a0 = (0.3, 0.2, 0.1) as the initial iteration, the average results also with continuous ten-time inversions are listed in Table 5, where ε, $\overline{a}^{inv}$, and *E*rr are all the same as those used in Table 3. It is important to note that we perform the inversion algorithm with convergent precision as eps = $1e{-}2$ for ε=1% and ε = 0.1% and eps = $1e-4$ for ε = 0.01% and ε = 0.001%, respectively.

Table 5. The inversion results using noisy data for S=2.

ε	$\overline{a}^{inv}$	$\overline{Err}$
1%	(0.853161, 0.762050, 0.411797)	$9.4438e-2$
0.1%	(0.878925, 0.716188, 0.520835)	$2.7124e-2$
0.01%	(0.900086, 0.699732, 0.500985)	$8.2268e-4$
0.001%	(0.900007, 0.699989, 0.500042)	$3.4895e-5$

From Tables 3 and 5, we can see that the inversion solutions with random noisy data approximate to the exact solutions as the noise level gets smaller. Although the inversion results for *S*=2 with large noises are not so good as that for *S*=1, they are still satisfactory and show a numerical stability for the inverse problem.

CONCLUSIONS

The numerical determination problem for the fractional orders in the multiterm TFDE is investigated using some measurements at the interior point of the space domain. The inversion problem is unique, and numerical inversions with random noisy data are performed successfully by using the homotopy regularization algorithm. The homotopy regularization algorithm can also be utilized to determine the multiple fractional orders in the multidimensional case as long as a solution to the forward problem can be worked out.

It is noted that numerical inversions can be implemented smoothly in the case that the fractional orders satisfy the order condition (2) given in Section 1, which is just a necessary condition for the uniqueness of the inverse problem. The inversion results are very satisfactory if coping with accurate data; however, they become a little bad in the case of $S=2$ when using noisy data with noises greater than 1%.Therefore, we can say that the stability of the inverse problem here with more fractional derivatives, that is, $S\geq 2$, could be severely ill-posed in spite of the fact that uniqueness is valid. We will deal with the inverse problem in the multidimensional case and we have to seek more effective inversion algorithms for the multiterm TFDE with $S>2$ in the future work.

CONFLICTS OF INTEREST

The authors declare that there are no conflicts of interest regarding the publication of this paper.

ACKNOWLEDGMENTS

This work is supported by the National Natural Science Foundation of China (nos. 11371231, 11071148).

REFERENCES

1. E. E. Adams and L. W. Gelhar, "Field study of dispersion in a heterogeneous aquifer: 2. Spatial moments analysis," Water Resources Research, vol. 28, no. 12, pp. 3293–3307, 1992.
2. D. A. Benson, The Fractional Advection-Dispersion Equation: Development and Application. Dissertation of Doctorial Degree [Dissertation, thesis], University of Nevada, Reno, Reno. USA, 1998.
3. B. Berkowitz, H. Scher, and S. E. Silliman, "Anomalous transport in laboratory-scale, heterogeneous porous media," Water Resources Research, vol. 36, no. 1, pp. 149–158, 2000.
4. R. Caponetto, G. Dongola, L. Fortuna, and I. Petras, Fractional Order Systems: Modeling and Control Applications, World Scientific, Singapore, 2010.
5. M. Giona, S. Cerbelli, and H. E. Roman, "Fractional diffusion equation and relaxation in complex viscoelastic materials," Physica A: Statistical Mechanics and Its Applications, vol. 191, no. 1–4, pp. 449–453, 1992.
6. Y. Hatano and N. Hatano, "Dispersive transport of ions in column experiments: An explanation of long-tailed profiles," Water Resources Research, vol. 34, no. 5, pp. 1027–1033, 1998.
7. F. Mainardi, Fractional Calculus and Waves in Linear Viscoelasticity: An Introduction to Mathematical Models, Imperial College Press, London, UK, 2010.View at: Publisher Site
8. L. Zhou and H. M. Selim, "Application of the fractional advection-dispersion equation in porous media," Soil Science Society of America Journal, vol. 67, no. 4, pp. 1079–1084, 2003.
9. C. F. Coimbra, "Mechanics with variable-order differential operators," Annalen der Physik, vol. 12, no. 11-12, pp. 692–703, 2003.
10. C. F. Lorenzo and T. T. Hartley, "Variable order and distributed order fractional operators," Nonlinear Dynamics, vol. 29, no. 1-4, pp. 57–98, 2002.
11. H. T. Pedro, M. H. Kobayashi, J. M. Pereira, and C. F. Coimbra, "Variable order modeling of diffusive-convective effects on the oscillatory flow past a sphere," Journal of Vibration and Control, vol. 14, no. 9-10, pp. 1659–1672, 2008.
12. I. Podlubny, Fractional Differential Equations, vol. 198, Academic Press, San Diego, Calif, USA, 1999.View at: MathSciNet

13. A. A. Kilbas, H. M. Srivastava, and J. J. Trujillo, Theory and Applications of Fractional Differential Equations, New York, NY, USA, Elsevier, 2006.View at: MathSciNet
14. V. Daftardar-Gejji and S. Bhalekar, "Boundary value problems for multi-term fractional differential equations," Journal of Mathematical Analysis and Applications, vol. 345, no. 2, pp. 754–765, 2008.
15. Y. Luchko, "Maximum principle for the generalized time-fractional diffusion equation," Journal of Mathematical Analysis and Applications, vol. 351, no. 1, pp. 218–223, 2009.
16. Y. Luchko, "Initial-boundary problems for the generalized multi-term time-fractional diffusion equation," Journal of Mathematical Analysis and Applications, vol. 374, no. 2, pp. 538–548, 2011.
17. H. Jiang, F. Liu, I. Turner, and K. Burrage, "Analytical solutions for the multi-term time-space Caputo-Riesz fractional advection-diffusion equations on a finite domain," Journal of Mathematical Analysis and Applications, vol. 389, no. 2, pp. 1117–1127, 2012.
18. X.-L. Ding and Y.-L. Jiang, "Analytical solutions for the multi-term time-space fractional advection-diffusion equations with mixed boundary conditions," Nonlinear Analysis. Real World Applications. An International Multidisciplinary Journal, vol. 14, no. 2, pp. 1026–1033, 2013.
19. X.-L. Ding and J. J. Nieto, "Analytical solutions for the multi-term time-space fractional reaction-diffusion equations on an infinite domain," Fractional Calculus and Applied Analysis, vol. 18, no. 3, pp. 697–716, 2015.
20. Z. Li, Y. Liu, and M. Yamamoto, "Initial-boundary value problems for multi-term time-fractional diffusion equations with positive constant coefficients," Applied Mathematics and Computation, vol. 257, pp. 381–397, 2015.
21. B. Jin, R. Lazarov, Y. Liu, and Z. Zhou, "The Galerkin finite element method for a multi-term time-fractional diffusion equation," Journal of Computational Physics, vol. 281, pp. 825–843, 2015.
22. G. Li, C. Sun, X. Jia, and D. Du, "Numerical solution to the multi-term time fractional diffusion equation in a finite domain," Numerical Mathematics. Theory, Methods and Applications, vol. 9, no. 3, pp. 337–357, 2016.

23. F. Liu, M. M. Meerschaert, R. J. McGough, P. Zhuang, and Q. Liu, "Numerical methods for solving the multi-term time-fractional wave-diffusion equation," Fractional Calculus and Applied Analysis, vol. 16, no. 1, pp. 9–25, 2013.
24. D. A. Murio, "Stable numerical solution of a fractional-diffusion inverse heat conduction problem," Computers & Mathematics with Applications, vol. 53, no. 10, pp. 1492–1501, 2007.
25. J. J. Liu and M. Yamamoto, "A backward problem for the time-fractional diffusion equation," Applicable Analysis. An International Journal, vol. 89, no. 11, pp. 1769–1788, 2010.
26. J. J. Liu, M. Yamamoto, and L. L. Yan, "On the reconstruction of unknown time-dependent boundary sources for time fractional diffusion process by distributing measurement," Inverse Problems, vol. 32, no. 1, article 015009, Article ID 015009, 2016.
27. K. Sakamoto and M. Yamamoto, "Initial value/boundary value problems for fractional diffusion-wave equations and applications to some inverse problems," Journal of Mathematical Analysis and Applications, vol. 382, no. 1, pp. 426–447, 2011.
28. V. K. Tuan, "Inverse problem for fractional diffusion equation," Fractional Calculus and Applied Analysis. An International Journal for Theory and Applications, vol. 14, no. 1, pp. 31–55, 2011.
29. G. Chi, G. Li, and X. Jia, "Numerical inversions of a source term in the FADE with a Dirichlet boundary condition using final observations," Computers & Mathematics with Applications, vol. 62, no. 4, pp. 1619–1626, 2011.
30. M. Yamamoto and Y. Zhang, "Conditional stability in determining a zeroth-order coefficient in a half-order fractional diffusion equation by a Carleman estimate," Inverse Problems, vol. 28, no. 10, article 105010, 2012.
31. Y. Luchko, W. Rundell, M. Yamamoto, and L. Zuo, "Uniqueness and reconstruction of an unknown semilinear term in a time-fractional reaction-diffusion equation," Inverse Problems, vol. 29, no. 6, article 065019, Article ID 065019, 2013.
32. T. Wei and J. Wang, "A modified quasi-boundary value method for an inverse source problem of the time-fractional diffusion equation," Applied Numerical Mathematics, vol. 78, pp. 95–111, 2014.

33. T. Wei, X. L. Li, and Y. S. Li, "An inverse time-dependent source problem for a time-fractional diffusion equation," Inverse Problems, vol. 32, no. 8, article 085003, Article ID 085003, 2016.
34. Y. Liu, W. Rundell, and M. Yamamoto, "Strong maximum principle for fractional diffusion equations and an application to an inverse source problem," Fractional Calculus and Applied Analysis, vol. 19, no. 4, pp. 888–906, 2016.
35. B. Jin and W. Rundell, "A tutorial on inverse problems for anomalous diffusion processes," Inverse Problems, vol. 31, no. 3, article 035003, 2015.
36. A. N. Bondarenko and D. S. Ivaschenko, "Numerical methods for solving inverse problems for time fractional diffusion equation with variable coefficient," Journal of Inverse and Ill-Posed Problems, vol. 17, no. 5, pp. 419–440, 2009.
37. J. Cheng, J. Nakagawa, M. Yamamoto, and T. Yamazaki, "Uniqueness in an inverse problem for a one-dimensional fractional diffusion equation," Inverse Problems. An International Journal on the Theory and Practice of Inverse Problems, Inverse Methods and Computerized Inversion of Data, vol. 25, article 115002, 2009.
38. G. Li, D. Zhang, X. Jia, and M. Yamamoto, "Simultaneous inversion for the space-dependent diffusion coefficient and the fractional order in the time-fractional diffusion equation," Inverse Problems, vol. 29, no. 6, article 065014, 2013.
39. Z. Li and M. Yamamoto, "Uniqueness for inverse problems of determining orders of multi-term time-fractional derivatives of diffusion equation," Applicable Analysis, vol. 94, no. 3, pp. 570–579, 2015.
40. S. Tatar and S. Ulusoy, "A uniqueness result for an inverse problem in a space-time fractional diffusion equation," Electronic Journal of Differential Equations, vol. 2013, no. 258, pp. 1–9, 2013.
41. S. Chen, F. Liu, X. Jiang, I. Turner, and K. Burrage, "Fast finite difference approximation for identifying parameters in a two-dimensional space-fractional nonlocal model with variable diffusivity coefficients," SIAM Journal on Numerical Analysis, vol. 54, no. 2, pp. 606–624, 2016.
42. Z. Li, O. Y. Imanuvilov, and M. Yamamoto, "Uniqueness in inverse boundary value problems for fractional diffusion equations," Inverse Problems, vol. 32, no. 1, article 015004, 2016.

43. C. Sun, G. Li, and X. Jia, "Simultaneous inversion for the diffusion and source coefficients in the multi-term TFDE," Inverse Problems in Science and Engineering, pp. 1–21, 2017.
44. D. Zhang, G. Li, X. Jia, and H. Li, "Simultaneous inversion for space-dependent diffusion coefficient and source magnitude in the time fractional diffusion equation," Journal of Mathematics Research, vol. 5, no. 2, pp. 65–78, 2013.

CHAPTER

7

Solutions of Direct and Inverse Even-Order Sturm-Liouville Problems Using Magnus Expansion

Upeksha Perera [1,3] and Christine Böckmann [2]

[1]Department of Mathematics, University of Kelaniya, Kelaniya 11600, Sri Lanka

[2]Institut für Mathematik, Universität Potsdam, 14476 Potsdam, Germany

[3]Current address: Institut für Mathematik, Universität Potsdam, 14476 Potsdam, Germany.

ABSTRACT

In this paper Lie group method in combination with Magnus expansion is utilized to develop a universal method applicable to solving a Sturm–Liouville problem (SLP) of any order with arbitrary boundary conditions. It is

Citation: Perera, U., & Böckmann, C. (2019). Solutions of Direct and Inverse Even-Order Sturm-Liouville Problems Using Magnus Expansion. Mathematics, 7(6), 544. (24 pages) DOI: https://doi.org/10.3390/math7060544

shown that the method has ability to solve direct regular (and some singular) SLPs of even orders (tested for up to eight), with a mix of (including non-separable and finite singular endpoints) boundary conditions, accurately and efficiently. The present technique is successfully applied to overcome the difficulties in finding suitable sets of eigenvalues so that the inverse SLP problem can be effectively solved. The inverse SLP algorithm proposed by Barcilon (1974) is utilized in combination with the Magnus method so that a direct SLP of any (even) order and an inverse SLP of order two can be solved effectively.

Keywords: higher-order Sturm–Liouville problems; inverse Sturm–Liouville problems; Magnus expansion

INTRODUCTION

Direct and inverse eigenvalue problems (EVP) of linear differential operators play an important role in all vibration problems in engineering and physics [**1**]. The theory of direct Sturm–Liouville problems (SLP) started around the 1830s in the independent works of Sturm and Liouville.The inverse Sturm–Liouville theory originated in 1929 [**2**].

In this paper, we consider the 2mth order, nonsingular, self-adjoint eigenvalue problem:

$$\begin{aligned}&(-1)^m(p_m(x)y^{(m)})^{(m)}+(-1)^{m-1}(p_{m-1}(x)y^{(m-1)})^{(m-1)}\\&+\ldots+(p_2(x)y'')''-(p_1(x)y')'+p_0(x)y=\lambda w(x)y,\qquad a<x<b\end{aligned} \tag{1}$$

along with y satisfying some general, separated boundary conditions at a and b (see Equation (2)). Usually, the functions p_k,($0\leq k\leq m$), and w(x) are continuous on the finite closed interval [a,b], and pm has continuous derivative. Further assumptions A1–A4 are placed on the coefficient functions in Section 2.1. For Equation (1), the direct eigenvalue problems is concerned with determining the λ given the coefficient information p_k, ($0\leq k\leq m$), and the inverse eigenvalue problem is concerned with reconstructing the unknown coefficient functions p_k, ($0\leq k\leq m$) from the knowledge of suitable spectral data satisfying the equation.

There is a variety of numerical solutions for the simplest case of Equation (1) known as the direct Sturm–Liouville problem with m=1, the most notable ones being Finite difference method (FDM) [3], Numerov's method (NM) [4], Finite element method (FEM) [5], Modified Numerov's method (MNM) [6], FEM with trigonometric hat functions using Simpson's rule (FEMS) and using trapezoidal rule (FEMT) [7], Lie-Group methods

(LGM) [8], MATSLISE [9], Magnus integrators (MI) [10], and Homotopy Perturbation Method (HPM) [11].

For a discussion on numerical methods for inverse SLP, refer to Freiling and Yurko [12]. McLaughlin [13] provided an overview of analytical methods for second- and fourth-order inverse problems. For the inverse SLP, some available methods are iterative solution (IS) [14] and Numerov's method (NM) [15].

Slightly fewer techniques are available for the direct eigenvalue problem of case m=2 known as fourth-order Sturm–Liouville problem (FSLP). For example, Adomian decomposition method (ADM) [16], Chebyshev spectral collocation method (CSCM) [17], Homotopy Perturbation Method (HPM) [11], homotopy analysis method (HAM) [18], Differential quadrature method (DQM) and Boubaker polynomials expansion scheme (BPES) [19], Chebychev method (CM) [20], Spectral parameter power series (SPPS) [21], Chebyshev differentiation matrices (CDM) [22], variational iteration method (VIM) [23], Matrix methods (MM) [24], and Lie Group method [25] are the prominent techniques available.

Among others Barcilon [**26**], and McLaughlin [**27**,**28**] proposed numerical algorithms for solving the inverse FSLP.

Several methods are available for m=3, or Sixth-order SLP (SSLP). They are shooting method [**29**], Chebyshev spectral collocation method (CSCM) [**17**], Adomian decomposition method (ADM) [**30**], and Chebyshev polynomials (CP) [**31**].

However, as the order of the direct problem increases, it becomes much harder to solve and also the accuracy of the eigenvalues decreases with the index since only a small portion of numerical eigenvalues are reliable [**32**].

All of the above-mentioned methods are applicable only to a particular order of the SLP and specific set(s) of boundary conditions.

In contrast, in this paper, the Lie group method in combination with the Magnus expansion is utilized to construct a method applicable to solving any order of SLP with arbitrary boundary conditions, which was initially applied for solving SLP by Moan [**8**] and Ledoux et al. [**10**] and FSLP by Mirzaei [**25**].

The proposed method is universal in the sense that it can be applied to solve a SLP of any (even) order with different types of boundary conditions (including some singular) subject to computational feasibility inherent to large matrix computations.

The inverse problem is much harder as restrictions on the spectral data should be placed to ensure the uniqueness. Barcilon [**33**] proved that, in general, m+1 spectra associated with m+1 distinct "admissible" boundary conditions are required to determine p_m's uniquely (with p_m=w=1). Later, McLaughlin and Rundell [**34**] proved that the measurement of a particular eigenvalue for an infinite set of different boundary conditions is sufficient to determine the unknown potential. Both these theorems an infinite number of accurate eigenvalue measurements, which are hard to obtain in practice as the higher eigenvalues are usually more expensive to compute than lower ones [**35**]. However, the present technique can be applied to address some of the above difficulties in finding suitable set(s) of eigenvalues so that the inverse problem can be effectively solved.

Section 2 describes the method constructed using the Lie group method in combination with the Magnus expansion and **Section 3** provides some numerical examples of direct Sturm–Liouville problems of different orders m=1,2,3,4 with a variety of (including one singular) boundary conditions. Furthermore, the efficiency and accuracy of the proposed technique is illustrated, using Example 1. **Section 4** presents two examples of Inverse SLP of order two with a symmetric and a non-symmetric potential. Finally, **Section 5** concludes with discussion and a summary.

MATERIALS AND METHODS

Notations

In this paper, Equation (**1**) is considered under the assumptions:

A1: All coefficient functions are real valued.

A2: Interval (a,b) is finite.

A3: The coefficient functions p_k, $(0 \leq k \leq m-1)$, w and $1/p_m$ are in $L^1(a,b)$.

A4: Infima of pm and *w* are both positive.

(The last two assumptions, A3 and A4, are relaxed in Example 3.) By Greenberg and Marletta [**29**], provided the above assumptions are true, one gets:

R1: The eigenvalues are bounded below.

R2: The eigenvalues can be ordered: $\lambda_0 \leq \lambda_1 \leq \lambda_2 \leq \ldots$.

R3: $\lim_{k\to\infty} \lambda_k = +\infty$.

Defining quasi-derivatives:

$$
\begin{aligned}
&u_k = y^{(k-1)}, \qquad 1 \le k \le m,\\
&v_1 = (-1)^m p_m y^{(m)},\\
&v_2 = (-1)^m (p_m y^{(m)})' + (-1)^{m-1}(p_{m-1} y^{(m-1)}),\\
&\quad\vdots\\
&v_k = (-1)^m (p_m y^{(m)})^{(k-1)} + (-1)^{m-1}(p_{m-1} y^{(m-1)})^{(k-2)} + \ldots + (-1)^{m-k+1}(p_{m-1} y^{(m-k+1)}),\\
&\quad\vdots\\
&v_2 = p_2 y'' - (p_3 y''')' + (p_4 y^{(4)})'' + \ldots + (-1)^{m-2}(p_m y^{(m)})^{(m-2)},\\
&\quad\vdots\\
&v_m = (-1)^{m-1}(p_m y^{(m)})^{(m-1)} + \ldots - (p_3 y''')'' + (p_2 y'')' - p_1 y',
\end{aligned}
$$

the general, separated, self-adjoint boundary conditions can be written in the form [**29**]:

$$A_1 u(a) + A_2 v(a) = 0, \qquad B_1 u(b) + B_2 v(b) = 0, \tag{2}$$

where

1. A_1,A_2,B_1,B_2 are m×m real matrices;
2. $A_1A^T_2=A_2A^T_1$ and $B_1B^T2=B_2B^T_1$; and
3. m×2m matrices $(A_1:A_2)$ and $(B_1:B_2)$ have rank *m*.

Equation (**1**) in the matrix form is:

$$U'=G(x)U \tag{3}$$

$$AU(a)+BU(b)=0 \tag{4}$$

with

$$
G(x) = \left(\begin{array}{ccccc|ccccc}
0 & 1 & 0 & 0 & 0 & 0 & 0 & 0 & 0 & 0\\
0 & 0 & 1 & 0 & 0 & 0 & 0 & 0 & 0 & 0\\
0 & 0 & 0 & \ddots & 0 & 0 & 0 & 0 & 0 & 0\\
0 & 0 & 0 & 0 & 1 & 0 & 0 & 0 & 0 & 0\\
0 & 0 & 0 & 0 & 0 & 1/p_m(x) & 0 & 0 & 0 & 0\\
\hline
0 & 0 & 0 & 0 & p_{m-1}(x) & 0 & 1 & 0 & 0 & 0\\
0 & 0 & 0 & \cdot^{\cdot^{\cdot}} & 0 & 0 & 0 & 1 & 0 & 0\\
0 & 0 & p_2(x) & 0 & 0 & 0 & 0 & 0 & \ddots & 0\\
0 & p_1(x) & 0 & 0 & 0 & 0 & 0 & 0 & 0 & 1\\
\lambda w(x) - p_0(x) & 0 & 0 & 0 & 0 & 0 & 0 & 0 & 0 & 0
\end{array}\right)_{2m\times 2m} \tag{5}
$$

$$U = [u_1, u_2, \ldots, u_m, v_1, v_2, \ldots, v_m],$$

$$A = \begin{pmatrix} A_1 & A_2 \\ 0_m & 0_m \end{pmatrix}, \qquad B = \begin{pmatrix} 0_m & 0_m \\ B_1 & B_2 \end{pmatrix}.$$

Lie Algebra

The Lie group (G,·) was introduced by Lie [**36**] as a differentiable manifold with the algebraic structure of a group. The tangent space at identity is called the Lie algebra of G and denoted by g. A binary operation $[\cdot,\cdot]:g\times g\rightarrow g$ is called a Lie bracket. Let $x,y\in g$, then the Lie bracket $[x,y]$ is defined as:

$$[x,y]=xy-yx. \quad (6)$$

Special Lie group is defined as

$$SL(F,n)=\{y\in F^{n\times n},\det(y)=1\}. \quad (7)$$

A system of linear ordinary differential equations on SL(F,n) has the form [**37**]:

$$y'=G(x)y,\mathrm{tr}(G(x))=0\Leftrightarrow\det(y)=1. \quad (8)$$

Letting $U(x)=Y(x)U(a)$, where $U(a)$ represents the initial data, $Y(x)$ satisfies the differential equation

$$Y'(x)=G(x)Y(x),Y(a)=I \quad (9)$$

where I stands for the n−dimensional identity matrix [**38**]. Since $\mathrm{tr}(G(x))=0$, we have $\det Y(x)=1$, so that Equation (**8**) is on the Lie Group SL(R4,4), hence can be solved by the Magnus expansion. Let $Y(x)$ be a solution of Equation (**9**) for $x\geq a$; the solution of the system in Equation (**3**) with initial condition $U(a)$ is

$$U(x)=Y(x)U(a).$$

Thus, $U(b)=Y(b)U(a)$ and using boundary conditions in Equation (**4**):

$$AU(a)+BY(b)U(a)=0\Rightarrow[A+BY(b)]U(a)=0.$$

Here, $U(a)\neq0$. Hence, the eigenvalues λ of the SLP in Equation (**1**) are the roots of the characteristic equation:

$$F(\lambda)=\det(A+BY(b))=0. \quad (10)$$

As the problem in Equation (**1**) is self-adjoint, the roots of Equation (**10**) are simple [**14**], hence can be calculated using the bisection method or any other root finding method. Although an explicit expression for $F(\lambda)$ cannot be found, we can find the roots by computing $F(\lambda)$ for a real number λ. Besides, one can plot the function $F(\lambda)$ to get an idea of the locations of the roots, and to specify a smaller bracket for the bisection method, thus increasing the efficiency of the method.

However, the slow convergence and high computational time requirements make it impossible to use the bisection method, especially when it comes to higher-order SLPs. As the explicit form of F (and its derivatives) is unknown, a Newton-like method cannot apply. To overcome these difficulties, we propose an alternative root finding method in the following paragraph.

Multisection Method

A new root finding method, named "Multisection method"—a variant of the bisection method—is proposed, which can converge to the desired root faster. The idea is to divide the root interval into *m* subsections (m=2 being the bisection method) and locate the sign changing interval. Then, this interval is again refined into *m* subsections and the sign changing interval is located. This continues until the desired accuracy or the maximum number of iterations is reached (See Example A1 in **Appendix A**, which illustrates the usage and the performance of multisection method.)

This method overcomes drawbacks in the bisection method, yet provides a simple and faster, derivative-free solution to the root-finding problem. This method is able to find even multiple roots in an interval, and, given a sufficiently small subsection length, it is able to find roots that are cluster closer to each other (see Example 4). Unlike the bisection method, this method does not require the bracketing interval to have different signs at the endpoints. In the present paper, the input to this method is a set of function values, as it does not require specifying the explicit functional form of *F*, which is another advantage of this method. Due to the faster convergence of the multisection method (even at the risk of more computational cost), it can be used to find the eigenvalues (or roots) of the characteristic function.

Magnus Expansion

Magnus [39] proposed a solution to Equation (**8**) as

$$y(x)=\exp\Omega(x)y(a), \qquad (11)$$

and a series expansion for the exponent

$$\Omega(x) = \sum_{k=1}^{\infty} \Omega_k(x), \qquad (12)$$

which is called the Magnus expansion. Its first term can be written as

$$\Omega_1(x) = \int_0^x G(t_1)dt_1$$

$$\Omega_2(x) = \frac{1}{2}\int_0^x dt_1 \int_0^{t_2} dt_2[G(t_1), G(t_2)]$$

$$\Omega_3(x) = \frac{1}{6}\int_0^x dt_1 \int_0^{t_1} dt_2 \int_0^{t_2} dt_3[G(t_1), [G(t_2), G(t_3)]] + [G(t_3), [G(t_2), G(t_1)]]$$

$$\vdots = \vdots$$

where [A,B]≡AB−BA is the matrix commutator of *A* and *B*.

Numerical Procedure

The Magnus series only converges locally, hence, to calculate Ω, the interval [a,b] should be divided into *N* steps such that the Magnus series converges in each subinterval $[x_{n-1},x_n]$, n=1,…,N, with x_N=b.

Then, the solution at x_N is represented by

A $$Y(x_N) = \prod_{n=1}^{N} \exp(\Omega(x_{n-1}, x_n))Y_0, \quad (13)$$

and the series $\Omega(x_{n-1},x_n)$ has to be appropriately truncated.

The three steps in the procedure are:

E1: Ω series is truncated at an appropriate order: For achieving an integration method of order 2s(s>1) only terms up to Ω_{2s-2} in the Ω series are required [**38**].

E2: The multivariate integrals in the truncated series $\Omega^{[p]}=\sum_{i=1}^{p}\Omega_p$ are replaced by conveniently chosen approximations: if their exact evaluation is not possible or is computationally expensive, a numerical quadrature may be used instead. Suppose that bi,ci,(i=1,…,k) are the weights and nodes of a particular quadrature rule, say *X* of order *p*, respectively. Then,

$$G(0) = \int_{t_n}^{t_n+h} G(t)dt = h\sum_{i=1}^{k} b_i G_i + \mathcal{O}(h^{p+1}),$$

with $G_i \equiv G(t_n+cih)$. By Jódar and Marletta [**40**], systems with matrices of general size require a λ−dependent step size restriction of the form $h \le O(|\lambda|^{-1/4})$ in order to be defined.

E3: The exponential of the matrix $\Omega^{[p]}$ has to be computed. This is the most expensive step of the method. For this, Matlab function expm() is used, which provides the value for machine accuracy.

The algorithm then provides an approximation for $Y(x_{n+1})$ starting from $Y_n \approx Y(x_n)$, with $x_n+1=x_n+h$.

In the present paper, Ω is truncated using a sixth-order Magnus series and the integrals are approximated using the three-point Gaussian integration method [**41**]. $\Omega^{[6]}$ is obtained in the following equations. Let,

$$P_1 = hG(c_1h), \quad P_2 = hG(c_2h), \quad P_2 = hG(c_3h),$$

$$Q_1 = P_2, \quad Q_2 = \frac{\sqrt{15}}{3}(P_3 - P_1), \quad Q_3 = \frac{10}{3}(P_3 - 2P_2 + P_1),$$

$$R_1 = [Q_1, Q_2], \quad R_2 = [Q_1, 2Q_3 + R_1], \quad R_3 = [-20Q_1 - Q_3 + R_1, Q_2 - \frac{1}{60}R_2].$$

Then,

$$\Omega^{[6]}(h) = Q_1 + \frac{1}{12}Q_3 + \frac{1}{240}R_3, \tag{14}$$

where $c_1 = \frac{1}{2} - \frac{\sqrt{15}}{10}, c_2 = \frac{1}{2}, c_3 = \frac{1}{2} + \frac{\sqrt{15}}{10}$ are the nodes of Gaussian quadrature.

Inverse SLP Algorithm

The Magnus method's ability to handle various types of boundary conditions, adaptability to extend to higher-order problems, and accuracy with regard to higher index eigenvalues naturally allow it to be useful in generating suitable spectra for constructing and testing different inverse SLPs.

The iterative solution presented by Barcilon [**14**] was selected as the inverse SLP algorithm for two reasons: First, there is no evidence of actual implementation of Barcilon's algorithm in the literature. Second, other numerical methods available for the inverse problem are not amenable to generalization for higher order equations or systems.

The following is the Barcilon's algorithm in brief:

Consider the Sturm–Liouville problem

$$y'' + (\sigma - q(x))y = 0, \quad x \in (-1,1) \qquad y(-1) = y(1) = 0, \tag{15}$$

for a symmetric and normalized potential q(x), i.e.,

$$q(-x) = q(x) \text{ and } \int_{-1}^{1} q(x)dx = 0. \tag{16}$$

Then, the spectrum $\{\sigma_n\}^{\infty}_1$ uniquely determines the symmetric potential q(x) [42]. The eigenvalue problem in Equation (15) is equivalent to the following pair of Sturm–Liouville problems:

$$u''+(\lambda-q(x))u=0, x\in(0,1) u(0)=u(1)=0, \tag{17}$$

$$v''+(\mu-q(x))v=0, x\in(0,1) v'(0)=v(1)=0. \tag{18}$$

In addition, the two spectra $\{\lambda_n\}^{\infty}_1$ and $\{\mu_n\}^{\infty}_1$ are interlaced,

$$0 < \mu_1 < \lambda_1 < \mu_2 < \lambda_2 < \mu_3 < \ldots. \tag{19}$$

Letting,

$$\begin{aligned} w_{2n}(x) &= u_n^2(x), \quad \nu_{2n} = 4\lambda_n \\ w_{2n-1}(x) &= v_n^2(x), \quad \nu_{2n-1} = 4\mu_n \end{aligned} \tag{20}$$

Equations (**17**) and (**18**) are combined into:

$$w''' - 4qw' - 2q'w = -\nu w', \qquad w'(0) = w(1) = w'(1) = 0. \tag{21}$$

Equation (**21**) is not self-adjoint and its adjoint is given by:

$$\omega''' - 4q\omega' - 2q'\omega = -\nu\omega', \qquad \omega(0) = \omega''(0) = \omega(1) = 0. \tag{22}$$

Then, the inverse SLP procedure presented in Algorithm 1 can be used to recover the unknown potential *q*.

For a symmetric potential, a fixed set of eigenfunctions can be used in the updating Equation (24) for $w\^{}_n(x)$, for example, eigenfunctions for the case $q \equiv 0$ [**14**]. This is possible because,

"if $q \in L^2([a,b])$, the *k*th eigenvalue $\lambda_k(q,a,b)$ behaves asymptotically as

$$\lambda_k(q,a,b) = \lambda_k(0,a,b) + \bar{q} + \delta_k(q,a,b), \tag{23}$$

where $\lambda_k(0,a,b)$ is the *k*th eigenvalue of the SLP with the same BCs and zero potential, q¯ is the mean value of *q* and $\delta_k(q,a,b)$ is the remainder for smooth potential. Moreover, the information that the given spectrum provides about the variation of the unknown potential are contained in the terms δk(q,a,b) and, in view of their behaviour, the first eigenvalues are the most important for the reconstruction of *q*."

([**43**] p. 2)

This implies that the required set can be restricted to the first few eigenvalues. In fact, when the number of input eigenvalues increases, the error also increases, which can be attributed to the approximation errors in the higher index eigenvalues (see Figure 7).

Implementation details of the steps of Algorithm 1 are as follows:

Line (Data:) Input eigenvalue sequence is limited to N=10 to get a finite sequence. Using the Magnus method on Equations (**17**) and (**18**), the truncated eigenvalue sequences $\{\lambda_n\}^N_1$, $\{\mu_n\}^N_1$ are obtained and, using Equation (**20**), they are combined to obtain $\{\nu\^{}_n\}^{2N}_1$.

Line (Result:) The output is a finite vector of function values: $q_0, q_1, \ldots, q_n$, where *n* is the number of subdivisions in the x−axis.

Line (1) Initial guess $q^{(0)}$ is set to q≡0.

Line (2) By setting q=0 in Equations (**17**) and (**18**), the following expressions are easily obtained: $\{u^{(0)}_n\}^\infty_1 = \{\sin(n\pi x)\}^\infty_1$, $\{\lambda^{(0)}_n\}^\infty_1 = \{(n\pi)^2\}^\infty_1$, $\{v^{(0)}_n\}^\infty_1 = \{\cos(n-1/2)\pi x\}^\infty_1$, $\{\mu^{(0)}_n\}^\infty_1 = \{((n-1/2)\pi)^2\}^\infty_1$, $\{w^{(0)}_n\}^\infty_1 = \{\cos(\nu^{(0)}_n ——\sqrt{\pi x})\}^\infty 1$, $\{\omega^{(0)}_n\}^\infty_1 = \{\sin(\nu^{(0)}_n ——\sqrt{\pi x})\}^\infty_1$, $\{\nu^{(0)}_n\}^\infty_1 = \{\lambda^{(0)n}\}^{\infty 1} \cup \{\mu^{(0)}_n\}^\infty_1$. For the rest of the algorithm, the eigenfunctions $\{w^{(k)}_n(x), \omega^{(k)}_n(x)\}2^N_1$ are kept fixed at $\{w^{(0)}_n(x), \omega^{(0)}_n(x)\}2^N_1$.

Line (3) The optimal while loop condition: $\hat{\nu}_n \neq \nu_n^{(k-1)}$,n=1,2,... may never reach due to various errors in the numerical procedure, and is relaxed to $\sum_{n=1}^{2N} |\hat{\nu}_n - \nu_n^{(k-1)}| < tol$, and/or k<kMAX where tol is a fixed tolerance and kMAX is a maximum number of iterations.

Line (4) The required derivatives and integrals in Equation (24) are approximated using the Matlab built-in functions gradient() and trapz(), respectively. Furthermore, the Matlab built-in function griddedInterpolant() with linear interpolation method is used to obtain an explicit functional form of q from the set of points $q_0, q_1, \ldots, q_n$. This is necessary because, in the next step, the explicit form of q should be the input to the Magnus method (which allows evaluating q(x) at an arbitrary point). Only one of the EVPs (Equation (**21**) or (**22**)), needs to be solved using the Magnus method since only the eigenvalue sequence $\{\nu^{(k)}_n\}^{2N}_1$ is needed (as the eigenfunctions are kept fixed) in the subsequent iterations.

Algorithm 1: Solving inverse SLP.

Data: Sequence $\{\hat{\nu}_n\}_1^{\infty}$, the combined eigenvalues of the two equations (17) and (18) using Equation (20)

Result: the unknown potential function $q(x)$

1 Set an initial guess $q^{(0)}$, $k = 1$;

2 Solve EVPs (21), (22), and get the eigensolutions: $\left\{\nu_n^{(0)}, w_n^{(0)}(x), \omega_n^{(0)}(x)\right\}_1^{\infty}$;

3 **while** $\sum_n |\hat{\nu}_n - \nu_n^{(k-1)}| \neq 0$ **do**

4 Set

$$\hat{w}_n(x) = \frac{\int_0^1 w_n^{(k-1)}(\xi)d\xi}{\int_0^1 w_n^{(k-1)}(\xi)\left(\frac{d\omega_n^{(k-1)}}{d\xi}\right)d\xi} \frac{d\omega_n^{(k-1)}}{dx} \qquad (24)$$

and

$$q^{(k)}(x) = q^{(k-1)}(x) + \frac{1}{4}\sum_{n=1}^{\infty}(\hat{\nu}_n - \nu_n^{(k-1)})\hat{w}_n(x) \qquad (25)$$

Solve EVPs (21), (22), and get the eigensolutions: $\left\{\nu_n^{(k)}, w_n^{(k)}(x), \omega_n^{(k)}(x)\right\}_1^{\infty}$;

5 $k = k + 1$;

6 **end**

RESULTS OF DIRECT STURM–LIOUVILLE PROBLEMS

In this section, some numerical examples of Sturm–Liouville problems of different orders are presented. Furthermore, the efficiency and accuracy of the proposed technique outlined in the previous section is investigated, numerically. Two examples are presented in the next section, illustrating the Magnus method's usefulness in solving inverse SLPs.

Here, we made use of the MATLAB 2014 package to code the three algorithms, namely: multi-section method (Listing S1), the Magnus method for solving SLP (Listing S2), and Barcilon's method for solving inverse SLP (Listing S3). The codes were executed on an Intel ® Core ™ i3 CPU with power 2.40 GHz, equipped with 8 GB RAM (see Supplementary Materials).

For the numerical results (unless otherwise specified), the parameter values were: m=100, n=500, and L=5 where *n* is the number of subdivisions in the interval [a,b] and *m* is the number of subdivisions in the interval $[\lambda_0,\lambda^*]$, λ^* being the maximum eigenvalue searching, and *L* is the number of multisection steps used to calculate each eigenvalue in the characteristic function.

The performance of the Magnus method was measured by the absolute error Ek, which is defined as

$$E_k = \left|\lambda_k^{(Exact)} - \lambda_k^{(Magnus)}\right|, \qquad k = 1,2,\ldots \tag{26}$$

where $\lambda^{(Magnus)}_k$ indicates the *k*th eigenvalue obtained by the Magnus method and $\lambda^{(Exact)}_k$ is the *k*th exact eigenvalue and the relative error ϵ_k which is defined as

$$\epsilon_k = \left|\frac{\lambda_k^{(Exact)} - \lambda_k^{(Magnus)}}{\lambda_k^{(Exact)}}\right|, \qquad k = 1,2,\ldots. \tag{27}$$

There are several parameters in the Magnus method, which affect the performance and the accuracy of the method:

1. eigenvalue index: *k* or eigenvalue: λ_k
2. number of subdivisions in λ: *m*
3. number of subdivisions in *x*: *n*
4. number of multisection iterations: *L*

The performance was measured by the computation time used. Here, the Matlab function timeit() was used. timeit measures the time required to run a function by calling the specified function multiple times, and computing the median of the measurements. It provides a robust measurement of the time required for function execution and the function provides a more vigorous estimate [**44**]. The accuracy was measured using the absolute and relative errors as defined in Equations (**26**) and (**27**), respectively. Example 1 illustrates the affect of these variables on a SLP.

Example 1.

Consider the eigenvalue problem:

$-y''=\lambda y, 0<x<1, y(0)=y(1)=0,$ (28)

having the exact eigenvalues $\lambda_k=(k\pi)^2$, k=1,2,….

In ***Figure*** *1a, the absolute error is increasing (as the eigenvalues are increasing quadratically), however, the relative error is (independent of λ) gradually decreasing as* $\lambda\to\infty$*.This shows the Magnus method's ability to find higher index eigenvalues with less relative error. (see* ***Table A1*** *in* ***Appendix A*** *for a detailed analysis of the code). In addition, the computation time is very slowly increasing linearly with the index of the eigenvalue (****Figure*** *1b). As expected, accuracy and computation time increase with* m*−the number of subdivisions of λ and* L*−the number of multisection iteration steps (****Figure 2****). After around* L=15*, the error no longer improves, thus it is sufficient to stop at 15 iterations.*

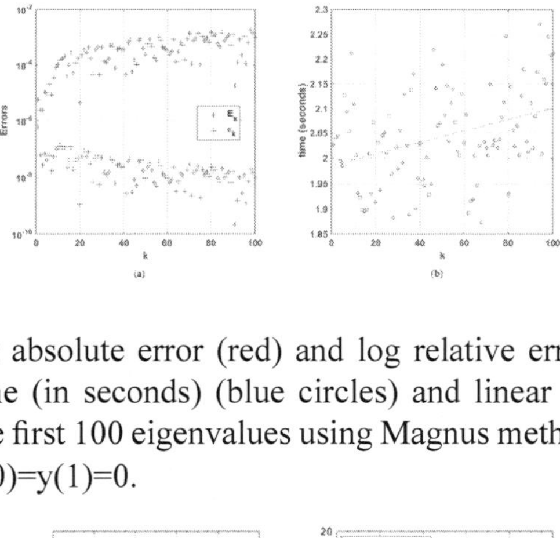

Figure 1. (**a**) Log absolute error (red) and log relative error (magenta); and (**b**) computing time (in seconds) (blue circles) and linear trend line (orange dashed line) for the first 100 eigenvalues using Magnus method for the problem $-y''=\lambda y, 0<x<1$, $y(0)=y(1)=0$.

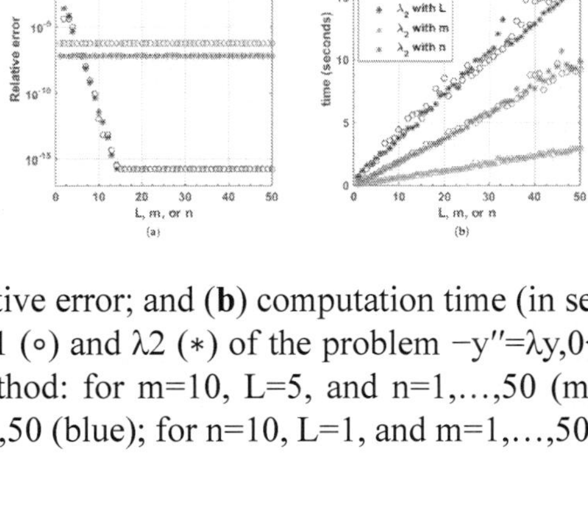

Figure 2. (**a**) Relative error; and (**b**) computation time (in seconds) for the first two eigenvalues λ1 (∘) and λ2 (∗) of the problem $-y''=\lambda y, 0<x<1$, $y(0)=y(1)=0$ using Magnus method: for m=10, L=5, and n=1,…,50 (magenta); for n=10, m=10, and L=1,…,50 (blue); for n=10, L=1, and m=1,…,50 (green).

*As **Figure 2**b shows, the computation time increases with the number of subdivisions of x:* n−, *but there is no change in the error (**Figure 2**a). This is obvious, since in Equation (**28**) all coefficient functions are constant, hence G is independent of x and the number of subdivisions n.*

Example 2

(SL Regular problem)**.** *This is an almost singular eigenvalue problem due to Paine et al. [**3**].*

$$-y''+(x+0.1)^{-2}y=\lambda y, 0<x<\pi, y(0)=y(\pi)=0. \quad (29)$$

***Table 1** and **Table 2** list the first four eigenvalues of this problem and compare the relative errors with FDM ([**3**], Table 4), NM ([**4**], Table 3), FEM ([**5**], Table 3), MNM ([**6**], Table 5), FEMT and FEMS ([**7**], Table 2), and LGM ([**8**], Table 4) methods. The Magnus method exhibits good relative errors compared with the other methods.*

Table 1. First four eigenvalues of $-y''+(x+0.1)-2y=\lambda y$, $0<x<\pi$, $y(0)=y(\pi)=0$.

k	Eigenvalue	
	λ_k [6], Table 4	Magnus
1	$1.5198658E+00$	$1.5198658206340E+00$
2	$4.9433098E+00$	$4.9433098197780E+00$
3	$1.0284663E+01$	$1.0284662639575E+01$
4	$1.7559958E+01$	$1.7559957737891E+01$

Table 2. Relative errors for the first four eigenvalues of $-y''+(x+0.1)^{-2}y=\lambda y$, $0<x<\pi$, $y(0)=y(\pi)=0$. (Minimum relative error for each eigenvalue is in bold.)

k	Relative Error							
	Magnus	FDM [3]	NM [4]	FEM [5]	MNM [6]	FEMT [7]	FEMS [7]	LGM [8]
1	$1.21E-08$	$\mathbf{0.00E+00}$	$2.63E-06$	$1.91E-04$	$3.95E-06$	$1.32E-04$	$1.41E-03$	$2.74E-11$
2	$4.00E-09$	$8.09E-05$	$4.86E-06$	$3.03E-04$	$8.50E-06$	$2.16E-04$	$2.02E-04$	$\mathbf{2.27E-11}$
3	$3.50E-08$	$8.75E-05$	$7.10E-06$	$3.51E-04$	$8.75E-06$	$2.78E-04$	$2.37E-04$	$\mathbf{5.26E-11}$
4	$1.49E-08$	$9.68E-05$	$9.23E-06$	$3.59E-04$	$1.73E-05$	$3.17E-04$	$2.55E-04$	$\mathbf{1.15E-09}$

***Figure 3**a shows the characteristic function for this problem. It can be seen that the roots (eigenvalues) are simple and also the eigenvalues are further apart as the index increases.*

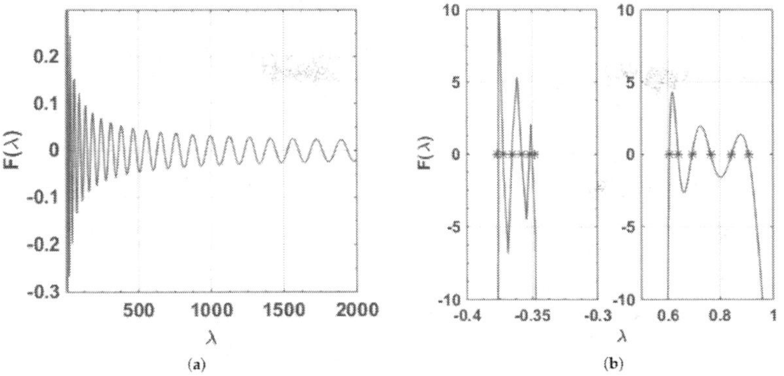

Figure 3. (**a**) Characteristic function of $-y''+(x+0.1)^{-2}y=\lambda y$, $0<x<\pi$, $y(0)=y(\pi)=0$. (**b**) Parts of the characteristic equation (blue) and first 12 eigenvalues (red) for the version of Mathieu Equation (**31**).

Example 3

(Singular SLP). *The Magnus method is robust and can handle even singular Sturm–Liouville Problems. For example, consider the Bessel equation of order* 12 *([**45**], Test problem 19):*

$$xy'' + \frac{1}{4x}y = \lambda xy, \quad 0 < x < 1, \qquad y(1) = 0. \tag{30}$$

Here, $p(0)=0$ *and the exact eigenvalues are* $\lambda_k=(k\pi)^2$. *The endpoint* $a=0$ *is singular and a limit-circle non-oscillatory (LCN) point (Note that Assumptions A3 and A4 mentioned in* **Section 2.1** *no longer hold). Taking Friedrich's BCs,* $f=1, g=\ln x$, *and writing the boundary condition coefficients as:*

$$\begin{pmatrix} 1 & \log(0) \\ 1 & \log(1) \end{pmatrix} \begin{pmatrix} y(0) & y(1) \\ y'(0) & y'(1) \end{pmatrix} = \begin{pmatrix} 0 \\ 0 \end{pmatrix}$$

the eigenvalues can be easily obtained as: $\lambda_1=9.864557319999999$, $\lambda_2=39.458190680000001$, $\lambda_3=88.780786309999996$, *and so on.*

Example 4.

*This example—a version of the Mathieu equation—is also due to (Pryce [**45**], Test problem 5).*

$$y'' + (\cos x)y = \lambda y, \quad 0 < x < 40, \qquad y(0) = y(40) = 0. \tag{31}$$

Here, the lower eigenvalues form clusters of 6 (see **Figure** *3b). Using a sufficiently small step size (<0.004: the minimum difference between* *two consecutive eigenvalues) for the subdivision in λ–dimension, the Magnus method is able to find even the lower eigenvalues with an accuracy comparable to Matslise's (see **Table 3**).*

Table 3. The first 17 eigenvalues and errors for the version of the Mathieu equation (Equation (**31**)) using Matslise and Magnus method.

k	Matslise [9]	Magnus	Absolute Difference
1	$-3.76845881566946E-01$	$-3.7684593940E-01$	$5.78330540124128E-08$
2	$-3.72222021386702E-01$	$-3.7222207135E-01$	$4.99632979988895E-08$
3	$-3.65517698755253E-01$	$-3.6551773700E-01$	$3.82447470359537E-08$
4	$-3.58145409533808E-01$	$-3.5814543445E-01$	$2.49161919985141E-08$
5	$-3.51818307563677E-01$	$-3.5181832065E-01$	$1.30863230252132E-08$
6	$-3.48153086495489E-01$	$-3.4815309260E-01$	$6.10451100779841E-09$
7	$6.06260772683456E-01$	$6.0626073600E-01$	$3.66834560505680E-08$
8	$6.39995069435221E-01$	$6.3999503360E-01$	$3.58352210128032E-08$
9	$6.94009291118601E-01$	$6.9400929280E-01$	$1.68139902001485E-09$
10	$7.64487943746816E-01$	$7.6448803840E-01$	$9.46531840684273E-08$
11	$8.43278584575029E-01$	$8.4327879680E-01$	$2.12224970930208E-07$
12	$9.07400354525007E-01$	$9.0740065280E-01$	$2.98274993038028E-07$
13	$1.27292510798947E+00$	$1.2729246720E+00$	$4.35989469860232E-07$
14	$1.38181949226729E+00$	$1.3818194944E+00$	$2.13271000859550E-09$
15	$1.52597349120834E+00$	$1.5259736064E+00$	$1.15191659988412E-07$
16	$1.69586866996862E+00$	$1.6958688256E+00$	$1.55631380005516E-07$
17	$1.88425137561091E+00$	$1.8842516480E+00$	$2.72389089950309E-07$

Example 5

(FSLP: Simple BCs). *This example is due to (Rattana and Böckmann [**24**], Example 1):*

$$y^{(4)} + y = \lambda y, \qquad y(0) = y''(0) = y(1) = y''(1) = 0. \tag{32}$$

*By Schueller [**46**], exact eigenvalues are:* $\lambda_k=(k\pi)^4+1, k=1,2,\ldots$

*From the results in **Table 4** and **Table 5** and **Figure 4**, it is clear that the Magnus method is more accurate. However, unlike the other methods, the accuracy decreases with the index of the eigenvalue due to the increasing magnitude of the eigenvalue. In contrast, for the second-order SLP using the Magnus method, the relative error decreases with the index (**Figure 1**a).*

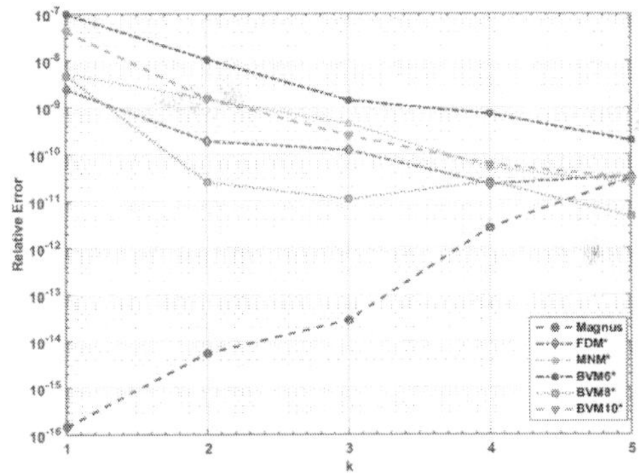

Figure 4. Log relative error of the first five eigenvalues of the problem in Equation (**32**).

Table 4. The first five eigenvalues for $y^{(4)}+y=\lambda y$, $y(0)=y''(0)=y(1)=y''(1)=0$.

k	Eigenvalue	
	λ_k	**Magnus**
1	$9.84090910340024E+01$	$9.84090910340024E+01$
2	$1.55954545654404E+03$	$1.55954545654403E+03$
3	$7.89113637375420E+03$	$7.89113637375442E+03$
4	$2.49377273047046E+04$	$2.49377273046367E+04$
5	$6.08816818962515E+04$	$6.08816818943245E+04$

Table 5. The relative errors of the first five eigenvalues for $y^{(4)}+y=\lambda y$, $y(0)=y''(0)=y(1)=y''(1)=0$. Relative errors for all the methods except Magnus method are calculated from the eigenvalues reported in ([**24**], Table 5). * denotes the methods with correction terms. (Minimum relative error for each eigenvalue is in bold.)

k	Relative Error					
	Magnus	FDM *	MNM *	BVM6 *	BVM8 *	BVM10 *
1	$\mathbf{1.44E-16}$	$2.38E-09$	$5.11E-09$	$9.73E-08$	$4.71E-09$	$4.14E-08$
2	$\mathbf{5.54E-15}$	$1.88E-10$	$1.49E-09$	$1.04E-08$	$2.57E-11$	$1.52E-09$
3	$\mathbf{2.84E-14}$	$1.22E-10$	$4.43E-10$	$1.40E-09$	$1.10E-11$	$2.47E-10$
4	$\mathbf{2.72E-12}$	$2.31E-11$	$5.16E-11$	$7.11E-10$	$2.60E-11$	$6.07E-11$
5	$3.17E-11$	$3.44E-11$	$2.45E-11$	$1.96E-10$	$\mathbf{4.73E-12}$	$3.15E-11$

Example 6 (FSLP: General BCs). *This example involving more general boundary conditions was taken from ([24], Example 4) to illustrate the Magnus method's versatility.*

$$\begin{aligned} & y^{(4)} + (\sin(x) + 2)y = \lambda y, \\ y(0) - y'(0) &= 0, \quad y(0) - y''(0) = 0, \\ y(1) - y'(1) &= 0, \quad y(1) - y''(1) = 0. \end{aligned} \tag{33}$$

*As Rattana and Böckmann [**24**], (pp. 153) claimed, the exact eigenvalues of this problem could not be computed and there is no other method except matrix methods (MM) proposed by them to approximate the eigenvalues.*

***Table 6** lists the first five eigenvalues of the problem and it can be seen that the values by the Magnus method are comparable with values computed by matrix methods.* **Figure 4** *we shows that the BVM8 method with correction terms performs better than the other methods considered (except Magnus) in that example. **Table 6** clearly shows that Magnus method's values are closer to the BVM8's values for this example.*

Table 6. The first five eigenvalues of FSLP in Equation (**33**). * Eigenvalues (except for the Magnus method) were computed using the codes provided by F. Fischer, PhD student at the Potsdam University, Germany. For each eigenvalue computed using Magnus method, the closest ones by other methods are in bold.

k	Magnus	FDM *	MNM *	BVM6 *	BVM8 *	BVM10 *
1	$3.5224017568E+00$	$3.444790E+00$	$2.604110E+00$	$\mathbf{3.522380E+00}$	$\mathbf{3.522380E+00}$	$\mathbf{3.522380E+00}$
2	$5.0300125581E+02$	$5.025637E+02$	$5.017770E+02$	$5.029616E+02$	$\mathbf{5.029632E+02}$	$5.029627E+02$
3	$3.8059852148E+03$	$3.804507E+03$	$3.804414E+03$	$3.805476E+03$	$\mathbf{3.805498E+03}$	$3.805491E+03$
4	$1.4620083327E+04$	$1.461435E+04$	$\mathbf{1.461949E+04}$	$1.461735E+04$	$1.461746E+04$	$1.461743E+04$
5	$3.9946254570E+04$	$3.992330E+04$	$\mathbf{3.995076E+04}$	$3.993665E+04$	$3.993704E+04$	$3.993692E+04$

Example 7. *This example illustrates Magnus method's superiority over some of the other existing methods. This problem describes the free lateral vibration of a uniform clamped-hinged beam (CH) [**17**]:*

$$y^{(4)} = \lambda y, \qquad y(0) = y'(0) = y(1) = y''(1) = 0. \tag{34}$$

*From [**17**], Equation (45), the eigenvalues are the solutions of the equation:*

$$\tanh \sqrt[4]{\lambda} = \tan \sqrt[4]{\lambda}. \tag{35}$$

*As **Table 7** and **Table 8** point out, the Magnus method has a comparable relative error to the other methods considered (CSCM ([**17**], Table 4), ADM [**16**], HAM [**18**], PDQ and FDQ ([**19**], Table 1), CM ([**20**], Tables 1 and 2), CDM ([**22**], Table 2), VIM ([**23**], Table 2), and SPSS [**21**]) with respect to the first five eigenvalues.*

Table 7. The first five eigenvalues of the problem in Equation (**34**).

k	**Eigenvalue**	
	λ_k	**Magnus**
1	$2.377210675300E+02$	$2.377210675300E+02$
2	$2.496487437860E+03$	$2.496487437860E+03$
3	$1.086758221698E+04$	$1.086758221697E+04$
4	$3.178009645408E+04$	$3.178009645380E+04$
5	$7.400084934916E+04$	$7.400084934040E+04$

Table 8. Relative error of the first five eigenvalues for the problem in Equation (**34**). (Minimum relative error for each eigenvalue is in bold.)

k	**Relative Error**									
	Magnus	**CSCM** [17]	**ADM** [16]	**HAM** [18]	**PDQ** [19]	**FDQ** [19]	**CM** [20]	**CDM** [22]	**VIM** [23]	**SPSS** [21]
1	$4.6E-12$	$2.4E-6$	$4.7E-12$	$\mathbf{0.0E+00}$	$7.6E-9$	$3.2E-9$	$4.7E-12$	$2.0E-09$	$6.7E-11$	$4.6E-12$
2	$1.4E-12$	$9.8E-6$	$4.4E-12$	$4.0E-12$	$4.4E-8$	$4.8E-8$	$3.0E-12$	$7.9E-10$	$4.0E-10$	$\mathbf{1.2E-12}$
3	$1.2E-12$	$9.3E-6$	$1.0E-06$	$\mathbf{9.2E-13}$	$1.7E-8$	$2.2E-8$	$5.1E-12$	$2.3E-10$	$1.8E-09$	$2.1E-12$
4	$8.8E-12$	$9.3E-6$	$9.6E-03$	$\mathbf{6.0E-12}$	$2.4E-8$	$1.7E-8$	$8.6E-09$	$2.0E-11$	$2.0E-09$	$4.2E-11$
5	$1.2E-10$	$9.3E-6$	-	$\mathbf{3.5E-11}$	$3.0E-8$	$1.9E-8$	-	$7.5E-11$	$3.5E-09$	$2.3E-10$

More examples on Magnus methods for FSLP can be found in Mirzaei [**25**].

Example 8 (SSLP).

$$-y^{(6)} = \lambda y, \quad 0 < x < \pi, \qquad y(0) = y''(0) = y^{(4)}(0) = y(\pi) = y''(\pi) = y^{(4)}(\pi) = 0. \qquad 36)$$

This example is due to Greenberg and Marletta [29] (Problem (1)), having exact eigenvalues: $\lambda_k = k^6$, k= *1,2, ...*

Again, as shown in ***Table 9****, the Magnus method shows excellent accuracy with regard to this sixth-order SLP compared to other methods, namely Shooting method ([****29****], Table 1), ADM ([****30****], Table 2), CP ([****31****], Table 2), and VIM ([****23****], Table 3).*

Table 9. Eigenvalues and relative error for the problem in Equation (**36**). (Minimum relative error for each eigenvalue is in bold.)

k	**Eigenvalue**		**Relative Error**				
	Exact	**Magnus**	**Magnus**	**Shooting** [29]	**ADM** [30]	**CP** [31]	**VIM** [23]
1	$1.00000E+00$	$9.99999999999844E-01$	$1.56E-13$	$2.00E-07$	$\mathbf{0.00E+00}$	$\mathbf{0.00E+00}$	$5.50E-09$
2	$6.40000E+01$	$6.40000000000014E+01$	$2.20E-14$	$1.72E-07$	$\mathbf{0.00E+00}$	$\mathbf{0.00E+00}$	$7.32E-09$
3	$7.29000E+02$	$7.28999999999632E+02$	$5.05E-13$	-	$\mathbf{0.00E+00}$	$\mathbf{0.00E+00}$	$5.88E-09$
4	$4.09600E+03$	$4.09599999998837E+03$	$2.84E-12$	$1.71E-07$	$\mathbf{0.00E+00}$	$\mathbf{0.00E+00}$	$1.61E-08$
5	$1.56250E+04$	$1.56249999999687E+04$	$2.00E-12$	-	$1.28E-13$	$\mathbf{6.40E-15}$	$2.82E-09$
6	$4.66560E+04$	$4.66559999797840E+04$	$4.33E-10$	$1.71E-07$	$1.00E-08$	$\mathbf{2.18E-15}$	$4.62E-09$
7	$1.17649E+05$	$1.17648998941287E+05$	$9.00E-09$	-	$1.34E-04$	$\mathbf{6.50E-11}$	$2.94E-08$
8	$2.62144E+05$	$2.62143988377187E+05$	$4.43E-08$	$1.53E-07$	$1.24E-01$	-	$\mathbf{9.50E-09}$
9	$5.31441E+05$	$5.31440822457814E+05$	$3.34E-07$	-	-	-	$\mathbf{1.25E-07}$
10	$1.00000E+06$	$1.00000000400001E+06$	$\mathbf{4.00E-09}$	$1.70E-07$			

To the authors' best knowledge, there exists no algorithms for solving eighth-order SLP in the literature. However, the proposed Magnus method can be successfully applied with literally no modification to the algorithm. Only the matrix G has to be modified using Equation (**36**).

Example 9 (Eighth-order Sturm–Liouville problem). *Squaring the fourth-order SLP:*

$$y^{(4)} = \lambda y, \qquad y(0) = y''(0) = y(1) = y''(1) = 0, \tag{37}$$

with exact eigenvalues $\lambda_k=(k\pi)^4, k=1,2,\ldots,$ *the following eighth-order SLP is obtained:*

$$y^{(8)} = \lambda^2 y, \qquad y(0) = y''(0) = y^{(4)}(0) = y^{(6)}(0) = y(1) = y''(1) = y^{(4)}(1) = y^{(6)}(1) = 0, \tag{38}$$

with exact eigenvalues $\lambda^2_k=(k\pi)^8, k=1,2, \ldots.$

*The first five eigenvalues have good accuracy, although the error is steadily increasing with the index of the eigenvalue (see **Table 10**). This is because the error also depends on the magnitude of the eigenvalue.*

Table 10. Eigenvalues and relative errors for the problem in Equation (**38**) using m=100, n=100, and L=10.

k	**Eigenvalue**		**Relative Error**
	λ_k	**Magnus**	
1	$9.48853101607057E+03$	$9.48853101607059E+03$	$1.91704005653253E-15$
2	$2.42906394011407E+06$	$2.42906394011394E+06$	$5.21434895376847E-14$
3	$6.22542519964390E+07$	$6.22542519964604E+07$	$3.43361538321867E-13$
4	$6.21840368669201E+08$	$6.21840368667400E+08$	$2.89626411740934E-12$
5	$3.70645742815257E+09$	$3.70645742800000E+09$	$4.11625897732872E-11$

Specifically, the error in a pth-order method grows as $O(h^{p+1}\lambda^{p/2-1})$*, and thus one expects poor approximations for large eigenvalues [38].*

It is clear that Magnus method has the potential to find the first eigenvalues of a SLP of arbitrary (even) order with good accuracy.

In addition, it is observed for the EVPs with constant coefficients (hence, constant and mostly sparse G matrix) the errors are smaller (n=2: Example 1; n=4: Examples 5 and 7; n=6: Example 8; and n=8: Example 9) irrespective of the order. However, when G is no longer a constant, the error is not as good even for the SLP of order two (see Example 2). This can be attributed to the matrix exponent calculation step and can be improved by using a more accurate method in place of the expm() function.

RESULTS OF INVERSE STURM–LIOUVILLE PROBLEMS

Example 10

(Symmetric potential). *Consider the EVP*

$u''+(\lambda-q(x))u=0, x\in(0,1)$

Suppose $q(x)=\cos(\pi x)$, which is symmetric and normalized. The truncated eigenvalue sequences at N=10, for Dirichlet boundary conditions (DDBCs): $u(0)=u(1)=0$ and Dirichlet– Neumann boundary conditions (DNBCs): $v'(0)=v(1)=0$ are calculated using the Magnus method (using a tolerance 1E−13), and combined into $\{v\hat{}n\}^{2N}_{1}$. This is the input to the algorithm. Now, using an initial guess, say q=0, the eigenvalue sequences for DDBCs and DNBCs are calculated again using the Magnus method and the sequence $\{v^{(0)}n\}^{2N}_{1}$ is constructed. Then, q is updated using Equations (24) and (25). This cycle is repeated again. The iteration stops when the desired accuracy is reached.

Figure 5 and Figure 6 show the exact and reconstructed potential q and the log absolute errors in the reconstructed potential and the eigenvalue sequences, respectively. The supremum norm difference between the reconstructed q and the actual q is 8.229454025521221E−05. It is clear that the error is larger at the end-points than in the middle. At the mid-point, q(x) is 0 and, from Equation (23), this implies a minimum error (see Figure 6a). The supremum norm difference between the reconstructed eigenvalues and the actual ones is 1.302601049246732E−07. In addition, after about 15 iterations, the differences in the two sets of eigenvalues no longer improves, and only oscillates, implying that more iterations may try to overfit (Figure 6b).

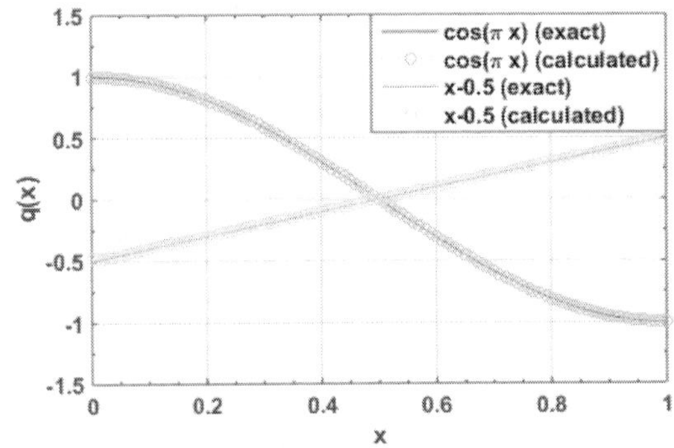

Figure 5. Exact and reconstructed potentials $q(x)=\cos(\pi x)$ and $q(x)=x-0.5$.

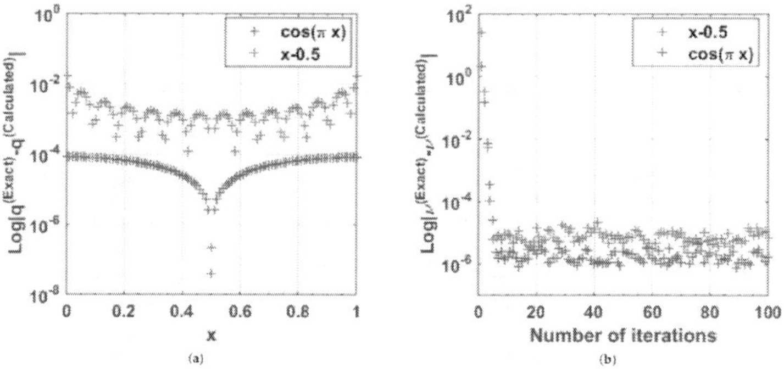

Figure 6. (**a**) Log absolute error in the reconstructed q; and (**b**) log absolute error sum of the difference of the eigenvalue sequences $\{\nu^{\wedge}_{n}\}$ and $\{\nu^{(k-1)}_{n}\}$ for reconstructing the potentials q(x)=cos(πx) (blue) and q(x)=x−0.5 (red).

For this potential, as the number of eigenvalues increases the error also increases, due to the errors in approximating higher index eigenvalues (see Figure 7). This agrees with Aceto et al. [43] who claimed that the first eigenvalues are the most important for the reconstruction of a symmetric, zero-mean potential q, thus the required set should be restricted to the first few eigenvalues. In addition, after about six iterations, the error no longer improves, and even increases for some values of N. Thus, it is advisable to stop the iteration procedure by checking the error.

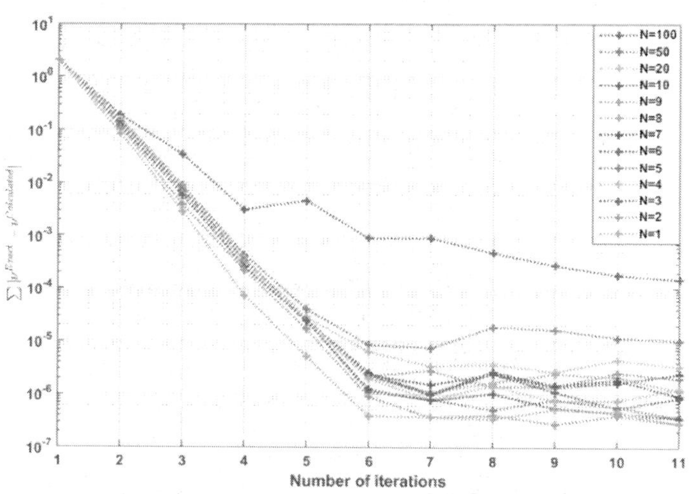

Figure 7. Sum of the absolute errors in the reconstructed eigenvalues: $\sum|\nu^{Exact}-\nu^{Calculated}||$ using the Magnus method vs. the iteration number, using eigenvalue sets of sizes N=1,…,10,20,50,100 for reconstructing the potential q(x)=cos(πx).

Example 11

(Non-symmetric potential). For a second example, consider $q(x)=x-0.5$, which is non-symmetric but normalized. Figure 5 and Figure 6 show the exact and reconstructed potential q and the log absolute errors in the reconstructed potential and the reconstructed eigenvalue sequence, respectively. The supremum norm difference between the reconstructed q and the actual q is 0.016647548176497. Although not as good as for the symmetric potential, this can be improved if the exact eigenvalues for the exact potential (here, we are calculating them using the Magnus method) are known. The supremum norm difference between the reconstructed eigenvalues sequence and the actual ones is 0.004796140075996. Similar to previous example, the error is larger at the end-points than in the middle (Figure 6a).

According to Andrew [**15**], there are three major sources of error with the inverse problem:

E4: attempting to compute q using only a finite (and often quite small) number of eigenvalues, even though the complete infinite set is required for the determination of q;

E5: only approximations of the eigenvalues are available and errors in the given eigenvalues may be especially serious for higher eigenvalues; and

E6: all errors in the numerical solution of the direct problem cause errors in the solution of the inverse problem.

The last one is the only error source that is affected by the choice of numerical method and which can be reduced by using a high accuracy method such as the Magnus method.

DISCUSSION AND CONCLUSIONS

There are several places where approximation errors occur in the Magnus method. For the direct problem, they are (see **Section 2.5** also):

E1: truncation of the Magnus series

E2: calculation of multivariate integrals

E3: computing the matrix exponent

In the case of the inverse problem (see also **Section 4**) they are:

E4: using a finite number of eigenvalues

E5: using approximate eigenvalues

E6: errors in the direct problem

E7: computing the explicit form of q by an interpolation method

E8: approximating derivatives and integrals using numerical methods

Thus, the accumulated error would be high for the direct and inverse SLPs. For the direct SLP, we can address E1 by using a higher order Magnus method, E2 by using higher order quadrature methods, and E3 by using a more accurate method from the methods proposed by Moler and Van Loan [**47**]. For the inverse SLP, E4 is no longer a problem as Aceto et al. [**43**] argued that the first eigenvalues are the most important for the reconstruction of the unknown potential. E5 and E6 can also be reduced by reducing the errors in the direct problem (E1–E3). E8 can be eliminated by using exact derivatives and integrals when possible. Thus, it turn outs that the error in the inverse problem solely depends on the errors in the direct problem and E7 (interpolating q from a set of points).

By construction, the Magnus method leads to a global error of order $O(h^p)$ if a pth-order Magnus method is applied, yet it turns out that the error also depends on the magnitude of the eigenvalue. Specifically, the error in a pth-order method grows as $O(h^{p+1}\lambda^{p/2-1})$, and thus one expects poor approximations for larger eigenvalues [**38**]. This method has the disadvantage that the cost of integration, while increasing very slowly as a function of λ, wouls be $O\left(\left(\begin{matrix}2n\\n\end{matrix}\right)^2\right)$ for a problem arising from a 2nth-order Sturm–Liouville problem [**40**].

The Magnus method was tested with finite interval SLPs of even orders (up to eight) along with regular and a finite singular endpoint BC, and good accuracy with regard to first few eigenvalues was obtained. One limitation of the Magnus method is that, although it is more accurate and efficient with constant coefficient BVPs irrespective of the order, when G is no longer a constant, the error is large even for the SLP of order 2. This needs to be addressed in a future work, as this would in turn effect the performance of the inverse SLP algorithm. It is an open problem whether this method can be extended to solve other types of EVPs such as different finite singular endpoints, infinite intervals, and multiple eigenvalues.

The present inverse SLP algorithm is less accurate with non-symmetric potentials, and it also should be improved to handle non-normalized, and/or non-smooth potentials. In a future work, the algorithm for inverse SLP may be modified to include general finite interval problems (i.e., problems in the domain (a,b)) for different sets of BCs (provided the spectra are interlaced), infinite BCS, and different classes of potential functions, such as non-smooth, discontinuous, oscillating, etc. In addition, a more suitable interpolation method for reconstructing q may be examined. Furthermore, Barcilon [**26**] generalized the inverse SLP algorithm presented in [**14**] to higher-order problems. This can be utilized in combination with Magnus method so that a direct or inverse SLP of any (even) order can be solved effectively.

In conclusion, the Magnus method is able to deal with general the kind of boundary conditions, since the boundary conditions are presented in a matrix form. Here, it is shown that the proposed technique has the ability to solve regular Sturm–Liouville problems of any even order, efficiently. By Ledoux et al. [**10**], a modified Magnus method can be used with Lobatto points to improve the error. Although in theory, this method can be used for solving higher order SLPs; for it to be effective in practice, the various errors in the approximation steps should be addressed adequately.

AUTHOR CONTRIBUTIONS

Conceptualization, C.B.; Data curation, U.P.; Funding acquisition, U.P.; Investigation, U.P.; Methodology, U.P.; Resources, C.B.; Software, U.P.; Supervision, C.B.; Validation, U.P.; Visualization, U.P.; Writing—original draft, U.P.; and Writing—review and editing, C.B.

FUNDING

This research was partially funded by University of Kelaniya, Sri Lanka.

ACKNOWLEDGMENTS

We would like to thank Florian Fischer, Ph.D. student at the Potsdam University, Germany, for providing the Matlab code for FDM, MNM and BVM allowing us to compute the eigenvalues in **Table 6**. We would like to thank the reviewers for their thoughtful comments and efforts towards improving our manuscript.

CONFLICTS OF INTEREST

The authors declare no conflict of interest.

ABBREVIATIONS

The following abbreviations are used in this manuscript:

ADM	Adomian decomposition method
FEMS	FEM with Simpson's rule
BC	Boundary conditions
FEMT	FEM with trapezoidal rule
BVM	Boundary Value Method
FSLP	Fourth-order Sturm–Liouville problem
CDM	Chebyshev differentiation matrices
HAM	homotopy analysis method
CM	Chebychev method
LGM	Lie-Group method
CP	Chebyshev polynomials
LCN	limit-circle non-oscillatory
CSCM	Chebyshev spectral collocation method
MNM	Modified Numerov's method
DDBCs	Dirichlet boundary conditions
NM	Numerov's method
DNBCs	Dirichlet–Neumann boundary conditions
PDQ	Polynomial-based differential quadrature
EVP	Eigenvalue problem
SLP	Sturm–Liouville problem
FDM	Finite difference method
SPPS	Spectral parameter power series
FDQ	Fourier expansion-based differential quadrature
SSLP	Sixth-order SLP
FEM	Finite element method
VIM	variational iteration method

APPENDIX A

Example A1

(Multi-section method). ***Figure A1*** *compares the log errors for the bisection and multisection methods for approximating the root of* $f(x)=x^3-x-2$ *in the interval* [1,2] *using* m=3,10,100 *subdivisions. Even after more than 20 iterations, the bisection method reports an error of* 1.5×10^{-8} *where as* *after only 7 iterations multisection method with* m=100 *reached an error* 4.5×10^{-14}. *Even* m=3 *multisection method converged within 19 iterations to the same accuracy. Obviously, the computation time increases and error decreases with the number of steps m and tolerance (**Figure A2**).*

Table A1 shows the profile summary of the Matlab implementation of Magnus method. It can be seen that the most time is spent on calculating Ω, and most function calls are for the functions G,p,w, and q (which are the coefficients matrix and the coefficients of the equation, respectively). The most self time is utilized by the function G. In addition, Problem (B) requires more computational time than for Problem (A) with other factors remaining fixed, as the latter has a constant G matrix.

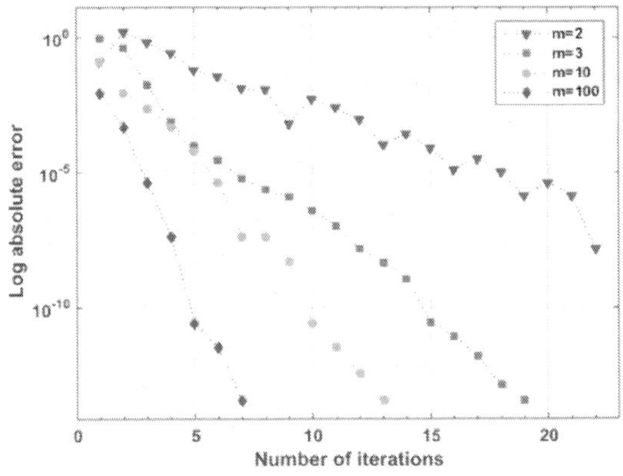

Figure A1. Log errors for the bisection (m=2) and multisection method for approximating the root of $f(x)=x^3-x-2$ in the interval [1,2] using m=3,10,100 subdivisions.

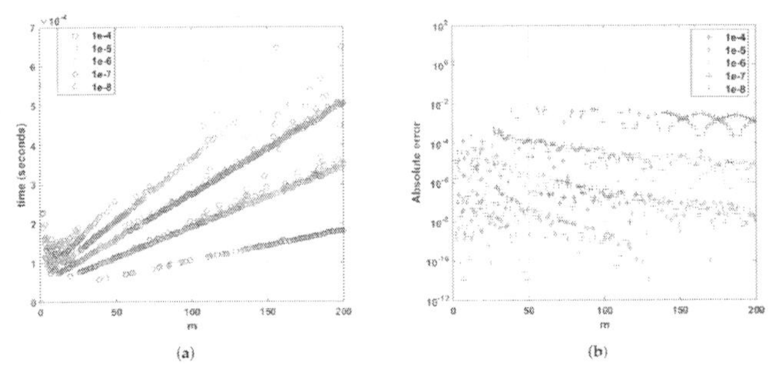

Figure A2. (**a**) Time; and (**b**) log absolute error for the multisection method for approximating the root of $f(x)=x^3-x-2$ in the interval [1,2] using m=2,…,200 subdivisions with tolerance in {1e−4,…,1e−8}.

Table A1. Profile summary of the slp function generated by Matlab with m=10, n=10, and L=5 for finding the eigenvalue in: (**A**) the interval [0,10] for $y''+\lambda y=0, y(0)=y(1)=0$; and (**B**) the interval [0,2] for $y''+(x+0.1)-2y=\lambda y, y(0)=y(\pi)=0$. * Self time is the time spent in a function excluding the time spent in its child functions. Self time also includes overhead resulting from the process of profiling. All time values are in seconds.

Function Name	Calls	Total Time		Self Time *	
		A	B	A	B
slp	1	2.442 s	**2.461 s**	0.028 s	**0.031 s**
Omega	660	**2.323 s**	2.292 s	0.024 s	**0.030 s**
R3	660	2.163 s	**2.200 s**	**0.056 s**	0.032 s
G	44,880	1.454 s	**1.480 s**	0.948 s	**1.014 s**
R2	1320	1.336 s	**1.404 s**	0.040 s	**0.063 s**
R1	3960	1.114 s	**1.307 s**	**0.096 s**	0.095 s
Q2	9240	0.843 s	**0.900 s**	**0.140 s**	0.124 s
Q3	4620	**0.736 s**	0.548 s	**0.096 s**	0.078 s
G2	17,160	**0.711 s**	0.653 s	**0.140 s**	0.110 s
G1	13,860	**0.602 s**	0.516 s	**0.104 s**	0.078 s
Q1	12,540	0.527 s	**0.623 s**	**0.076 s**	0.048 s
G3	13,860	0.481 s	**0.654 s**	0.096 s	**0.156 s**
p	44,880	**0.216 s**	0.171 s	**0.216 s**	0.171 s
w	44,880	**0.180 s**	0.141 s	**0.180 s**	0.141 s
q	44,880	0.110 s	**0.155 s**	0.110 s	**0.155 s**
expm	660	0.091 s	**0.139 s**	0.008 s	**0.031 s**
expm>PadeApproximantOfDegree	660	0.067 s	**0.108 s**	0.067 s	**0.092 s**
expm>expmchk	660	**0.016 s**	0.000 s	**0.016 s**	0.000 s
expm>getPadeCoefficients	660	0.000 s	**0.016 s**	0.000 s	**0.016 s**

REFERENCES

1. Lanczos, C. *An Iteration Method for the Solution of the Eigenvalue Problem of Linear Differential and Integral Operators*; United States Government Press Office: Los Angeles, CA, USA, 1950.
2. Ambarzumian, V. Über eine frage der eigenwerttheorie. *Zeitschrift für Physik A Hadrons and Nuclei* **1929**, *53*, 690–695.
3. Paine, J.W.; de Hoog, F.R.; Anderssen, R.S. On the correction of finite difference eigenvalue approximations for Sturm-Liouville problems. *Computing* **1981**, *26*, 123–139.
4. Andrew, A.L.; Paine, J.W. Correction of Numerov's eigenvalue estimates. *Num. Math.* **1985**, *47*, 289–300.
5. Andrew, A.L.; Paine, J.W. Correction of finite element estimates for Sturm-Liouville eigenvalues. *Num. Math.* **1986**, *50*, 205–215.
6. Vanden Berghe, G.; De Meyer, H. A modified Numerov method for higher Sturm-Liouville eigenvalues. *Int. J. Comput. Math.* **1990**, *37*, 63–77.
7. Berghe, G.V.; De Meyer, H. A finite-element estimate with trigonometric hat functions for Sturm–Liouville eigenvalues. *J. Comput. Appl. Math.* **1994**, *53*, 389–396.
8. Moan, P.C. *Efficient Approximation of Sturm-Liouville Problems Using Lie-Group Methods*; Report–DAMPT NA; Department of Applied Mathematics & Theoretical Physics, University of Cambridge: Cambridge, UK, 1998.
9. Ledoux, V.; Daele, M.V.; Berghe, G.V. MATSLISE: A MATLAB package for the numerical solution of Sturm-Liouville and Schrödinger equations. *ACM Trans. Math. Softw.* **2005**, *31*, 532–554.
10. Ledoux, V.; Daele, M.V.; Berghe, G.V. Efficient numerical solution of the one-dimensional Schrodinger eigenvalue problem using Magnus integrators. *IMA J. Numer. Anal.* **2009**, *30*, 751–776.
11. Atay, M.T.; Kartal, S. Computation of Eigenvalues of Sturm-Liouville Problems using Homotopy Perturbation Method. *Int. J. Nonlinear Sci. Numer. Simul.* **2010**, *11*, 105–112.
12. Freiling, G.; Yurko, V.A. *Inverse Sturm-Liouville Problems and Their Applications*; NOVA Science Publishers: New York, NY, USA, 2001.
13. McLaughlin, J.R. Analytical methods for recovering coefficients in differential equations from spectral data. *SIAM Rev.* **1986**, *28*, 53–72.

14. Barcilon, V. Iterative solution of the inverse Sturm-Liouville problem. *J. Math. Phys.* **1974**, *15*, 429–436.
15. Andrew, A.L. Numerov's method for inverse Sturm–Liouville problems. *Inverse Probl.* **2005**, *21*, 223–238.
16. Attili, B.S.; Lesnic, D. An efficient method for computing eigenelements of Sturm-Liouville fourth-order boundary value problems. *Appl. Math. Comput.* **2006**, *182*, 1247–1254.
17. Mai-Duy, N. An effective spectral collocation method for the direct solution of high-order ODEs. *Commun. Numer. Methods Eng.* **2006**, *22*, 627–642.
18. Abbasbandy, S.; Shirzadi, A. A new application of the homotopy analysis method: Solving the Sturm–Liouville problems. *Commun. Nonlinear Sci. Numer. Simul.* **2011**, *16*, 112–126.
19. Yücel, U.; Boubaker, K. Differential quadrature method (DQM) and Boubaker polynomials expansion scheme (BPES) for efficient computation of the eigenvalues of fourth-order Sturm–Liouville problems. *Appl. Math. Model.* **2012**, *36*, 158–167.
20. El-Gamel, M.; Sameeh, M. An Efficient Technique for Finding the Eigenvalues of Fourth-Order Sturm-Liouville Problems. *Appl. Math.* **2012**, *3*, 920–925.
21. Khmelnytskaya, K.V.; Kravchenko, V.V.; Baldenebro-Obeso, J.A. Spectral parameter power series for fourth-order Sturm–Liouville problems. *Appl. Math. Comput.* **2012**, *219*, 3610–3624.
22. Taher, A.S.; Malek, A.; Momeni-Masuleh, S. Chebyshev differentiation matrices for efficient computation of the eigenvalues of fourth-order Sturm–Liouville problems. *Appl. Math. Model.* **2013**, *37*, 4634–4642.
23. Taher, A.H.S.; Malek, A. An efficient algorithm for solving high order Sturm-Liouville problems using variational iteration method. *Fixed Point Theory* **2013**, *14*, 193–210.
24. Rattana, A.; Böckmann, C. Matrix methods for computing eigenvalues of Sturm–Liouville problems of order four. *J. Comput. Appl. Math.* **2013**, *249*, 144–156.
25. Mirzaei, H. Computing the eigenvalues of fourth order Sturm-Liouville problems with Lie Group method. *Iran. J. Numer. Anal. Optim.* **2017**, *7*, 1–12.
26. Barcilon, V. On the solution of inverse eigenvalue problems of high orders. *Geophys. J. Int.* **1974**, *39*, 143–154.

27. McLaughlin, J.R. An Inverse Eigenvalue Problem of Order Four. *SIAM J. Math. Anal.* **1976**, *7*, 646–661.
28. McLaughlin, J.R. An Inverse Eigenvalue Problem of Order Four—An Infinite Case. *SIAM J. Math. Anal.* **1978**, *9*, 395–413.
29. Greenberg, L.; Marletta, M. Oscillation Theory and Numerical Solution of Sixth Order Sturm–Liouville Problems. *SIAM J. Numer. Anal.* **1998**, *35*, 2070–2098.
30. Lesnic, D.; Attili, B.S. An efficient method for sixth-order Sturm-Liouville problems. *Int. J. Sci. Technol.* **2007**, *2*, 109–114.
31. Allame, M.; Ghasemi, H.; Kajani, M.T. An appropriate method for eigenvalues approximation of sixth-order Sturm-Liouville problems by using integral operation matrix over the Chebyshev polynomials. In *AIP Conference Proceedings*; AIP Publishing: College Park, MD, USA, 2015; Volume 1684, p. 090001.
32. Zhang, Z. How Many Numerical Eigenvalues Can We Trust? *J. Sci. Comput.* **2015**, *65*, 455–466.
33. Barcilon, V. On the uniqueness of inverse eigenvalue problems. *Geophys. J. R. Astron. Soc.* **1974**, *38*, 287–298.
34. McLaughlin, J.R.; Rundell, W. A uniqueness theorem for an inverse Sturm–Liouville problem. *J. Math. Phys.* **1987**, *28*, 1471–1472.
35. Bailey, P.B.; Gordon, M.K.; Shampine, L.F. Automatic solution of the Sturm-Liouville problem. *ACM Trans. Math. Softw.* **1978**, *4*, 193–208.
36. Lie, S. *Theorie der Transformationsgruppen*; Ripol Classic: Moscow, Russia, 1970; Volume 1.
37. Iserles, A.; Norsett, S. On the solution of linear differential equations in Lie groups. *Philos. Trans. R. Soc. Lond. Ser. A Math. Phys. Eng. Sci.* **1999**, *357*, 983–1019.
38. Blanes, S.; Casas, F.; Oteo, J.; Ros, J. The Magnus expansion and some of its applications. *Phys. Rep.* **2009**, *470*, 151–238.
39. Magnus, W. On the exponential solution of differential equations for a linear operator. *Commun. Pure Appl. Math.* **1954**, *7*, 649–673.
40. Jódar, L.; Marletta, M. Solving ODEs arising from non-selfadjoint Hamiltonian eigenproblems. *Adv. Comput. Math.* **2000**, *13*, 231–256.
41. Iserles, A. Magnus expansions and beyond. *Comb. Phys.* **2008**, *539*, 171–186.
42. Borg, G. Eine Umkehrung der Sturm-Liouvilleschen Eigenwertaufgabe:

Bestimmung der Differentialgleichung durch die Eigenwerte. *Acta Math.* **1946**, *78*, 1–96.

43. Aceto, L.; Ghelardoni, P.; Magherini, C. Boundary value methods for the reconstruction of Sturm–Liouville potentials. *Appl. Math. Comput.* **2012**, *219*, 2960–2974.
44. McKeeman, B. *MATLAB Performance Measurement—File Exchange—MATLAB Central*; MathWorks: Natick, MA, USA, 2008.
45. Pryce, J.D. A Test Package for Sturm-Liouville Solvers. *ACM Trans. Math. Softw.* **1999**, *25*, 21–57.
46. Schueller, A. Eigenvalue Asymptotics for Self-Adjoint, Fourth-Order, Ordinary Differential Operators. Ph.D. Thesis, University of Kentucky, Lexington, KY, USA, 1997.
47. Moler, C.; Van Loan, C. Nineteen dubious ways to compute the exponential of a matrix, twenty-five years later. *SIAM Rev.* **2003**, *45*, 3–49.

CHAPTER

8

Numerical Solution of Direct and Inverse Problems for Time-Dependent Volterra Integro-Differential Equation Using Finite Integration Method with Shifted Chebyshev Polynomials

Ratinan Boonklurb, Ampol Duangpan and Phansphitcha Gugaew

Department of Mathematics and Computer Science, Faculty of Science, Chulalongkorn University, Bangkok 10330, Thailand

ABSTRACT

In this article, the direct and inverse problems for the one-dimensional time-dependent Volterra integro-differential equation involving two integration

Citation: Boonklurb, R., Duangpan, A., & Gugaew, P. (2020). Numerical Solution of Direct and Inverse Problems for Time-Dependent Volterra Integro-Differential Equation Using Finite Integration Method with Shifted Chebyshev Polynomials. Symmetry, 12(4), 497.(19 pages) DOI: https://doi.org/10.3390/sym12040497

terms of the unknown function (i.e., with respect to time and space) are considered. In order to acquire accurate numerical results, we apply the finite integration method based on shifted Chebyshev polynomials (FIM-SCP) to handle the spatial variable. These shifted Chebyshev polynomials are symmetric (either with respect to the point x=L/2 or the vertical line x=L/2 depending on their degree) over [0,L] , and their zeros in the interval are distributed symmetrically. We use these zeros to construct the main tool of FIM-SCP: the Chebyshev integration matrix. The forward difference quotient is used to deal with the temporal variable. Then, we obtain efficient numerical algorithms for solving both the direct and inverse problems. However, the ill-posedness of the inverse problem causes instability in the solution and, so, the Tikhonov regularization method is utilized to stabilize the solution. Furthermore, several direct and inverse numerical experiments are illustrated. Evidently, our proposed algorithms for both the direct and inverse problems give a highly accurate result with low computational cost, due to the small number of iterations and discretization.

Keywords: finite integration method; shifted Chebyshev polynomial; direct and inverse problems; Volterra integro-differential equation; Tikhonov regularization method

INTRODUCTION

An integro-differential equation (IDE) is an equation which contains both derivatives and integrals of an unknown function. Several situations in the branches of science and engineering can be demonstrated by developing mathematical models which are often in the form of IDEs, such as in RLC circuit analysis, the activity of interacting inhibitory and excitatory neurons, the Wilson–Cowan model, and so on; see Reference [**1**] for more applications. In fact, many of these problems cannot be directly solved, because we may not know all necessary information or an incomplete system may be provided. This has led to the study of both direct and inverse problems for a certain type of one-dimensional IDE involving time, which is called the one-dimensional time-dependent Volterra IDE (TVIDE). Hence, in this study, we investigate the TVIDE of the following form

$$v_t(x,t) + \mathcal{L}v(x,t) = \int_0^t \kappa_1(x,\eta)v(x,\eta)d\eta + \int_0^x \kappa_2(\xi,t)v(\xi,t)d\xi + F(x,t), \tag{1}$$

for all (x,t)∈(0,L)×(0,T], where *x* and *t* represent space and time variables, respectively; L is the spatial linear differential operator of order *n*; κ_1(x,t) and κ_2(x,t) are the given continuously integrable kernel functions; and v(x,t) is an unknown function, which is to be determined subject to prescribed initial and boundary conditions. We remark that, if a forcing term F(x,t) of (**1**) is given, then this problem has only one unknown $v(x,t)\in C^{n,1}([0,L]\times[0,T])$ to be solved, and it is called a direct problem. In contrast, if the forcing term F(x,t) is missing, then this problem has two unknowns $F(x,t)\in C([0,L]\times[0,T])$ and $v(x,t)\in C^{n,1}([0,L]\times[0,T])$ to be solved, and it is called an inverse problem. However, for the inverse problem in this paper, we specifically define the forcing term $F(x,t):=\beta(t)f(x,t)$, where $\beta(t)$ is a missing source function to be retrieved and f(x,t) is the given function. We note that (**1**) has both $\int_0^t \kappa_1(x,\eta) v(x,\eta)d\eta$ and $\int_0^x \kappa_2(\xi,t)v(\xi,t)d\xi$, while several studies in the literature have considered similar problems containing only one of these two terms.

The Volterra IDE containing only an integration term with respect to time arises in many applications, including the compression of poro-viscoelastic media, blow-up problems, analysis of space–time-dependent nuclear reactor dynamics, and so on; see Reference [**2**]. The existence, uniqueness, and asymptotic behavior of its solution have been discussed in Reference [**3**]. There are many authors who have studied the numerical solution of this type of problem by using techniques such as the finite element method [**2**], finite difference method (FDM) [**4**], collocation methods in polynomial spline [**5**], the implicit Runge–Kutta–Nyström method [**6**], the Legendre collocation method [**7**], and so on.

On the other hand, the Volterra IDE containing only an integration term with respect to space has also been studied in various areas, such as for the one-dimensional viscoelastic problem and one-dimensional heat flow in materials with memory [**8**], modeling heat/mass diffusion processes, biological species coexisting together with increasing and decreasing rates of growth, electromagnetism, and ocean circulation, among others [**9**]. Moreover, the existence and uniqueness for this type of Volterra IDE were shown in Reference [**8**]. Consequently, abundant numerical methods have appeared for finding solutions to this type of Volterra IDE using, for example, spline collocation method [**10**], collocation method with implicit Runge–Kutta method [**11**], decomposition method [**12**], and so on.

However, our problem deals with a Volterra IDE involving both temporal and spatial integrations. There have been no results in the literature regarding the existence and uniqueness of solutions to this type of problem. In this paper, we concentrate on providing a decent numerical procedure to find approximate solutions for both the direct and inverse problems of the proposed TVIDE (**1**).

Generally, it is well-known that the classification of problems involving differential equations was defined by Hadamard [**13**] in 1902. Mathematical problems involving differential equations are well-posed if the following conditions hold: existence, uniqueness, and stability. Otherwise, the problem is called ill-posed; this normally occurs in the inverse problem. Even though the initial and boundary conditions are prescribed, it is not sufficient to guarantee that our inverse problem (**1**) has unique solutions $\beta(t)$ and $v(x,t)$. Hence, additional conditions (e.g., the observation or measurement of data) need to be involved. In practice, there are many kinds of additional conditions; for example, a fixed point of the system, an average time of the system, or an integral of the system. After the additional conditions has been added as an auxiliary condition in our inverse problem (**1**), we can obtain the existence and uniqueness of $\beta(t)$ and $v(x,t)$. However, the additional condition may contains measurement or observation errors, which may cause the instability in the solutions; namely, a small perturbation in the input data can produce a considerable error, especially for $\beta(t)$. Thus, some regularization techniques are required to overcome the ill-posedness and stabilize the solution.

There exist many schemes which are generally used to solve both direct and inverse problems of Volterra IDEs, such as the above-mentioned methods. However, those methods utilize the process of approximating differentiation. It is well-known that numerical differentiation is very sensitive to rounding errors, as its manipulation task involves division by a small step-size. On the other hand, the process of numerical integration involves multiplication by a small step-size and, so, it is very insensitive to rounding errors. In recent years, the finite integration method (FIM) has been developed to find approximate solutions of linear boundary value problems for partial differential equations (PDEs). The concept of FIM is to transform a given PDE into an equivalent integral equation, following which a numerical integration method, such as the trapezoid, Simpson, or Newton–Cotes methods (see References [**14,15,16**]), are applied. In 2018, Boonklurb et al. [**17**] modified the traditional FIM by using Chebyshev polynomials to solve one- and two-dimensional linear PDEs and obtained a more accurate result compared to the traditional FIMs and FDM. However, their technique [**17**] has not yet been utilized to overcome the direct and inverse problems of TVIDE, which are the major focuses of this work.

In this paper, we formulate numerical algorithms for solving the direct and inverse problems of TVIDE (**1**). We manipulate the idea of FIM in Reference [**17**] by using shifted Chebyshev polynomials, which we call the FIM with shifted Chebyshev polynomials (FIM-SCP), to deal with the spatial variable and use the forward difference quotient to estimate the time derivative. We further apply the Tikhonov regularization method to stabilize our ill-posed problem (**1**). The rest of the paper is organized as follows. In **Section 2**, the definition and some basic properties concerning the shifted Chebyshev polynomial are given to construct the shifted Chebyshev integration matrices. The Tikhonov regularization method is also presented in **Section 2**. In **Section 3**, we use the FIM-SCP and the forward difference quotient to devise efficient numerical algorithms to find approximate solutions to the direct and inverse problems of (**1**). Then, we implement our proposed algorithms through several examples, in order to demonstrate their efficiency compared with their analytical solutions. Furthermore, we also display the time convergence rate and CPU time (s) in **Section 4**. Finally, the conclusion and some directions for future work are given in **Section 5**.

PRELIMINARIES

In this section, we introduce some necessary tools for solving the direct and inverse problems of TVIDE (**1**): the FIM-SCP and the Tikhonov regularization method.

Shifted Chebyshev Integration Matrices

We first introduce the definition and some basic properties of shifted Chebyshev polynomials [**18**], which are used to establish the first- and higher-order shifted Chebyshev integration matrices based on the idea of constructing integration matrices in Reference [**17**]. However, we slightly modify this idea by instead using a shifted Chebyshev expansion suitable for solving our problem (**1**) without domain transformation. We give the definition and properties as follows.

Definition 1. *The shifted Chebyshev polynomial of degree* n≥0 *is defined by*

$$S_n(x) = \cos\left(n \arccos\left(\frac{2x}{L} - 1\right)\right) \quad \text{for } x \in [0, L].$$

Note that this shifted Chebyshev polynomial is symmetric, either with respect to the point x=L/2 or the vertical line x=L/2 over [0,L], depending on its degree. Next, we provide some important properties of the shifted Chebyshev polynomial, which we use to constructing the shifted Chebyshev integration matrix, as follows.

Lemma 1. (i) *For* n∈N, *the zeros of* $S_n(x)$ *are symmetrically distributed over* [0,L] *and given by*

$$x_k = \frac{L}{2}\left(\cos\left(\frac{2k-1}{2n}\pi\right)+1\right),\ k \in \{1,2,3,...,n\}. \tag{2}$$

(ii) *For* r∈N, *the* rth-*order derivatives of* $S_n(x)$ *at the endpoint* b∈{0,L} *are*

$$\frac{d^r}{dx^r}S_n(x)\bigg|_{x=b} = \prod_{k=0}^{r-1}\left(\frac{n^2-k^2}{2k+1}\right)\left(\frac{2b}{L}-1\right)^{n+r}. \tag{3}$$

(iii) *For* x∈[0,L], *the single-layer integrations of shifted Chebyshev polynomial* $S_n(x)$ *are*

$$\bar{S}_0(x) = \int_0^x S_0(\xi)\,d\xi = x,$$

$$\bar{S}_1(x) = \int_0^x S_1(\xi)\,d\xi = \frac{x^2}{L} - x,$$

$$\bar{S}_n(x) = \int_0^x S_n(\xi)\,d\xi = \frac{L}{4}\left(\frac{S_{n+1}(x)}{n+1} - \frac{S_{n-1}(x)}{n-1} - \frac{2(-1)^n}{n^2-1}\right),\ n \in \{2,3,4,...\}.$$

(iv) *Let* $\{x_k\}^n_{k=1}$ *be a set of zeros of* $S_n(x)$, *the shifted Chebyshev matrix* S *is defined by*

$$\mathbf{S} = \begin{bmatrix} S_0(x_1) & S_1(x_1) & \cdots & S_{n-1}(x_1) \\ S_0(x_2) & S_1(x_2) & \cdots & S_{n-1}(x_2) \\ \vdots & \vdots & \ddots & \vdots \\ S_0(x_n) & S_1(x_n) & \cdots & S_{n-1}(x_n) \end{bmatrix}.$$

Then, it has the multiplicative inverse S^{-1}=1/n diag(1,2,2,...,2)ST.

Next, we use the above definition and properties of shifted Chebyshev polynomials to construct the shifted Chebyshev integration matrices. First, let *N* be a positive integer and *L* be a positive real number. Define an approximate solution u(x) of a certain differential equation by a linear combination of shifted Chebyshev polynomials $S_n(x)$; that is,

$$u(x) = \sum_{n=0}^{N-1} c_n S_n(x) \text{ for } x \in [0,L]. \tag{4}$$

Let xk for k∈{1,2,3,...,N} be the interpolated points which are meshed by the zeros of $S_N(x)$ defined in (**2**).

$$\begin{bmatrix} u(x_1) \\ u(x_2) \\ \vdots \\ u(x_N) \end{bmatrix} = \begin{bmatrix} S_0(x_1) & S_1(x_1) & \cdots & S_{N-1}(x_1) \\ S_0(x_2) & S_1(x_2) & \cdots & S_{N-1}(x_2) \\ \vdots & \vdots & \ddots & \vdots \\ S_0(x_N) & S_1(x_N) & \cdots & S_{N-1}(x_N) \end{bmatrix} \begin{bmatrix} c_0 \\ c_1 \\ \vdots \\ c_{N-1} \end{bmatrix},$$

which is denoted by u=Sc. The unknown coefficient vector can be performed by $c=S^{-1}u$. Let us consider the single-layer integration of u(x) from 0 to x_k, which is denoted by $U^{(1)}(x_k)$; we obtain

$$U^{(1)}(x_k) = \int_0^{x_k} u(\xi)\, d\xi = \sum_{n=0}^{N-1} c_n \int_0^{x_k} S_n(\xi)\, d\xi = \sum_{n=0}^{N-1} c_n \bar{S}_n(x_k)$$

for k∈{1,2,3,...,N} or, in matrix form,

$$\begin{bmatrix} U^{(1)}(x_1) \\ U^{(1)}(x_2) \\ \vdots \\ U^{(1)}(x_N) \end{bmatrix} = \begin{bmatrix} \bar{S}_0(x_1) & \bar{S}_1(x_1) & \cdots & \bar{S}_{N-1}(x_1) \\ \bar{S}_0(x_2) & \bar{S}_1(x_2) & \cdots & \bar{S}_{N-1}(x_2) \\ \vdots & \vdots & \ddots & \vdots \\ \bar{S}_0(x_N) & \bar{S}_1(x_N) & \cdots & \bar{S}_{N-1}(x_N) \end{bmatrix} \begin{bmatrix} c_0 \\ c_1 \\ \vdots \\ c_{N-1} \end{bmatrix}.$$

We denote the above matrix by $U^{(1)}=\bar{S}c=\bar{S}S^{-1}u:=Au$, where $A=\bar{S}S^{-1}:=[a_{ki}]_{N\times N}$ is called the first-order shifted Chebyshev integration matrix for the FIM-SCP; that is,

$$U^{(1)}(x_k) = \int_0^{x_k} u(\xi)\, d\xi = \sum_{i=1}^{N} a_{ki} u(x_i).$$

Next, consider the double-layer integration of u(x) from 0 to x_k, which denoted by $U^{(2)}(x_k)$. We have

$$U^{(2)}(x_k) = \int_0^{x_k} \int_0^{\xi_2} u(\xi_1)\, d\xi_1 d\xi_2 = \sum_{i=1}^{N} a_{ki} \int_0^{x_i} u(\xi_1)\, d\xi_1 = \sum_{i=1}^{N} \sum_{j=1}^{N} a_{ki} a_{ij} u(x_j)$$

for k∈{1,2,3,...,N}. It can be written, in matrix form, as $U^{(2)}=A^2u$. Similarly, we can calculate the *n*-layer integration of u(x) from 0 to x_k, which is denoted by $U^{(n)}(x_k)$. Then, we have

$$U^{(n)}(x_k) = \int_0^{x_k} \cdots \int_0^{\xi_2} u(\xi_1)\, d\xi_1 \cdots d\xi_n = \sum_{i_n=1}^{N} \cdots \sum_{j=1}^{N} a_{ki_n} \cdots a_{i_1 j} u(x_j)$$

for k∈{1,2,3,...,N}, which can be expressed, in matrix form, as $U^{(n)}=A^n u$.

Tikhonov Regularization Method

In this section, we briefly present the idea of the Tikhonov regularization method [19], which is usually applied to stabilize ill-posed problems, such as our inverse problem. Normally, the considered inverse problem can be represented by the system of m linear equations with n unknowns, as

$$Ax=b^{\epsilon}, \quad (5)$$

where bϵ is the vector in the right-hand side, which is perturbed by some noise ϵ, and x is the solution of the system (**5**) after perturbation. Tikhonov regularization replaces the inverse problem (**5**) by a minimization problem to obtain an efficiently approximate solution, which can be described as

$$\arg\min_{\mathbf{x}\in\mathbb{R}^n}\left\{\|\mathbf{Ax}-\mathbf{b}^{\epsilon}\|^2+\lambda\|\mathbf{x}\|^2\right\}, \quad (6)$$

where $\lambda>0$ is a regularization parameter balancing the weighting between the two terms of the function and $\|\cdot\|$ is the standard Euclidean norm. To reformulate the above minimization problem (**6**), we obtain

$$\arg\min_{\mathbf{x}\in\mathbb{R}^n}\left\{\left\|\begin{bmatrix}\mathbf{A}\\ \sqrt{\lambda}\mathbf{I}\end{bmatrix}\mathbf{x}-\begin{bmatrix}\mathbf{b}^{\epsilon}\\ \mathbf{0}\end{bmatrix}\right\|^2\right\}.$$

Clearly, this is a linear least-square problem in x. Then, the above problem turns out to be the normal equation of the form

$$\begin{bmatrix}\mathbf{A}\\ \sqrt{\lambda}\mathbf{I}\end{bmatrix}^{\top}\begin{bmatrix}\mathbf{A}\\ \sqrt{\lambda}\mathbf{I}\end{bmatrix}\mathbf{x}=\begin{bmatrix}\mathbf{A}\\ \sqrt{\lambda}\mathbf{I}\end{bmatrix}^{\top}\begin{bmatrix}\mathbf{b}^{\epsilon}\\ \mathbf{0}\end{bmatrix}.$$

To simplify the above equation, the solution x under the regularization parameter λ (denoted by $x\lambda$) can be computed by

$$\mathbf{x}_{\lambda}=(\mathbf{A}^{\top}\mathbf{A}+\lambda\mathbf{I})^{-1}\mathbf{A}^{\top}\mathbf{b}^{\epsilon}. \quad (7)$$

We can see that the accuracy of $x\lambda$ in (**7**) depends on the regularization parameter λ, which plays an important role in the calculation: A large regularization parameter may over-smoothen the solution, while a small regularization parameter may lose the ability to stabilize the solution. Therefore, a suitable choice of the regularization parameter λ is very significant for finding a stable approximate solution. There are many approaches for choosing a value of the parameter λ, such as the discrepancy principle criterion, the generalized cross-validation, the L-curve method, and so on. Nevertheless, the regularization parameter λ in this paper is chosen according to Morozov's discrepancy principle combined with Newton's method, which has been proposed in Reference [**20**]. We provide the procedure for calculating the optimal regularization parameter λ below, which can be carried out by the following steps:

Step 1: Set n=0 and give an initial regularization parameter $\lambda_0 > 0$.

Step 2: Compute $\mathbf{x}_{\lambda_n} = (\mathbf{A}^\top \mathbf{A} + \lambda_n \mathbf{I})^{-1} \mathbf{A}^\top \mathbf{b}^\epsilon$.

Step 3: Compute $\nabla \mathbf{x}_{\lambda_n} = -(\mathbf{A}^\top \mathbf{A} + \lambda_n \mathbf{I})^{-1} \mathbf{x}_{\lambda_n}$.

Step 4: Compute $G(\lambda_n) = \|\mathbf{A}\mathbf{x}_{\lambda_n} - \mathbf{b}^\epsilon\|^2 - \epsilon^2$.

Step 5: Compute $G'(\lambda_n) = 2\lambda_n \|\mathbf{A}\nabla \mathbf{x}_{\lambda_n}\|^2 + 2\lambda_n^2 \|\nabla \mathbf{x}_{\lambda_n}\|^2$.

Step 6: Compute $\lambda_{n+1} = \lambda_n - \frac{G(\lambda_n)}{G'(\lambda_n)}$.

Step 7: If $\|\lambda_{n+1} - \lambda_n\| < \delta$ for a tolerance δ, end. Else, set n=n+1 and return to Step 2.

Therefore, we receive the optimal regularization parameter λ, which is the terminal value λn obtained from the above procedure. When the regularization parameter λ is fixed as the mentioned optimal value, we can directly obtain the corresponding regularized solution by (**7**).

NUMERICAL ALGORITHMS FOR DIRECT AND INVERSE PROBLEMS OF TVIDE

In this section, we apply the FIM-SCP described in **Section 2.1** to devise the numerical algorithms for solving both the direct and inverse TVIDE problems (**1**), in order to obtain accurate approximate results. Let *u* be an approximate solution of *v* in (**1**). Then, we have the following linear TVIDE over the domain Ω=(0,L)×(0,T]:

$$u_t(x,t) + \mathcal{L}u(x,t) = \int_0^t \kappa_1(x,\eta)u(x,\eta)d\eta + \int_0^x \kappa_2(\xi,t)u(\xi,t)d\xi + F(x,t), \tag{8}$$

subject to the initial condition

$$u(x,0)=\phi(x), x\in[0,L], \tag{9}$$

and the boundary conditions

$$u^{(r)}(b,t)=\psi r(t), t\in[0,T], \tag{10}$$

for b∈{0,L} and r∈{0,1,2,...,n−1}, where *t* and *x* represent time and space variables, respectively. Additionally, κ_1, κ_2, *F*, ϕ, and ψr are given continuous functions and L is the spatial linear differential operator of order *n* defined by $\mathcal{L} := \sum_{i=0}^{n} p_i(x,t)\frac{d^i}{dx^i}$, where $p_i(x,t)$ is given and sufficiently smooth.

Procedure for Solving the Direct TVIDE Problem

First, we linearize (**8**) by uniformly discretizing the temporal domain into *M* subintervals with time step τ. Then, we specify (**8**) at a time t_m=mτ for m∈N and use the first-order forward difference quotient to estimate the time derivative term ut. Next, we replace each *x* by x_k for k∈{1,2,3,...,N} as generated by the zeros of the shifted Chebyshev polynomial SN(x) defined in (**2**). Thus, we have

$$\frac{u^{\langle m\rangle}-u^{\langle m-1\rangle}}{\tau}+\mathcal{L}u^{\langle m\rangle}=\int_0^{t_m}\kappa_1(x_k,\eta)u(x_k,\eta)d\eta+\int_0^{x_k}\kappa_2(\xi,t_m)u(\xi,t_m)d\xi+F^{\langle m\rangle}, \tag{11}$$

where $u^{\langle m\rangle}=u^{\langle m\rangle}(x_k)=u(x_k,t_m)$ and $F^{\langle m\rangle}=F^{\langle m\rangle}(x_k)=F(x_k,t_m)$. Next, consider the first integral term with respect to time by letting it be $J^{\langle m\rangle}1(x_k)$, we approximate $J^{\langle m\rangle}1(x_k)$ by using the trapezoidal rule. Thus, we approximate $J^{\langle m\rangle}{}_1(x_k)$ as

$$\begin{aligned}
J_1^{\langle m\rangle}(x_k) &:= \int_0^{t_m}\kappa_1(x_k,\eta)u(x_k,\eta)d\eta \\
&= \sum_{i=0}^{m-1}\int_{t_i}^{t_{i+1}}\kappa_1(x_k,\eta)u(x_k,\eta)d\eta \\
&\approx \sum_{i=0}^{m-1}\frac{\tau}{2}\left(\kappa_1^{\langle i\rangle}(x_k)u^{\langle i\rangle}(x_k)+\kappa_1^{\langle i+1\rangle}(x_k)u^{\langle i+1\rangle}(x_k)\right) \\
&= \frac{\tau}{2}\kappa_1^{\langle 0\rangle}(x_k)u^{\langle 0\rangle}(x_k)+\tau\sum_{i=1}^{m-1}\kappa_1^{\langle i\rangle}(x_k)u^{\langle i\rangle}(x_k)+\frac{\tau}{2}\kappa_1^{\langle m\rangle}(x_k)u^{\langle m\rangle}(x_k)
\end{aligned}$$

for each $x_k\in\{x_1,x_2,x_3,...,x_N\}$. The above equation can be written, in matrix form, as

$$\mathbf{J}_1^{\langle m\rangle}=\frac{\tau}{2}\mathbf{K}_1^{\langle 0\rangle}\mathbf{u}^{\langle 0\rangle}+\tau\sum_{i=1}^{m-1}\mathbf{K}_1^{\langle i\rangle}\mathbf{u}^{\langle i\rangle}+\frac{\tau}{2}\mathbf{K}_1^{\langle m\rangle}\mathbf{u}^{\langle m\rangle}, \tag{12}$$

where each parameter in (**12**) can be defined as follows:

$$\begin{aligned}
\mathbf{J}_1^{\langle m\rangle} &= \left[J_1^{\langle m\rangle}(x_1),J_1^{\langle m\rangle}(x_2),J_1^{\langle m\rangle}(x_3),...,J_1^{\langle m\rangle}(x_N)\right]^\top, \\
\mathbf{u}^{\langle i\rangle} &= \left[u^{\langle i\rangle}(x_1),u^{\langle i\rangle}(x_2),u^{\langle i\rangle}(x_3),...,u^{\langle i\rangle}(x_N)\right]^\top, \\
\mathbf{K}_1^{\langle i\rangle} &= \operatorname{diag}\left(\kappa_1^{\langle i\rangle}(x_1),\kappa_1^{\langle i\rangle}(x_2),\kappa_1^{\langle i\rangle}(x_3),...,\kappa_1^{\langle i\rangle}(x_N)\right).
\end{aligned}$$

Then, we consider the second integral term with respect to space by letting it be J⟨m⟩2(xk) and using the idea of FIM-SCP (as described in **Section 2.1**) to approximate it. Then, we obtain

$$J_2^{\langle m\rangle}(x_k):=\int_0^{x_k}\kappa_2(\xi,t_m)u(\xi,t_m)d\xi=\int_0^{x_k}\kappa_2^{\langle m\rangle}(\xi)u^{\langle m\rangle}(\xi)d\xi\approx\sum_{i=1}^{N}a_{ki}\kappa_2^{\langle m\rangle}(x_i)u^{\langle m\rangle}(x_i)$$

for each $x_k \in \{x_1, x_2, x_3, \ldots, x_N\}$. The above equation can be written, in matrix form, as

$$\mathbf{J}_2^{\langle m \rangle} = \mathbf{A}\mathbf{K}_2^{\langle m \rangle}\mathbf{u}^{\langle m \rangle}, \tag{13}$$

where $A = S\overline{\ }S^{-1}$ is the shifted Chebyshev integration matrix defined in **Section 2.1**,

$$\begin{aligned}
\mathbf{J}_2^{\langle m \rangle} &= \left[J_2^{\langle m \rangle}(x_1), J_2^{\langle m \rangle}(x_2), J_2^{\langle m \rangle}(x_3), \ldots, J_2^{\langle m \rangle}(x_N)\right]^{\top}, \\
\mathbf{u}^{\langle m \rangle} &= \left[u^{\langle m \rangle}(x_1), u^{\langle m \rangle}(x_2), u^{\langle m \rangle}(x_3), \ldots, u^{\langle m \rangle}(x_N)\right]^{\top}, \\
\mathbf{K}_2^{\langle m \rangle} &= \operatorname{diag}\left(\kappa_2^{\langle m \rangle}(x_1), \kappa_2^{\langle m \rangle}(x_2), \kappa_2^{\langle m \rangle}(x_3), \ldots, \kappa_2^{\langle m \rangle}(x_N)\right).
\end{aligned}$$

Then, we apply the FIM-SCP (described in **Section 2.1**) to eliminate all spatial derivatives from (**11**) by taking the *n*-layer integral on both sides of (**11**), to obtain the following equation at the shifted Chebyshev node x_k, as defined in (**2**), as

$$\int_0^{x_k} \cdots \int_0^{\xi_2} \left(\frac{u^{\langle m \rangle} - u^{\langle m-1 \rangle}}{\tau} + \mathcal{L}u^{\langle m \rangle}\right) d\xi_1 \ldots d\xi_n = \int_0^{x_k} \cdots \int_0^{\xi_2} \left(J_1^{\langle m \rangle} + J_2^{\langle m \rangle} + F^{\langle m \rangle}\right) d\xi_1 \ldots d\xi_n. \tag{14}$$

To simplify the *n*-layer integration of the spatial derivative terms of $Lu^{\langle m \rangle}$, by letting it be $Q^{\langle m \rangle}(x_k)$ and using the technique of integration by parts, we have

$$\begin{aligned}
Q^{\langle m \rangle}(x_k) :&= \int_0^{x_k} \cdots \int_0^{\xi_2} \mathcal{L}u^{\langle m \rangle}(\xi_1)\, d\xi_1 \ldots d\xi_n \\
&= \int_0^{x_k} \cdots \int_0^{\xi_2} \sum_{i=0}^{n} p_i(\xi_1, t_m) \frac{d^i}{dx^i} u^{\langle m \rangle}(\xi_1)\, d\xi_1 \ldots d\xi_n \\
&= \sum_{i=0}^{n} (-1)^i \binom{n}{i} \int_0^{x_k} \cdots \int_0^{\eta_2} p_n^{(i)}(\eta_1, t_m) u^{\langle m \rangle}(\eta_1) d\eta_1 \ldots d\eta_i \\
&+ \int_0^{x_k} \left(\sum_{i=0}^{n-1} (-1)^i \binom{n-1}{i} \int_0^{\xi_n} \cdots \int_0^{\eta_2} p_{n-1}^{(i)}(\eta_1, t_m) u^{\langle m \rangle}(\eta_1) d\eta_1 \ldots d\eta_i\right) d\xi_n \\
&+ \int_0^{x_k} \int_0^{\xi_n} \left(\sum_{i=0}^{n-2} (-1)^i \binom{n-2}{i} \int_0^{\xi_{n-1}} \cdots \int_0^{\eta_2} p_{n-2}^{(i)}(\eta_1, t_m) u^{\langle m \rangle}(\eta_1) d\eta_1 \ldots d\eta_i\right) d\xi_{n-1} d\xi_n \\
&\vdots \\
&+ \int_0^{x_k} \cdots \int_0^{\xi_2} p_0(\xi_1, t_m) u^{\langle m \rangle}(\xi_1) d\xi_1 \ldots d\xi_n + d_1 \frac{x_k^{n-1}}{(n-1)!} + d_2 \frac{x_k^{n-2}}{(n-2)!} + d_3 \frac{x_k^{n-3}}{(n-3)!} + \ldots + d_n,
\end{aligned}$$

where $d_1, d_2, d_3, \ldots, d_n$ are the arbitrary constants which emerge from the process of integration by parts. Then, we substitute each $x_k \in \{x_1, x_2, x_3, \ldots, x_N\}$ into the above equation and utilize the idea of FIM-SCP. Thus, we can express it, in matrix form, by

$$
\begin{aligned}
\mathbf{Q}^{\langle m \rangle} &= \sum_{i=0}^{n}(-1)^i\binom{n}{i}\mathbf{A}^i\mathbf{P}_n^{(i)}\mathbf{u}^{\langle m \rangle} + \sum_{i=0}^{n-1}(-1)^i\binom{n-1}{i}\mathbf{A}^{i+1}\mathbf{P}_{n-1}^{(i)}\mathbf{u}^{\langle m \rangle} \\
&+ \sum_{i=0}^{n-2}(-1)^i\binom{n-2}{i}\mathbf{A}^{i+2}\mathbf{P}_{n-2}^{(i)}\mathbf{u}^{\langle m \rangle} + \cdots + \mathbf{A}^n\mathbf{P}_0^{(0)}\mathbf{u}^{\langle m \rangle} + \mathbf{X}_n\mathbf{d} \\
&= \sum_{j=0}^{n}\sum_{i=0}^{n-j}(-1)^i\binom{n-j}{i}\mathbf{A}^{i+j}\mathbf{P}_{n-j}^{(i)}\mathbf{u}^{\langle m \rangle} + \mathbf{X}_n\mathbf{d},
\end{aligned}
\tag{15}
$$

where $A=S\overline{\quad}S^{-1}$ is the shifted Chebyshev integration matrix, $d=[d_1,d_2,d_3,\ldots,d_N]^T$,

$$
\begin{aligned}
\mathbf{Q}^{\langle m \rangle} &= \left[Q^{\langle m \rangle}(x_1), Q^{\langle m \rangle}(x_2), Q^{\langle m \rangle}(x_3), \ldots, Q^{\langle m \rangle}(x_N)\right]^{\top}, \\
\mathbf{X}_n &= \left[\mathbf{x}_{n-1}, \mathbf{x}_{n-2}, \mathbf{x}_{n-3}, \ldots, \mathbf{x}_0\right] \text{ for each } \mathbf{x}_i = \tfrac{1}{i!}\left[x_1^i, x_2^i, x_3^i, \ldots, x_N^i\right]^{\top}, \\
\mathbf{P}_{n-j}^{(i)} &= \operatorname{diag}\left(p_{n-j}^{(i)}(x_1,t_m), p_{n-j}^{(i)}(x_2,t_m), p_{n-j}^{(i)}(x_3,t_m), \ldots, p_{n-j}^{(i)}(x_N,t_m)\right).
\end{aligned}
$$

Finally, we vary all points $x_k \in \{x_1,x_2,x_3,\ldots,x_N\}$ in (**14**) and rearrange them into matrix form by using the FIM-SCP with the derived matrix equations (**12**), (**13**), and (**15**); thus, we obtain

$$
\frac{\mathbf{A}^n\mathbf{u}^{\langle m \rangle} - \mathbf{A}^n\mathbf{u}^{\langle m-1 \rangle}}{\tau} + \mathbf{Q}^{\langle m \rangle} = \mathbf{A}^n\mathbf{J}_1^{\langle m \rangle} + \mathbf{A}^n\mathbf{J}_2^{\langle m \rangle} + \mathbf{A}^n\mathbf{F}^{\langle m \rangle}
$$

or, factorizing the unknown solution $u^{\langle m \rangle}$ explicitly, as

$$
\begin{aligned}
&\left(\mathbf{A}^n + \tau\textstyle\sum_{j=0}^{n}\sum_{i=0}^{n-j}(-1)^i\binom{n-j}{i}\mathbf{A}^{i+j}\mathbf{P}_{n-j}^{(i)} - \frac{\tau^2}{2}\mathbf{A}^n\mathbf{K}_1^{\langle m \rangle} - \tau\mathbf{A}^{n+1}\mathbf{K}_2^{\langle m \rangle}\right)\mathbf{u}^{\langle m \rangle} \\
&+\mathbf{X}_n\mathbf{d} = \tfrac{\tau^2}{2}\mathbf{A}^n\mathbf{K}_1^{\langle 0 \rangle}\mathbf{u}^{\langle 0 \rangle} + \tau^2\textstyle\sum_{i=1}^{m-1}\mathbf{A}^n\mathbf{K}_1^{\langle i \rangle}\mathbf{u}^{\langle i \rangle} + \mathbf{A}^n\mathbf{u}^{\langle m-1 \rangle} + \tau\mathbf{A}^n\mathbf{F}^{\langle m \rangle}.
\end{aligned}
\tag{16}
$$

Next, consider the given boundary conditions (**10**) at the endpoints $b \in \{0,L\}$. We can convert them into matrix form by using the linear combination of shifted Chebyshev polynomial (**4**) in term of the rth-order derivative of *u* at the iteration time t_m and using (**3**). Then, we have

$$
\frac{d^r}{dx^r}u^{\langle m \rangle}(x)\Big|_{x=b} = \sum_{n=0}^{N-1} c_n^{\langle m \rangle}\frac{d^r}{dx^r}S_n(x)\Big|_{x=b} = \psi_r(t_m)
$$

for all $r \in \{0,1,2,\ldots,n-1\}$. We can express the above equation, in matrix form, as

$$\begin{bmatrix} S_0(b) & S_1(b) & \cdots & S_{N-1}(b) \\ S_0'(b) & S_1'(b) & \cdots & S_{N-1}'(b) \\ \vdots & \vdots & \ddots & \vdots \\ S_0^{(n-1)}(b) & S_1^{(n-1)}(b) & \cdots & S_{N-1}^{(n-1)}(b) \end{bmatrix} \begin{bmatrix} c_0^{\langle m \rangle} \\ c_1^{\langle m \rangle} \\ \vdots \\ c_{N-1}^{\langle m \rangle} \end{bmatrix} = \begin{bmatrix} \psi_0(t_m) \\ \psi_1(t_m) \\ \vdots \\ \psi_{n-1}(t_m) \end{bmatrix}, \tag{17}$$

which can be denoted by $Bc^{\langle m \rangle}=\Psi^{\langle m \rangle}$ or $BS-1u^{\langle m \rangle}=\Psi^{\langle m \rangle}$. Finally, we can construct the system of mth iterative linear equations from (**16**) and (**17**), which has N+n unknowns containing $u^{\langle m \rangle}$ and d, as follows:

$$\begin{bmatrix} \mathbf{H}^{\langle m \rangle} & \mathbf{X}_n \\ \mathbf{BS}^{-1} & \mathbf{0} \end{bmatrix} \begin{bmatrix} \mathbf{u}^{\langle m \rangle} \\ \mathbf{d} \end{bmatrix} = \begin{bmatrix} \mathbf{E}_1^{\langle m \rangle} \\ \mathbf{\Psi}^{\langle m \rangle} \end{bmatrix}, \tag{18}$$

where $H^{\langle m \rangle}$ is the coefficient matrix of $u^{\langle m \rangle}$ in (**16**) and $E^{\langle m \rangle}_1$ is the right-hand side column vector of (**16**). Consequently, the solution $u^{\langle m \rangle}$ can be approximated by solving the system (**18**) starting from the given initial condition (**9**); that is, $u\langle 0 \rangle=[\phi(x_1),\phi(x_2),\phi(x_3),\ldots,\phi(x_N)]^T$. Note that, when we would like to find a numerical solution u(x,t) at any point x∈[0,L] for the terminal time *T*, we can calculate it by the following formula:

$$u(x,T) = \sum_{n=0}^{N-1} c_n^{\langle m \rangle} S_n(x) = \mathbf{s}(x)\mathbf{c}^{\langle m \rangle} = \mathbf{s}(x)\mathbf{S}^{-1}\mathbf{u}^{\langle m \rangle},$$

where $s(x)=[S_0(x),S_1(x),S_2(x),\ldots,S_{N-1}(x)]$ and $u^{\langle m \rangle}$ is the final mth iterative solution of (**18**).

Procedure for Solving Inverse Problem of TVIDE

For the inverse problem in this paper, we specifically define the forcing term F(x,t):=β(t)f(x,t), where β(t) is a missing source function to be retrieved and f(x,t) is the given function. Thus, our considered time-dependent inverse TVIDE problem (**1**) becomes

$$u_t(x,t) + \mathcal{L}u(x,t) = \int_0^t \kappa_1(x,\eta)u(x,\eta)d\eta + \int_0^x \kappa_2(\xi,t)u(\xi,t)d\xi + \beta(t)f(x,t), \tag{19}$$

where *u* is an approximate solution of *v* and the other parameters are defined as in (**8**). The initial and boundary conditions of (**19**) are (**9**) and (**10**), which satisfy the compatibility conditions. Now, we remove all spatial derivatives from (**19**) and use the shifted Chebyshev integration matrix (as explained in **Section 2.1**). Then, we obtain the following matrix equation, based on the same process as in (**16**), as

$$\left(\mathbf{A}^n + \tau\sum_{j=0}^{n}\sum_{i=0}^{n-j}(-1)^i\binom{n-j}{i}\mathbf{A}^{i+j}\mathbf{P}_{n-j}^{(i)} - \frac{\tau^2}{2}\mathbf{A}^n\mathbf{K}_1^{\langle m\rangle} - \tau\mathbf{A}^{n+1}\mathbf{K}_2^{\langle m\rangle}\right)\mathbf{u}^{\langle m\rangle}$$
$$+\mathbf{X}_n\mathbf{d} - \tau\mathbf{A}^n\mathbf{f}^{\langle m\rangle}\beta^{\langle m\rangle} = \frac{\tau^2}{2}\mathbf{A}^n\mathbf{K}_1^{\langle 0\rangle}\mathbf{u}^{\langle 0\rangle} + \tau^2\sum_{i=1}^{m-1}\mathbf{A}^n\mathbf{K}_1^{\langle i\rangle}\mathbf{u}^{\langle i\rangle} + \mathbf{A}^n\mathbf{u}^{\langle m-1\rangle}, \tag{20}$$

where $\beta^{\langle m\rangle}=\beta(tm)$, $f^{\langle m\rangle}=[f(x_1,t_m),f(x_2,t_m),f(x_3,t_m),\ldots,f(x_N,t_m)]^T$ and the other parameters in (**20**) are as defined in **Section 3.1**. However, the occurrence of missing data is caused by the given conditions being insufficient to ensure a unique solution to our inverse problem. Hence, an additional condition or observed data needs to be involved. Thus, we use an additional condition, regarding the aggregated solution of the system, in the following form:

$$\int_0^L u(\xi,t)\,d\xi = g(t), \quad t \in [0,T], \tag{21}$$

where g(t) is the measured data at time *t*, which probably contains measurement errors. In order to illustrate the realistic phenomena of this problem, we assume that the measurement data of the aggregated solution g(t) involves some noise ϵ, which is denoted by $g\epsilon(t)$ (where $\|g\epsilon(t)-g(t)\|\leq\epsilon$) and define the noisy value ϵ by a random variable generated by the Gaussian normal distribution with mean $\mu=0$ and standard deviation $\sigma=p|g(t)|$, where *p* is the percentage of the noise to be input. Then, the additional condition (**21**) becomes

$$\int_0^L u(\xi,t)\,d\xi = g^{\epsilon}(t), \quad t \in [0,T]. \tag{22}$$

Using the concept of FIM-SCP, the additional condition (**22**) at time tm can be written, in vector form, as

$$\int_0^L u^{(m)}(\xi)\,d\xi = \sum_{n=0}^{N-1} c_n^{(m)}\int_0^L S_n(\xi)\,d\xi = \sum_{n=0}^{N-1} c_n^{(m)}\bar{S}_n(L)\,d\xi := \mathrm{z}c^{(m)} = \mathrm{z}S^{-1}\mathrm{u}^{(m)} = g^{\epsilon}(t_m), \tag{23}$$

where $z=[\bar{S}_0(L),\bar{S}_1(L),\bar{S}_2(L),\ldots,\bar{S}_{N-1}(L)]$ and each $\bar{S}_n$ (L) is as defined in Lemma 1(iii). Finally, we can establish the following system of mth iterative linear equations for the inverse TVIDE problem (**19**) by utilizing (**20**) and (**23**), which has N+n+1 unknown variables including $u^{\langle m\rangle}$, d, and $\beta^{\langle m\rangle}$, as

$$\begin{bmatrix} \mathbf{H}^{\langle m\rangle} & \mathbf{X}_n & -\tau\mathbf{A}^n\mathbf{f}^{\langle m\rangle} \\ \mathbf{B}\mathbf{S}^{-1} & 0 & 0 \\ \mathbf{z}\mathbf{S}^{-1} & 0 & 0 \end{bmatrix} \begin{bmatrix} \mathbf{u}^{\langle m\rangle} \\ \mathbf{d} \\ \beta^{\langle m\rangle} \end{bmatrix} = \begin{bmatrix} \mathbf{E}_2^{\langle m\rangle} \\ \mathbf{\Psi}^{\langle m\rangle} \\ g^{\epsilon}(t_m) \end{bmatrix}, \tag{24}$$

where $H^{\langle m\rangle}$ is the coefficient matrix of $u^{\langle m\rangle}$ defined in (**20**) and $E^{\langle m\rangle}_2$ is the right-hand side column vector of (**20**). Before seeking an approximate solution $u^{(m)}$ and source term $\beta^{(m)}$, as we have mentioned, we must address that our inverse problem is ill-posed. When a noisy value is input into the system, it may cause a significant error. Hence, we need to stabilize the solution of (**24**) by employing the Tikhonov regularization method. We denote the linear system (**24**) by the simplified matrix equation as

$$Ry=b^{\epsilon}. \qquad (25)$$

Applying the Tikhonov regularization method (**6**) in order to filter out the noise in the corresponding perturbed data, we can stabilize the numerical solution (**25**) by using (**7**). Thus, we have

$$\mathbf{y}_{\lambda} = (\mathbf{R}^{\top}\mathbf{R} + \lambda\mathbf{I})^{-1}\mathbf{R}^{\top}\mathbf{b}^{\epsilon}. \qquad (26)$$

Finally, we can receive the optimal regularization parameter λ by using Morozov's discrepancy principle combined with Newton's method, as described in **Section 2.2**. Thus, we can directly obtain the corresponding regularized solution by (**26**).

Algorithms for Solving the Direct and Inverse TVIDE Problems

For computational convenience, we summarize the aforementioned procedures for finding approximate solutions to the direct (**8**) and inverse (**19**) TVIDE problems in **Section 3.1** and **Section 3.2**, respectively, as the numerical Algorithms 1 and 2, which are in the form of pseudocode.

Algorithm 1 Numerical algorithm for solving the direct TVIDE problem via FIM-SCP

Input: $x, \tau, L, T, N, \phi(x), \psi_r(t), p_i(x,t), \kappa_1(x,t), \kappa_2(x,t)$, and $F(x,t)$.

Output: An approximate solution $u(x,T)$.

Set $x_k = \frac{L}{2}\left(\cos\left(\frac{2k-1}{2N}\pi\right)+1\right)$ for $k \in \{1,2,3,...,N\}$ in descending order.
Compute $\mathbf{A}, \mathbf{B}, \mathbf{S}, \bar{\mathbf{S}}, \mathbf{S}^{-1}, \mathbf{X}_n$, and $\mathbf{u}^{(0)}$.
Set $m = 1$ and $t_1 = \tau$.
while $t_m \leq T$ **do**
    Compute $\mathbf{K}_1^{(m)}, \mathbf{K}_2^{(m)}, \mathbf{F}^{(m)}, \mathbf{H}^{(m)}, \mathbf{\Psi}^{(m)}$, and $\mathbf{E}_1^{(m)}$.
    Find $\mathbf{u}^{(m)}$ by solving the linear system (18).
    Update $m = m + 1$.
    Compute $t_m = m\tau$.
end while
return Find $u(x,T) = \mathbf{s}(x)\mathbf{S}^{-1}\mathbf{u}^{(m)}$.

Algorithm 2 Numerical algorithm for solving the inverse TVIDE problem via FIM-SCP

Input: $x, p, \tau, \delta, L, T, N, \lambda_0, \phi(x), g(t), \psi_r(t), p_i(x,t), \kappa_1(x,t), \kappa_2(x,t)$, and $f(x,t)$.

Output: An approximate solution $u(x,T)$ and the source terms $\beta(t_m)$ at all discretized times.

```
Set x_k = (L/2)(cos((2k-1)/(2N) π) + 1) for k ∈ {1,2,3,...,N} in descending order.
Compute A, B, S, S̃, S^{-1}, X_n, A_N, and u^(0).
Set m = 1 and t_1 = τ.
while t_m ≤ T do
    Set the measurement data g^ε(t_m) = g(t_m) + ε, where ε ~ N(0, p^2 |g(t_m)|^2).
    Compute K_1^(m), K_2^(m), f^(m), H^(m), Ψ^(m), E_2^(m), R, and b^ε.
    Set n = 0.
    do
        Compute y_{λ_n} = (R^T R + λ_n I)^{-1} R^T b^ε.
        Compute ∇y_{λ_n} = -(R^T R + λ_n I)^{-1} y_{λ_n}.
        Compute G(λ_n) = ||R y_{λ_n} - b^ε||^2 - ε^2.
        Compute G'(λ_n) = 2λ_n ||R ∇y_{λ_n}||^2 + 2λ_n^2 ||∇y_{λ_n}||^2.
        Compute λ_{n+1} = λ_n - G(λ_n)/G'(λ_n).
        Update n = n + 1.
    while ||λ_n - λ_{n-1}|| ≥ δ
    Set the optimal regularization parameter λ = λ_n.
    Find u^(m) and β^(m) by explicitly solving y_λ using the matrix equation (7).
    Update m = m + 1.
    Compute t_m = mτ.
end while
return Find u(x,T) = s(x) S^{-1} u^(m).
```

NUMERICAL EXPERIMENTS

In this section, we implement our devised numerical algorithms for solving the direct and inverse TVIDE problems through several examples, in order to demonstrate the efficiency and accuracy of the solutions obtained by proposed methods. Examples 1 and 2 are used to examine Algorithm 1 for the direct TVIDE problems (**8**). Examples 3 and 4 are inverse TVIDE problems (**19**), as solved by Algorithm 2. Additionally, time convergence rates and CPU times(s) for each example are presented to indicate the computational cost and time. The time convergence rate is defined by $\text{Rate} = \lim_{t_m \to T} \frac{\|\mathbf{u}^*(t_{m+1}) - \mathbf{u}(t_{m+1})\|_\infty}{\|\mathbf{u}^*(t_m) - \mathbf{u}(t_m)\|_\infty}$, where T is the terminal time, tm is a partitioned time contained in [0,T], u∗(tm) is the exact solution at time t_m, $u(t_m)$ is the numerical solution at time tm, and $\|\cdot\|^\infty$ is the l^∞ norm. Graphical solutions of each example are also depicted. Our numerical algorithms were implemented using the MatLab R2016a software, run on a Intel(R) Core(TM) i7-6700 CPU @ 3.40 GHz computer system.

Example 1. *Consider the following direct TVIDE problem, which consists of a second-order derivative with constant coefficient for* x∈(0,1) *and* t∈(0,T]*:*

$$u_t + u_{xx} + u = \int_0^t 2e^{-x} u(x,y)\,dy + \int_0^x (t+\xi)\,u(\xi,t)\,d\xi + F(x,t), \tag{27}$$

where

F(x,t)=−t−e^x(t^2−3t+tx−1),

subject to the homogeneous initial condition u(x,0)=0 for x∈[0,1] and the Dirichlet boundary conditions u(0,t)=t and u(1,t)=te for t∈[0,T]. The analytical solution of this problem is u∗(x,t)=tex.

In the numerical testing based on Algorithm 1, we first took the double-layer integral of both sides of (**27**) and transformed it into matrix form (**16**). Then, we obtained the approximate solutions u(x,T) for this problem (**27**) by applying the numerical Algorithm 1. The accuracy of our obtained approximate results was measured by the mean absolute error, which compared it to the analytical solution at different values of x∈{0.1,0.3,0.5,0.7,0.9} and the terminal time T=1, as shown in **Table 1**. From **Table 1**, we observe that, when the partitioning number of the temporal domain *M* was fixed and nodal numbers *N* were increasingly varied, then the accuracy was significantly improved. Similarly, for a fixed nodal number *N* but various time partitioning numbers *M*, the accuracy results were also significantly improved. Moreover, the convergence rates with respect to the time in Algorithm 1 were estimated for various numbers of the time partition (M∈{5,10,15,20,25}) for the spatial points N=10, as shown in **Table 2**. We can notice, from **Table 2**, that these time convergence rates for the $\ell\infty$ norm indeed approached linear convergence for T∈{5,10,15}. The computational cost, in terms of CPU times (s), is also displayed in **Table 2**. Finally, a graph of our approximate solutions u(x,t) for different times *t* and the surface plot of the solution under the parameters N=20, M=20, and T=1 are depicted in **Figure 1**.

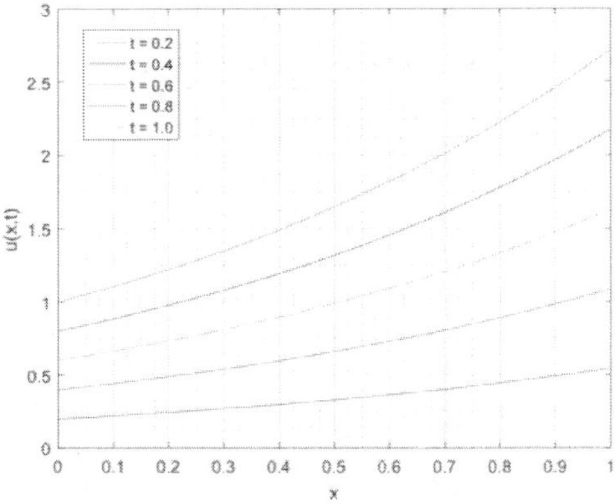

(a) $u(x,t)$ at different times t

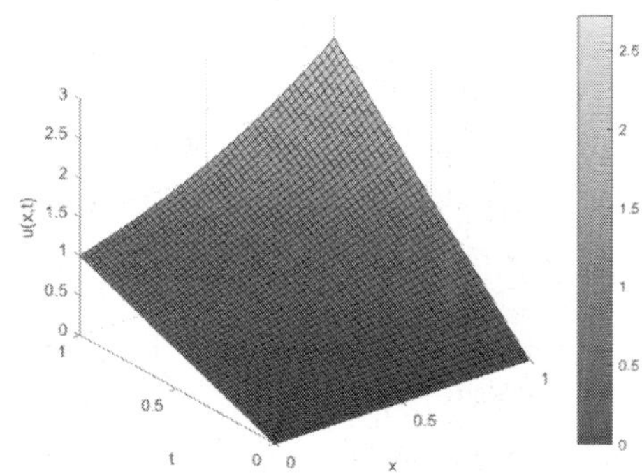

(b) Surface plot of $u(x,t)$

Figure 1. The graphical results of Example 1 for N=20, M=20, and T=1.

Table 1. Mean absolute errors between exact and numerical solutions of u(x,1) for Example 1.

x	$M = 20$			$N = 12$		
	$N = 8$	$N = 10$	$N = 12$	$M = 11$	$M = 13$	$M = 15$
0.1	1.6855×10^{-5}	1.1723×10^{-8}	1.0208×10^{-10}	2.3823×10^{-7}	5.0060×10^{-9}	3.3268×10^{-11}
0.3	4.3851×10^{-5}	3.0501×10^{-8}	2.6554×10^{-10}	6.1976×10^{-7}	1.3024×10^{-8}	8.6539×10^{-11}
0.5	5.3554×10^{-5}	3.7247×10^{-8}	3.2429×10^{-10}	7.5684×10^{-7}	1.5904×10^{-8}	1.0567×10^{-10}
0.7	4.2555×10^{-5}	2.9599×10^{-8}	2.5767×10^{-10}	6.0140×10^{-7}	1.2638×10^{-8}	8.3964×10^{-11}
0.9	1.5864×10^{-5}	1.1034×10^{-8}	9.6049×10^{-11}	2.2419×10^{-7}	4.7112×10^{-9}	3.1289×10^{-11}

Table 2. Time convergence rates and CPU times (s) for Example 1 by Algorithm 1 with N=10.

M	T = 5			T = 10			T = 15		
	$\|u^* - u\|_\infty$	Rate	Time(s)	$\|u^* - u\|_\infty$	Rate	Time(s)	$\|u^* - u\|_\infty$	Rate	Time(s)
5	4.298×10^{-12}	1.4584	0.0465	8.565×10^{-12}	1.5076	0.0456	1.262×10^{-11}	1.5112	0.0469
10	4.318×10^{-12}	1.2499	0.0487	8.576×10^{-12}	1.2863	0.0469	1.276×10^{-11}	1.3040	0.0485
15	4.309×10^{-12}	1.1723	0.0495	8.547×10^{-12}	1.2008	0.0481	1.275×10^{-11}	1.2135	0.0501
20	4.311×10^{-12}	1.1327	0.0506	8.533×10^{-12}	1.1555	0.0516	1.277×10^{-11}	1.1657	0.0535
25	1.135×10^{-12}	1.1353	0.0521	8.540×10^{-12}	1.1272	0.0538	1.277×10^{-11}	1.1365	0.0553

Example 2. *Consider the following direct TVIDE problem, which consists of a third-order derivative with variable coefficient for* x∈(0,1) *and* t∈(0,T]*:*

$$u_t + tu_{xxx} + \cos(x)u_{xx} = \int_0^t \frac{2}{x-1} u(x,\eta)d\eta + \int_0^x \frac{6t}{\xi - 1} u(\xi,t)d\xi + F(x,t), \tag{28}$$

where

$F(x,t)=x-x^2+xt^2(3x+1)-2t\cos(x)$,

subject to the initial condition u(x,0)=0 for x∈[0,1] and the boundary conditions u(0,t)=0, u(1,t)=0, and u′(0,t)=t for t∈[0,T]. The analytical solution of this problem is $u^*(x,t)=(x-x^2)t$.

We test the efficiency and accuracy of the proposed Algorithm 1 via the problem (**28**). First, we took a triple-layer integral on both sides of (**28**) and utilized the shifted Chebyshev integration matrix to transform it into matrix form (**16**). Next, we implemented Algorithm 1 to obtain numerical solutions u(x,T) for this problem (**28**). **Table 3** shows the precision of our obtained approximate results at different values of x∈{0.1,0.3,0.5,0.7,0.9} and at the terminal time T=1, through the mean absolute error. We can see that the accuracy was significantly improved according to an increase in the number of both the partitioning space and time domains. However, we observe that, in the case of fixed *N*, when *M* was increased, the mean absolute errors provide accurate results with a lower computational number *M*. Furthermore, the time convergence rates concerning the ℓ^∞ norm and CPU times (s) are demonstrated in **Table 4**, under various values of *M* (M∈{5,10,15,20,25}) and final times *T* (T∈{5,10,15}). The graphical solutions for u(x,t) in both one and two dimensions are shown in **Figure 2**.

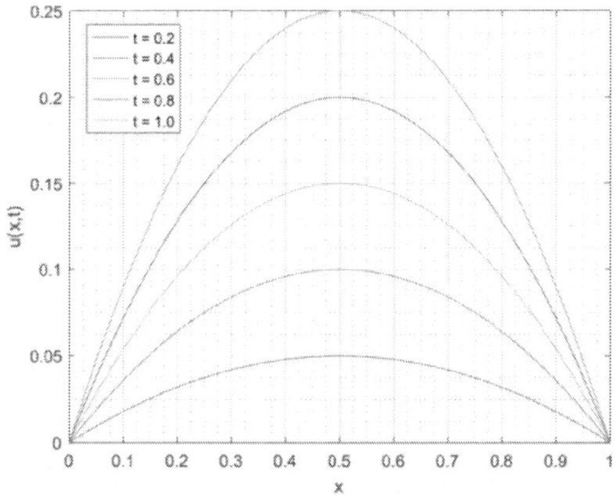

(**a**) $u(x,t)$ at different times t

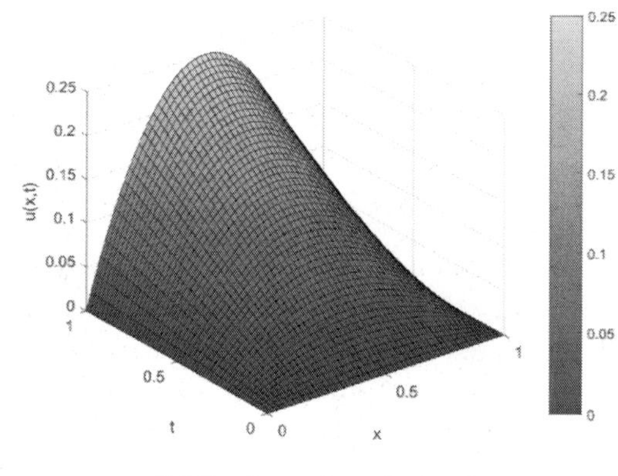

(**b**) Surface plot of $u(x,t)$

Figure 2. The graphical results of Example 2 for N=20, M=20, and T=1.

Table 3. Mean absolute errors between exact and numerical solutions of u(x,1) for Example 2.

x	$M = 10$			$N = 10$		
	$N = 8$	$N = 10$	$N = 12$	$M = 5$	$M = 10$	$M = 15$
0.1	3.5098×10^{-10}	5.0498×10^{-13}	1.6695×10^{-14}	5.1849×10^{-13}	5.0498×10^{-13}	4.9435×10^{-13}
0.3	1.0060×10^{-9}	1.1285×10^{-12}	4.1411×10^{-14}	1.1544×10^{-12}	1.1285×10^{-12}	1.0850×10^{-12}
0.5	1.0780×10^{-9}	1.4543×10^{-12}	5.1958×10^{-14}	1.4672×10^{-12}	1.4543×10^{-12}	1.3845×10^{-12}
0.7	1.0237×10^{-9}	1.1625×10^{-12}	4.5908×10^{-14}	1.1572×10^{-12}	1.1625×10^{-12}	1.0923×10^{-12}
0.9	3.1567×10^{-10}	5.2400×10^{-13}	2.0983×10^{-14}	5.1567×10^{-13}	5.2400×10^{-13}	4.9050×10^{-13}

Table 4. Time convergence rates and CPU times (s) for Example 2 by Algorithm 1 with N=10.

M	$T = 5$			$T = 10$			$T = 15$		
	$\|u^* - u\|_\infty$	Rate	Time(s)	$\|u^* - u\|_\infty$	Rate	Time(s)	$\|u^* - u\|_\infty$	Rate	Time(s)
5	1.426×10^{-12}	1.0241	0.0524	1.620×10^{-12}	1.0489	0.0531	3.549×10^{-12}	1.3720	0.0535
10	1.533×10^{-12}	1.0334	0.0577	1.635×10^{-12}	1.0278	0.0576	1.874×10^{-12}	1.1035	0.0576
15	1.426×10^{-12}	1.0208	0.0597	1.664×10^{-12}	1.0243	0.0585	1.806×10^{-12}	1.0294	0.0598
20	1.476×10^{-12}	1.0223	0.0609	1.609×10^{-12}	1.0210	0.0610	1.537×10^{-12}	1.0203	0.0620
25	1.496×10^{-12}	1.0165	0.0619	1.488×10^{-12}	1.0182	0.0638	1.276×10^{-12}	1.0079	0.0641

Example 3. *Consider the following inverse TVIDE problem, which consists of a second-order derivative with constant coefficient and a continuous forcing function* f(x,t) *for* x∈(0,1) *and* t∈(0,T]*:*

$$u_t - u_{xx} + 2u = \int_0^t 2\ln(x)u(x,\eta)d\eta + \int_0^x e^{-\xi}u(\xi,t)d\xi + \beta(t)f(x,t), \tag{29}$$

where

$$f(x,t) = e^{2t}\left[1 + t - x + e^x + te^{-x} - (2e^x + t)t\ln x\right],$$

subject to the initial condition u(x,0)=ex *for* x∈[0,1] *and the boundary conditions* u(0,t)=t+1 *and* u(1,t)=t+e *for* t∈[0,T]. *The additional condition, in terms of the aggregated solution of the system, is* g(t)=t+e−1. *The analytical solutions of this problem are* $u^*(x,t)=t+e^x$ *and* β∗(t)=e−2t.

Implementing the numerical Algorithm 2 by taking the double-layer integral of both sides of (29) and transforming it into matrix form (24), we obtained the approximate solutions u(x,1) and β(t) for this problem (29). As the additional condition was measurement data, there may be an error in the measurement. Therefore, we perturbed the additional condition g(t) with a percentage *p* of the noise (p∈{0%,1%,3%,5%}). In Table 5, we show the accuracy of the solutions u(x,1) and β(t), in terms of the mean absolute error, respectively, denoted by $\mathcal{E}_u = \frac{1}{N}\sum_{i=1}^{N}|u_i^* - u_i|$ and $\mathcal{E}_\beta = \frac{1}{M}\sum_{j=1}^{M}|\beta_j^* - \beta_j|$, and the values of the optimal regularization parameters λ at time t=1 with various M=N∈{5,10,15,20}. From Table 5, we can observe that the optimal regularization parameters λ were close to zero and the mean absolute errors for both E_u and E_β significantly increased with an increasing percentage p of the perturbation. Furthermore, we used the regularization parameter λ=0 to explore the rates of convergence with respect to the ℓ∞ norm and CPU times (s) for M=N∈{5,10,15,20} with the final times T∈{1,2,3} as shown in Table 6. The graphical solutions of the perturbed functions u(x,1) and β(t) for p∈{1%,3%,5%} are depicted in Figure 3.

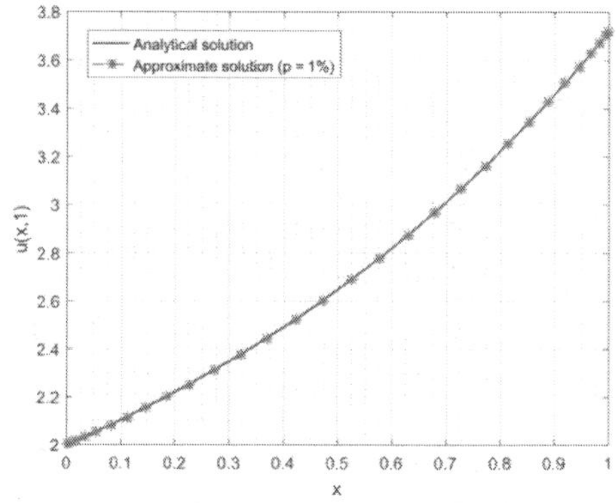

(**a**) $u(x, 1)$ with $p = 1\%$

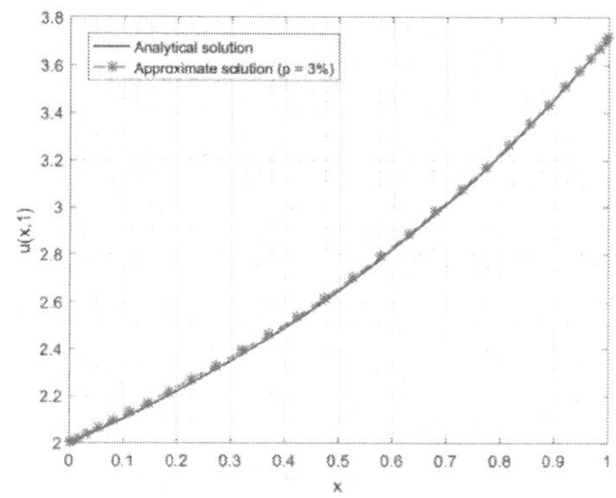

(**b**) $u(x, 1)$ with $p = 3\%$

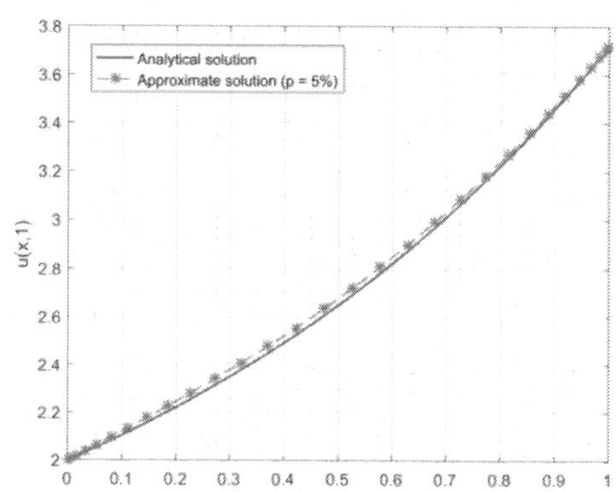

(**c**) $u(x, 1)$ with $p = 5\%$

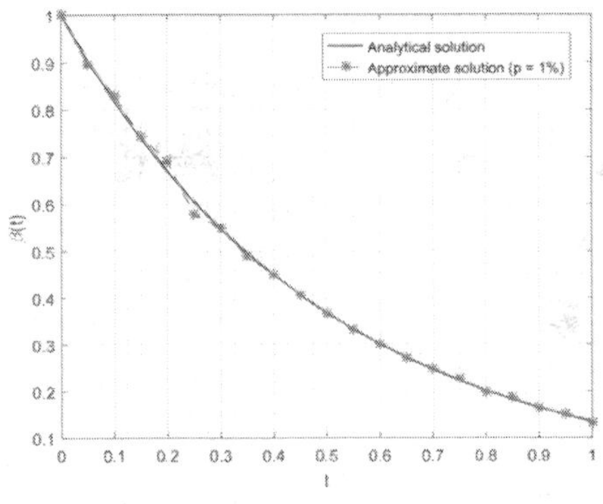

(d) $\beta(t)$ with $p = 1\%$

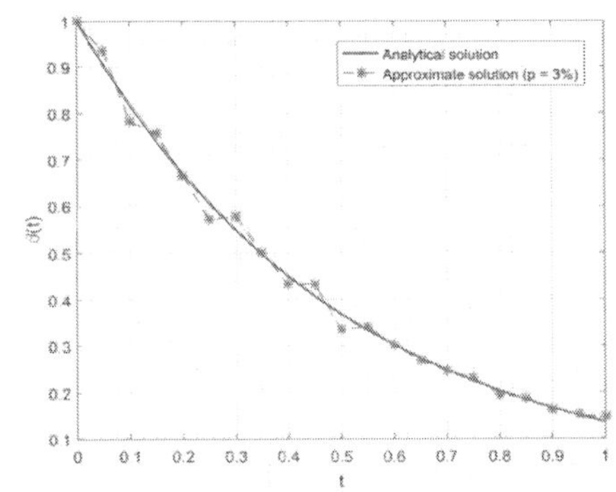

(e) $\beta(t)$ with $p = 3\%$

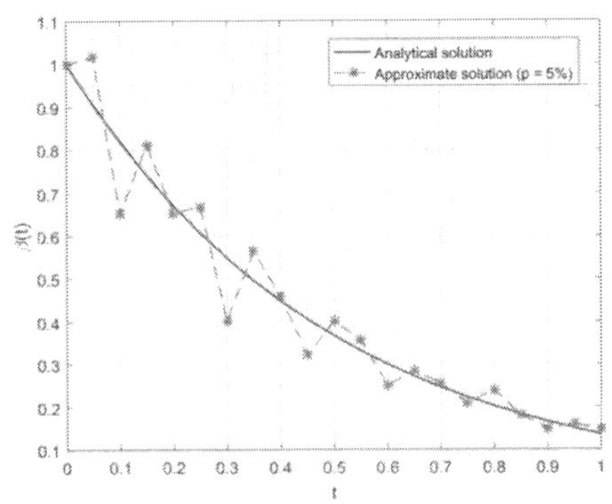

(f) $\beta(t)$ with $p = 5\%$

Figure 3. The graphical results of u(x,1) and β(t) for Example 3 with N=30 and M=20.

Table 5. Mean absolute errors of u(x,1) and β(t) for optimal regularization parameter λ of Example 3.

$M=N$	$p=0\%$			$p=1\%$		
	λ	$\mathcal{E}_u$	$\mathcal{E}_\beta$	λ	$\mathcal{E}_u$	$\mathcal{E}_\beta$
5	6.22×10^{-14}	1.6609×10^{-5}	7.9997×10^{-7}	3.11×10^{-12}	1.1372×10^{-4}	4.1098×10^{-4}
10	2.38×10^{-18}	3.9844×10^{-13}	1.0459×10^{-12}	2.20×10^{-13}	3.4011×10^{-4}	8.8853×10^{-4}
15	1.02×10^{-17}	9.9950×10^{-14}	1.6384×10^{-13}	9.17×10^{-12}	7.8857×10^{-4}	7.0288×10^{-4}
20	2.14×10^{-18}	2.9774×10^{-13}	1.7125×10^{-13}	4.33×10^{-14}	1.9024×10^{-4}	1.3201×10^{-3}
$M=N$	$p=3\%$			$p=5\%$		
	λ	$\mathcal{E}_u$	$\mathcal{E}_\beta$	λ	$\mathcal{E}_u$	$\mathcal{E}_\beta$
5	8.80×10^{-11}	1.2533×10^{-3}	8.3870×10^{-3}	8.61×10^{-12}	2.2805×10^{-3}	1.0964×10^{-2}
10	6.64×10^{-11}	3.7067×10^{-3}	9.5414×10^{-3}	1.11×10^{-11}	5.0047×10^{-3}	2.7003×10^{-2}
15	1.09×10^{-12}	8.4361×10^{-3}	7.0094×10^{-3}	8.79×10^{-12}	3.2925×10^{-2}	3.2201×10^{-2}
20	8.61×10^{-12}	3.3382×10^{-3}	1.3582×10^{-2}	1.40×10^{-13}	1.0214×10^{-2}	3.8774×10^{-2}

Table 6. Time convergence rates and CPU times (s) for Example 3 by Algorithm 2 with N=10.

M	$T=1$			$T=2$			$T=3$		
	$\|u^*-u\|_\infty$	Rate	Time(s)	$\|u^*-u\|_\infty$	Rate	Time(s)	$\|u^*-u\|_\infty$	Rate	Time(s)
5	8.495×10^{-13}	1.0046	0.0667	9.130×10^{-13}	1.0203	0.0655	3.602×10^{-12}	1.6644	0.0662
10	8.131×10^{-13}	0.9970	0.0679	8.659×10^{-13}	1.0042	0.0667	2.456×10^{-12}	1.2409	0.0673
15	7.851×10^{-13}	0.9965	0.0684	8.362×10^{-13}	1.0001	0.0675	1.731×10^{-12}	1.1355	0.0693
20	8.344×10^{-13}	1.0003	0.0716	7.829×10^{-13}	0.9967	0.0722	2.928×10^{-12}	1.0956	0.0720
25	8.362×10^{-13}	1.0003	0.0776	8.686×10^{-13}	1.0022	0.0766	2.134×10^{-12}	1.0699	0.0751

Example 4. *Consider the following inverse TVIDE problem, which consists of a second-order derivative with variable coefficient and the piecewise forcing function* f(x,t) *for* x∈(0,1) *and* t∈(0,T]*:*

$$u_t + u_{xx} + u_x - \cos(xt)u = \int_0^t 2\sin(x)u(x,\eta)d\eta - \int_0^x 3t\cos(\xi)u(\xi,t)d\xi + \beta(t)f(x,t), \tag{30}$$

where

$$f(x,t) = \begin{cases} \frac{1}{2}\left[2t\cos(2x) + t\sin(2x) + (t\cos(xt)+1)\sin^2 x\right], & 0 < t \le \frac{T}{3}, \\ \frac{1}{3}\left[2t\cos(2x) + t\sin(2x) + (t\cos(xt)+1)\sin^2 x\right], & \frac{T}{3} < t \le \frac{2T}{3}, \\ \frac{1}{4}\left[2t\cos(2x) + t\sin(2x) + (t\cos(xt)+1)\sin^2 x\right], & \frac{2T}{3} < t \le T, \end{cases}$$

subject to the initial condition u(x,0)=0 *for* x∈[0,1] *and the Dirichlet boundary conditions* u(0,t)=0 *and* u(1,t)=t$\sin^2$(1) *for* t∈[0,T]. *The additional condition, in terms of the aggregated solution of the system, is* g(t)=t/4(2−sin(2))+e^t. *The analytical solutions of this problem are* u∗(x,t)=t$\sin^2$(x) *and*

$$\beta^*(t) = \begin{cases} 2, & 0 < t \le \frac{T}{3}, \\ 3, & \frac{T}{3} < t \le \frac{2T}{3}, \\ 4, & \frac{2T}{3} < t \le T. \end{cases}$$

Based on the numerical Algorithm 2, we took the double-layer integral to both sides of (**30**) and transformed it into matrix form (**24**). We obtained the approximate solutions u(x,1) and β(t) for (**29**) by implementing Algorithm 2. **Table 7** shows the accuracy of the solutions u(x,1) and β(t) obtained by our numerical algorithm, in terms of the mean absolute errors E_u and E_β, as well as the values of the optimal regularization parameter λ at time t=1 with the noisy percentage p∈{0%,1%,5%,10%} under various M=N∈{6,9,12,15}. Although this problem had the piecewise forcing term f(x,t), our Algorithm 2 perfectly performed in providing accurate results, as shown in **Table 7**. The time convergence rates concerning the ℓ^∞ norm and CPU times (s) are shown in **Table 8**, under various numbers of M∈{6,9,12,15,18} with the final times T∈{1,2,3}. The graphical perturbed solutions u(x,1) and β(t) for p∈{1%,5%,10%} are shown in **Figure 4**.

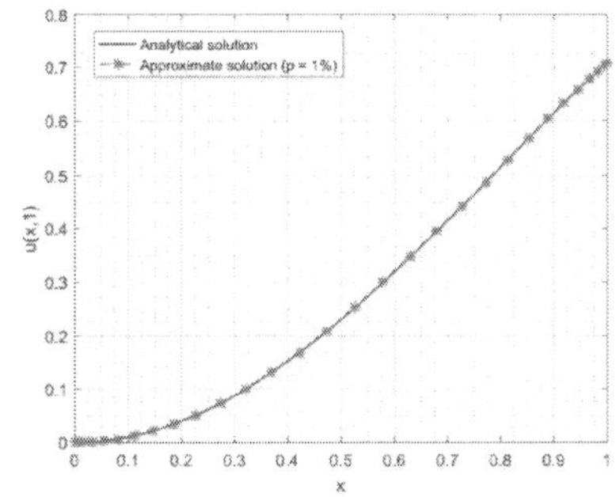

(**a**) $u(x, 1)$ with $p = 1\%$

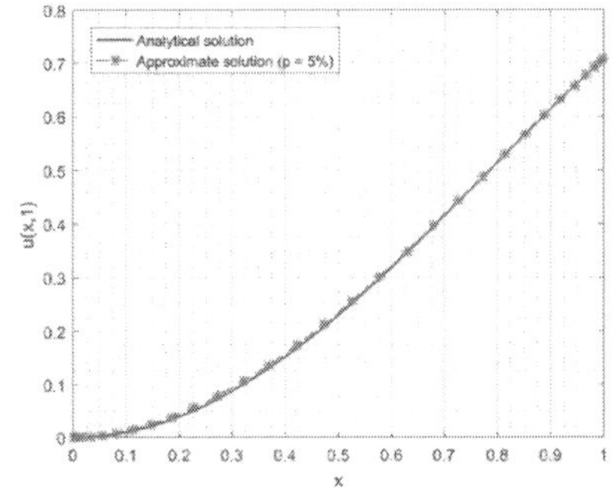

(**b**) $u(x, 1)$ with $p = 5\%$

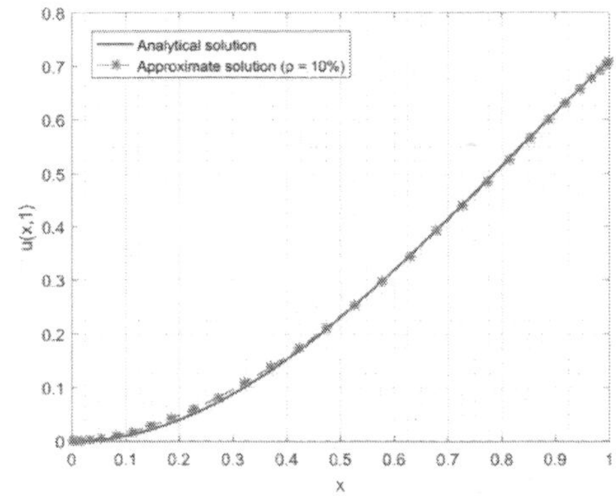

(**c**) $u(x, 1)$ with $p = 10\%$

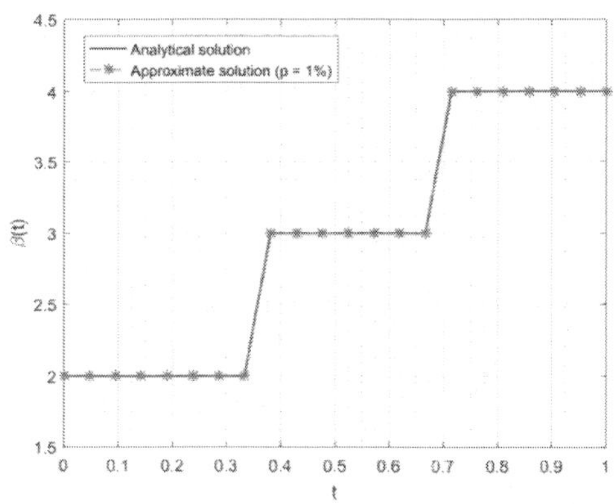

(**d**) $\beta(t)$ with $p = 1\%$

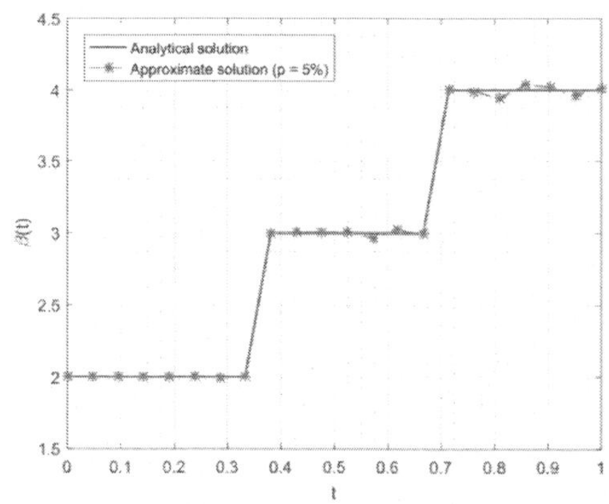

(**e**) $\beta(t)$ with $p = 5\%$

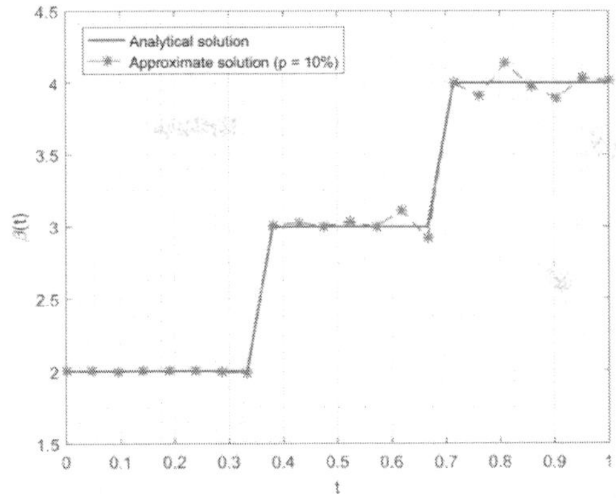

(f) $\beta(t)$ with $p = 10\%$

Figure 4. The graphical results u(x,T) and β(t) for Example 4 with N=30 and M=21.

Table 7. Mean absolute errors of u(x,1) and β(t) for optimal regularization parameter λ of Example 4.

$M = N$	$p = 0\%$			$p = 1\%$		
	λ	$\mathcal{E}_u$	$\mathcal{E}_\beta$	λ	$\mathcal{E}_u$	$\mathcal{E}_\beta$
6	1.25×10^{-14}	6.4987×10^{-6}	18123×10^{-4}	2.60×10^{-12}	7.8604×10^{-6}	1.8959×10^{-4}
9	7.41×10^{-17}	2.8057×10^{-9}	7.1782×10^{-9}	3.45×10^{-10}	1.2932×10^{-7}	4.0319×10^{-5}
12	2.65×10^{-20}	1.4031×10^{-13}	4.5672×10^{-12}	6.02×10^{-11}	2.6701×10^{-7}	5.1531×10^{-5}
15	6.11×10^{-21}	5.4903×10^{-14}	2.7330×10^{-13}	8.41×10^{-12}	2.9871×10^{-6}	6.8381×10^{-5}
$M = N$	$p = 5\%$			$p = 10\%$		
	λ	$\mathcal{E}_u$	$\mathcal{E}_\beta$	λ	$\mathcal{E}_u$	$\mathcal{E}_\beta$
6	4.19×10^{-12}	8.8690×10^{-5}	5.1802×10^{-4}	5.51×10^{-11}	6.5680×10^{-4}	3.7335×10^{-3}
9	6.20×10^{-13}	1.8419×10^{-5}	7.0830×10^{-4}	7.96×10^{-11}	4.4035×10^{-4}	2.3292×10^{-3}
12	4.11×10^{-12}	7.0910×10^{-5}	1.6889×10^{-3}	8.65×10^{-12}	6.4815×10^{-4}	6.3981×10^{-3}
15	1.01×10^{-13}	2.7507×10^{-4}	1.7821×10^{-3}	5.64×10^{-12}	5.9709×10^{-4}	6.1579×10^{-3}

Table 8. Time convergence rates and CPU times (s) for Example 4 by Algorithm 2 with N=12.

M	T = 1			T = 2			T = 3		
	$\lVert u^* - u \rVert_\infty$	Rate	Time(s)	$\lVert u^* - u \rVert_\infty$	Rate	Time(s)	$\lVert u^* - u \rVert_\infty$	Rate	Time(s)
6	2.681×10^{-13}	1.4574	0.0704	5.338×10^{-13}	1.4564	0.0726	8.024×10^{-13}	1.4563	0.0728
9	2.677×10^{-13}	1.3415	0.0727	5.353×10^{-13}	1.3406	0.0737	8.006×10^{-13}	1.3395	0.0746
12	2.681×10^{-13}	1.2758	0.0735	5.338×10^{-13}	1.2744	0.0753	8.011×10^{-13}	1.2743	0.0767
15	2.666×10^{-13}	1.2312	0.0749	5.360×10^{-13}	1.2332	0.0783	8.015×10^{-13}	1.2337	0.0807
18	2.682×10^{-13}	1.2028	0.0798	5.351×10^{-13}	1.2031	0.0799	7.989×10^{-13}	1.2022	0.0828

CONCLUSIONS AND DISCUSSION

In this paper, we utilized FIM-SCP combined with the forward difference quotient to create efficient and accurate numerical algorithms for solving the considered direct and inverse TVIDE problems. According to the numerical examples in **Section 4**, we have demonstrated the performance of our proposed Algorithm 1 for seeking the approximate solutions of direct TVIDE problems in Examples 1 and 2. We can see that, for Example 1—which involved a second-order derivative with constant coefficients—Algorithm 1 provided an accurate result. Furthermore, for a problem involving a higher-order derivative with variable coefficients, it still provided high accuracy, in terms of solutions, as demonstrated in Example 2. Moreover, we handled inverse TVIDE problems using Algorithm 2, the effectiveness of which was illustrated in Examples 3 and 4. We used the Tikhonov regularization method to deal with the instability of the inverse problem; it can be seen that, in the examples, the regularization parameter λ was close to zero. Algorithm 2 could handle both continuous and piecewise-defined forcing terms with high accuracy, as demonstrated in Examples 3 and 4. Furthermore, when we perturbed the problems by adding noisy values, our Algorithm 2 still overcame the noise and provided approximate results that approached the analytical solutions. We further notice that our presented methods provide high accuracy, even when using only a small number of nodal points. Evidently, when we decrease the time step, they will furnish more accurate results. The rates of convergence with respect to time (based on the ℓ_∞ norm) of our methods were observed to be linear. Finally, we also depicted the computational times for each example. However, we realize that there exist no theoretical error analysis results for the proposed numerical algorithms. Thus, our future research will study the error analysis, in order to find theories for order of accuracy and rate of convergence for our method. Another interesting direction for our future work is to extend our techniques to solve other types of IDEs and non-linear IDEs.

AUTHOR CONTRIBUTIONS

Conceptualization, R.B., A.D., and P.G.; methodology, R.B. and A.D.; software, A.D. and P.G.; validation, R.B., A.D., and P.G.; formal analysis, R.B.; investigation, A.D. and P.G.; writing—original draft preparation, A.D. and P.G.; writing—review and editing, R.B.; visualization, A.D. and P.G.; supervision, R.B.; project administration, R.B.; funding acquisition, R.B. All authors have read and agreed to the published version of the manuscript.

ACKNOWLEDGMENTS

The authors would like to thank the reviewers for their thoughtful comments and efforts towards improving our manuscript.

CONFLICTS OF INTEREST

The authors declare no conflict of interest.

ABBREVIATIONS

The following abbreviations are used in this manuscript:

FDM	finite difference method
FIM	finite integration method
FIM-SCP	finite integration method with shifted Chebyshev polynomial
IDE	integro-differential equation
PDE	partial differential equation
TVIDE	time-dependent Volterra integro-differential equation

REFERENCES

1. Zill, D.G.; Wright, W.S.; Cullen, M.R. *Differential Equations with Boundary-Value Problem*, 8th ed.; Brooks/Cole, Cengang Learning: Boston, MA, USA, 2013.
2. Yanik, E.G.; Fairweather, G. Finite element methods for parabolic and hyperbolic partial integro–differential equations. *Nonlinear Anal.* **1988**, *12*, 785–809.
3. Engle, H. *On Some Parabolic Integro–Differential Equations: Existence and Asymptotics of Solution*; Lecture Notes in Mathematics, Springer: Berlin, Germany, 1983.
4. Tang, T. A finite difference scheme for partial integro–differential equations with a weakly singular kernel. *Appl. Numer. Math.* **1993**, *11*, 309–319.
5. Aguilar, M.; Brunner, H. Collocation methods for second–order Volterra integro–differential equations. *Appl. Numer. Math.* **1988**, *4*, 455–470.
6. Brunner, H. Implicit Runge–Kutta–Nyström methods for general second–order Volterra integro–differential equations. *Comput. Math. Appl.* **1987**, *14*, 549–559.
7. Jiang, Y.J. On spectral methods for Volterra-type integro–differential equations. *J. Comput. Appl. Math.* **2009**, *230*, 333–340.
8. Burton, T.A. *Volterra Integral and Differential Equations*; Academic Press: New York, NY, USA, 1983.
9. Rahman, M. *Integral Equations and Their Applications*; WIT Press: Southampton, UK, 2007.
10. Hu, Q. Stieltjes derivatives and beta–polynomial spline collocation for Volterra integro–differential equations with singularities. *SIAM J. Numer.* **1996**, *33*, 208–220.
11. Brunner, H. Superconvergence in collocation and implicit Runge–Kutta methods for Volterra–type integral equations of the second kind. *Internet Schriftenreihe Numer. Math.* **1980**, *53*, 54–72.
12. El-Sayed, S.M.; Kaya, D.; Zarea, S. The decomposition method applied to solve high–order linear Volterra–Fredholm integro–differential equations. *Internet J. Nonlinear Sci. Numer. Simulat.* **2004**, *5*, 105–112.
13. Kabanikhin, S.I. Definitions and examples of inverse and ill–posed problems. *J. Inverse Ill-Pose Probl.* **2008**, *16*, 317–357.

14. Wen, P.H.; Hon, Y.C.; Li, M.; Korakianitis, T. Finite integration method for partial differential equations. *Appl. Math. Model.* **2013**, *37*, 10092–10106.

15. Li, M.; Chen, C.S.; Hon, Y.C.; Wen, P.H. Finite integration method for solving multi–dimensional partial differential equations. *Appl. Math. Model.* **2015**, *39*, 4979–4994.

16. Li, M.; Tian, Z.L.; Hon, Y.C.; Chen, C.S.; Wen, P.H. Improved finite integration method for partial differential equations. *Eng. Anal. Bound. Elem.* **2016**, *64*, 230–236.

17. Boonklurb, R.; Duangpan, A.; Treeyaprasert, T. Modified finite integration method using Chebyshev polynomial for solving linear differential equations. *J. Numer. Ind. Appl. Math.* **2018**, *12*, 1–19.

18. Rivlin, T.J. *Chebyshev Polynomials, From Approximation Theory to Algebra and Number Theory*, 2nd ed.; John Wiley and Sons: New York, NY, USA, 1990.

19. Tikhonov, A.N.; Goncharsky, A.V.; Stepanov, V.V.; Yagola, A.G. *Numerical Methods for the Solution of Ill–Posed Problems*; Springer: Dordrecht, The Netherlands, 1995.

20. Sun, Y. Indirect boundary integral equation method for the Cauchy problem of the Laplace equation. *J. Sci. Comput.* **2017**, *71*, 469–498.

CHAPTER 9

Direct and Inverse Problem for Geometric Perturbation of the Laplace Operator in a Strip

I. Khélifa and L. Chorfi

Department of Mathematics, LMA, Badji Mokhtar University, Annaba, Algeria

ABSTRACT

This paper is concerned with the inverse problem of determining geometric shape of a part γ of the boundary of a perturbed strip Ω from a pair of Cauchy data of a harmonic function u in Ω. This leads to the study of the direct problem. Using the variational method, we show that is well posed, and by the integral equation method we seek the solution in the form of combined double- and single-layer potential. For the identification of γ we prove a uniqueness result, that is, a pair of Cauchy data on the accessible part Γ_0

Citation: Khélifa, I., & Chorfi, L. (2019). Direct and inverse problem for geometric perturbation of the Laplace operator in a strip. Boundary Value Problems, 2019(1), 1-19. (19 pages) DOI: https://doi.org/10.1186/s13661-019-1245-6

uniquely determines the missing part γ of the boundary, and we derive a system of nonlinear integral equations equivalent to our inverse problem. We present numerical examples for both the direct and inverse problems.

INTRODUCTION

We consider the perturbed strip $\Omega \subset R^2$ as follows:

$$\Omega = \{(x,y) \in R^2 : 0 < y < 1 - h(x)\}, \quad (1)$$

where $h:R \to [0,1]$ is a continuous function which is a parametrization of a local perturbation of a strip $\Omega_0=\{(x,y)\in R^2:0<y<1\}$. We assume that there exist $a>0$ and $0<b<1$ such that $h\in C^2[-a,a]$ and it satisfies

(1) $h(x) = 0$ if $|x| \geq a$; (2) $0 \leq h(x) \leq b$ if $|x| \leq a$.

The boundary Γ of Ω is decomposed as $\Gamma=\Gamma_0\cup\Gamma_1$ with $\Gamma_0\cap\Gamma_1=\emptyset$ and $\Gamma_1=\Gamma^-_1\cup\gamma\cup\Gamma^+_1$ (see Fig. 1) where γ is the arc

$$\gamma = \{(x,y) \in R^2 : y = 1 - h(x), |x| \leq a\}.$$

For a given function $f\in H^{3/2}(R)$ consider the Dirichlet problem for the Laplace equation:

$$\Delta u = 0 \quad \text{in } \Omega \quad (2)$$

subject to the boundary condition

$$u = f \quad \text{on } \Gamma_0, \qquad u = 0 \quad \text{on } \Gamma_1. \quad (3)$$

The inverse problem we are concerned with consists in recovering the shape of $\gamma: y=1-h(x)$, $-a\leq x\leq a$, from the Cauchy data $f:=u|\Gamma_0$ and $g:=\partial u/\partial y|\Gamma_0$.

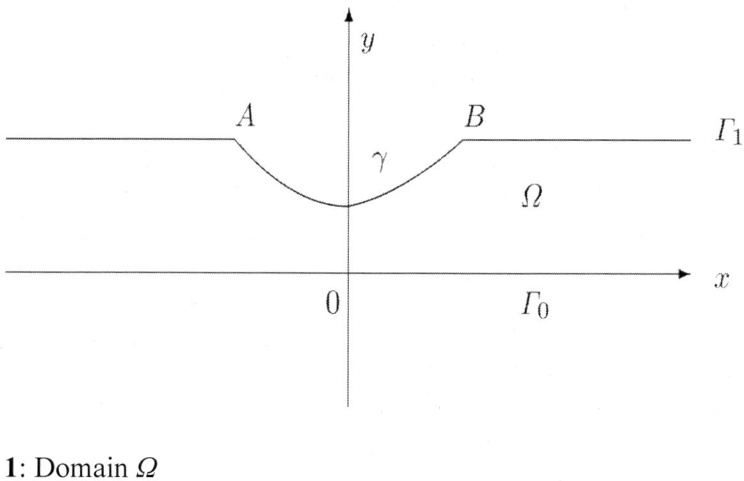

Figure 1: Domain Ω

This problem arises in electrostatic or thermal imaging methods, detecting a corrosion surface in nondestructive testing. This could be the case for an electrically conducting specimen, which is subject to wear by corrosion, causing material loss or cracks. In practice, it often happens that the unknown part of the boundary that has suffered corrosion is not accessible to direct inspection. The aim is to detect the presence of such defects by nondestructive methods from the knowledge of the imposed voltage $f:=u|\Gamma_0$ and the measured resulting current $g:=\partial u/\partial y|\Gamma_0$ on the accessible part Γ_0 of the boundary Γ. Various application related to the above model problem are described, for example, in [1, 11] (see also the references therein). Let us mention the articles [2, 5, 10, 15, 17] on the inverse boundary value problems for the Laplace equation in a bounded domain. Our problem presents additional difficulties related to the unbounded nature of the domain and, to our knowledge, it was not considered in the literature. In this work we generalize the potential method used in the precursor works of Kress and Rundel [9, 10].

The paper is organized as follows. In Sect. 2 we will be concerned with the direct problem, we proceed by showing the existence and uniqueness of the solution. Using Green's function, we obtain a representation of the solution in the form of potentials and we give a numerical test. Section 3 is devoted to the study of the inverse problem, a uniqueness theorem is obtained, we derive the two-by-two system of integral equations and describe the proposed iteration scheme. The paper concludes with some numerical examples.

THE DIRECT PROBLEM

Existence and Uniqueness

The following notations are used throughout this text. For $s \geq 0$, we say that $f \in Hs(R)$ if the norm

$$\|f\|_{H^s(\mathrm{R})} := \left(\int_{-\infty}^{+\infty} (1 + |\xi|^2)^s \left|\widehat{f}(\xi)\right|^2 d\xi \right)^{\frac{1}{2}}$$

is finite. For s=0, we denote $\|f\|_{L2(R)} := \|f\|_0$. Here the Fourier transform pair of f is defined by the formulae

$$\widehat{f}(\xi) \equiv \mathcal{F}(f)(\xi) := \frac{1}{\sqrt{2\pi}} \int_{-\infty}^{+\infty} e^{-ix\xi} f(x)\, dx,$$

$$f(x) = \frac{1}{\sqrt{2\pi}} \int_{-\infty}^{+\infty} e^{ix\xi} \widehat{f}(\xi)\, d\xi. \tag{4}$$

We will show that the direct problem (2)–(3) is well posed.

Theorem 1

Suppose that $f \in H^{3/2}(R)$. *Then*

1. *There exists a recovery function* $u_f \in H^2(\Omega)$ *such that* $u_f(x,0)=f(x)$.
2. *The problem* (2)–(3) *has a unique solution* $u \in H^1(\Omega)$.
3. The solution $u \in H^2_{loc}(\Omega)$.

Proof

1. Using Fourier transform of function *f*, we define uf by

$$u_f(x,y) = w_0(x,y)\varphi_0(y),$$

where

$$w_0(x,y) = \mathcal{F}^{-1}\left(e^{-y|\xi|}\hat{f}(\xi)\right)$$

and φ_0 is a truncation function of class $C^2[0,1]$ such that

$$\varphi_0(y) = \begin{cases} 1 & \text{if } 0 < y < b/2, \\ 0 & \text{if } b < y < 1. \end{cases} \tag{5}$$

It is easy to see that $u_f \in H^2(\Omega)$ and $u_f(x,0)=f(x)$.

2. We put $v=u-u_f$ and $F=-\Delta u_f \in L^2(\Omega)$. Then *v* solves the homogeneous Dirichlet problem:

$$\begin{cases} \Delta v = F & \text{in } \Omega; \\ v = 0 & \text{on } \Gamma. \end{cases} \tag{6}$$

The bilinear form $a(v,w)=\int_\Omega \nabla v \nabla w\, dxdy$, $v,w \in H^1_0(\Omega)$, is coercive since Ω is bounded in the *y* direction (see [3]). By the Lax–Milgram theorem, there exists a unique weak solution $v \in H^1_0(\Omega)$.

3. Now we show that $v \in H^2_{loc}(\Omega)$. The corners *A* and *B* of Γ_1 have an angle less that π, then from the regularity theory it follows that $v \in H^2(\Omega R)$, $\Omega R=\{(x,y) \in \Omega; |x| \leq R\}$, for any $R \geq a$ (see [6]).

This completes the proof.

Theorem 2

The Dirichlet-to-Neumann map $f \mapsto g = \partial u/\partial y(x,0)$ *where u is the solution of the Dirichlet problem* (2)–(3) *is continuous from* $H^{3/2}(R)$ *to* $L^2(R)$.

Proof

From the previous theorem, $g \in L^{loc2}(R)$. It suffices to prove that *g* is in $L^2(R_0,+\infty)$ for $R_0 \geq a+1$. Let $v \in H_0^{\,1}(\Omega)$ be the solution of (6). We put $w=\theta(x)v$ where $\theta \in C^2(R)$ is a truncation function such that $\theta(x)=0$ for $x \leq a$ and $\theta(x)=1$ for $x \geq R0$. Then *w* satisfies the equation

$$\begin{cases} \Delta w = F_1 & \text{in } \Omega_a^+ =]a,+\infty[\times]0,1[; \\ w = 0 & \text{on } \Gamma_a^+ = \partial\Omega_a^+, \end{cases} \tag{7}$$

with $F_1 = \theta F + v\theta'' + 2\frac{\partial v}{\partial x}\theta' \in L^2(\Omega_a^+)$. We can assume in (7) that a=0 (after a translation in the direction of *x*) and use sine-Fourier transform

$$\widehat{w}(\xi,y) = \sqrt{\frac{2}{\pi}}\int_0^\infty w(x,y)\sin(\xi x)\,dx, \qquad \widehat{F}_1(\xi,y) = \sqrt{\frac{2}{\pi}}\int_0^\infty F_1(x,y)\sin(\xi x)\,dx,$$

to obtain the second order differential equation in *y*:

$$\begin{cases} \widehat{w}_{yy} - \xi^2\widehat{w} = \widehat{F}_1(\xi,y), & y \in]0,1[, \\ \widehat{w}(\xi,0) = 0, \qquad \widehat{w}(\xi,1) = 0, \end{cases} \tag{8}$$

with the solution

$$\widehat{w}(\xi,y) = \int_0^1 G(\xi,y,z)\widehat{F}_1(\xi,z)\,dz$$

and $G(\xi,y,z)$ being the Green's function

$$G(\xi,y,z) = \frac{1}{\xi\cosh\xi}\begin{cases} \sinh(\xi z)\sinh(\xi(1-y)), & z \leq y; \\ \sinh(\xi y)\sinh(\xi(1-z)), & z \geq y. \end{cases} \tag{9}$$

Then

$$\frac{\partial\widehat{w}}{\partial y}(\xi,0) = \frac{-1}{\cosh\xi}\int_0^1 \sinh\big(\xi(1-z)\big)\widehat{F}_1(\xi,z)\,dz,$$

which leads to

$$\forall \xi \geq 0, \quad \left| \frac{\partial \widehat{w}}{\partial y}(\xi, 0) \right|^2 \leq \int_0^1 \left| \widehat{F}_1(\xi, z) \right|^2 dz$$

and then

$$\left\| \frac{\partial w}{\partial y}(\cdot, 0) \right\|_{L^2(0,+\infty)} \leq \|F_1\|_{L^2((0,+\infty)\times]0,1[)}.$$

On the other hand, we have

$$\|F_1\|_{L^2((R_0,+\infty)\times]0,1[)} \leq 2\|\theta\|_{C^2}\left(\|F\|_{L^2(\Omega)} + \|v\|_{H^1(\Omega)}\right)$$
$$\leq C_1\|F\|_{L^2(\Omega)} \leq C_2\|f\|_{H^{3/2}(\mathbb{R})},$$

since w=v, for $x \geq R_0$, and v=u−uf, so we get

$$g(x) = \frac{\partial w}{\partial y}(x, 0) + \frac{\partial u_f}{\partial y}(x, 0), \quad x \geq R_0,$$

and deduce that

$$\|g\|_{L^2} \leq C_2\|f\|_{3/2} + \|f'\|_0 \leq C_3\|f\|_{3/2},$$

ending the proof.

Theorem 3

(Uniqueness)

Suppose that u∈Hloc1(Ω) *satisfies*

$$\begin{cases} \Delta u = 0 & in\ \Omega, \\ \gamma_0(u) = 0 & on\ \Gamma, \\ \int_0^1 |u(x,y)|^2\, dy \to 0 & as\ |x| \to +\infty. \end{cases} \quad (10)$$

Then u=0.

Proof

1. In the semi-strip Ω^+_a=]a,+∞[×]0,1[, u has the representation

$$u(x,y) = \sum_{n=1}^{\infty} \sin(n\pi y)\left(A_n e^{-n\pi(x-a)} + B_n e^{n\pi(x-a)}\right),$$

and then, for R>a,

$$\int_0^1 |u(R,y)|^2\, dy = \frac{2}{\pi}\sum_{n=1}^{\infty}\left(A_n e^{-n\pi(R-a)} + B_n e^{n\pi(R-a)}\right)^2.$$

If $R\to+\infty$, we deduce that for all $n\in N*$,

$$\lim_{R\to+\infty}\left[A_n e^{-n\pi(R-a)}+B_n e^{n\pi(R-a)}\right]=0,$$

thus $\forall n\geq 1$, $B_n=0$ and $A_n=\frac{2}{\pi}\int_0^1 u(a,y)\sin(n\pi y)\,dy$.

Then we deduce the following estimate, for all $|x|\geq R$ and $y\in[0,1]$:

$$|u(x,y)|+\left|\frac{\partial u}{\partial x}(x,y)\right|\leq C(a)\|u\|_{H^1(\Omega_a)}e^{-\pi R}. \tag{11}$$

2. Suppose (10) holds. Applying the first Green's identity in the set $\Omega R=\{(x,y)\in\Omega;|x|<R\}$, for $R\geq a$, we have

$$\int_{\Omega_R}|\nabla u|^2\,dx\,dy=\int_0^1\left(u\frac{\partial u}{\partial x}\bigg|_{x=R}+u\frac{\partial u}{\partial x}\bigg|_{x=-R}\right)dy.$$

Inequality (11) implies that

$$\int_{\Omega_R}|\nabla u|^2\,dx\,dy=O(e^{-2\pi R}).$$

Letting $R\to+\infty$, we deduce that $\nabla u=0$ in Ω, and thus u is constant in Ω. Since $\gamma_0(u)=0$, it follows that $u=0$ in Ω.

Unperturbed Problem

By an unperturbed problem we mean the problem on the strip $[\Omega_0=R\times]0,1[$:

$$(P_0)\quad\begin{cases}\Delta u_0=0 & \text{in } \Omega_0,\\ u_0(x,0)=f(x), \quad u_0(x,1)=0, & \forall x\in R.\end{cases}$$

Using the partial Fourier transform with respect to x, we obtain the solution

$$u_0(x,y)=\frac{1}{\sqrt{2\pi}}\int_{-\infty}^{+\infty}a(\xi,y)\hat{f}(\xi)e^{ix\xi}\,d\xi, \tag{12}$$

where

$$a(\xi,y)=\frac{\sinh(\xi(1-y))}{\sinh\xi}. \tag{13}$$

Proposition 1

Assume that $f\in H3/2(R)$. *Then* $u_0\in H2(\Omega_0)$ *and it satisfies*

(i) $\|u_0\|_{H^2(\Omega_0)} \leq C\|f\|_{H^{3/2}(\mathrm{R})}$, (ii) $\left\|\frac{\partial u_0}{\partial y}(x,0)\right\|_{L^2(\mathrm{R})} \leq C\|f\|_{L^2(\mathrm{R})}$.

Proof

(i) We use the asymptotic behavior of a(ξ,y), that is,

$$a(\xi,y) \simeq 1-y \quad \text{if } \xi \to 0 \quad \text{and} \quad a(\xi,y) \simeq e^{-y|\xi|} \quad \text{if } |\xi| \to +\infty.$$

Indeed, for example, we have

$$\int_{\Omega_0}\left|\frac{\partial^2 u_0}{\partial x^2}\right|^2 dx\,dy = \int_{\mathrm{R}} \xi^4 |\hat{f}(\xi)|^2 \left(\int_0^1 |a(\xi,y)|^2\, dy\right) d\xi$$

and

$$\int_0^1 |a(\xi,y)|^2\, dy \leq C\int_0^1 e^{-2y|\xi|}\, dy \leq \frac{1}{2|\xi|}.$$

Then

$$\int_{\Omega_0}\left|\frac{\partial^2 u_0}{\partial x^2}\right|^2 dx\,dy \leq \int_{\mathrm{R}} |\xi|^3 |\hat{f}(\xi)|^2\, d\xi \leq \|f\|_{H^{3/2}(\mathrm{R})}.$$

(ii) We have $\widehat{\frac{\partial u_0}{\partial y}}(\xi,0) = -(\xi \coth \xi)\widehat{f}(\xi)$. Since the function ξ↦ξcothξ is bounded on R,

$$\left\|\frac{\partial u_0}{\partial y}(x,0)\right\|_{L^2(\mathrm{R})} \leq C\|f\|_0.$$

Green's function of the strip Ω_0

The Green's function of the Dirichlet problem for the Laplace equation$2XTÿÿ in the strip Ω_0 is defined by (see [13]) by

$$G((x,y),(\xi,\eta)) = \frac{1}{4\pi}\log\left[\frac{\cosh \pi(x-\xi) - \cos \pi(y+\eta)}{\cosh \pi(x-\xi) - \cos \pi(y-\eta)}\right],$$

which satisfies in the sense of the distribution

$$\begin{cases} \Delta G = \delta(x-\xi, y-\eta) & \text{in } \Omega_0, \\ G|_{y=0} = 0, \\ G|_{y=1} = 0. \end{cases} \tag{14}$$

For a fixed point source (ξ,η)∈Ω, *G* exhibits the following 9+6:

$$G((x,y),(\xi,\eta)) = -\frac{1}{2\pi}\log r + \frac{1}{4\pi}\log\left(\frac{4\sin\pi\eta}{\pi}\right) + O(r) \tag{15}$$

as $r = \sqrt{(x-\xi)^2 + (y-\eta)^2} \to 0$. Also

$$G((x,y),(\xi,\eta)) = O\left(\frac{1}{|x|}\right) \quad \text{as } |x| \to +\infty \text{ (uniformly in } y \in [0,1]) \tag{16}$$

and

$$\frac{\partial G}{\partial x}((x,y),(\xi,\eta)) = O\left(e^{-\pi|x|}\right) \quad \text{as } |x| \to +\infty \text{ (uniformly in } y \in [0,1]). \tag{17}$$

Integral Equation Method

Let u be the solution of problem (2)–(3), and u0u0 the solution of (P_0). We put $v_0=u-u_0$, and then

$$\begin{cases} \Delta v_0 = 0 & \text{in } \Omega, \\ v_0 = 0 & \text{on } \Gamma_0, \\ v_0 = -u_0 & \text{on } \gamma, \\ v_0 = 0 & \text{on } \Gamma_1 \setminus \gamma. \end{cases} \tag{18}$$

We introduce the boundary integral equation formulation for (18) as follows. For this we apply Green's second theorem to the harmonic functions v_0 and $G(\cdot,(\xi,\eta))$ in the domain $\Omega R,\epsilon=\Omega R\setminus B((\xi,\eta),\epsilon)$, $\Omega R=\{(x,y)\in\Omega:|x|<R\}$ to obtain

$$\int_{\partial\Omega_{R,\epsilon}} \left(v_0\frac{\partial G}{\partial n} - G\frac{\partial v_0}{\partial n}\right) ds = 0.$$

The function $v_0 \in H^1(\Omega)$ satisfies the decay property in (10). Then from the proof of Theorem 3 (part 1, estimate (11)), we deduce that

$$\lim_{R\to+\infty}\int_0^1 \left(\left|v_0(\pm R,y)\right|^2 + \left|\frac{\partial v_0}{\partial x}(\pm R,y)\right|^2\right) dy = 0,$$

and (16)–(17) lead to the limit

$$\lim_{R\to+\infty}\int_0^1 \left(v_0\frac{\partial G}{\partial x} - G\frac{\partial v_0}{\partial x}\right)\bigg|_{x=\pm R} dy = 0. \tag{19}$$

Then, as $R\to+\infty$ and $\epsilon\to 0$, we obtain the integral representation formula

$$v_0(\xi,\eta) = \int_\gamma \left(v_0 \frac{\partial G}{\partial n} - G \frac{\partial v_0}{\partial n} \right) ds, \quad (\xi,\eta) \in \Omega. \tag{20}$$

We put $\psi = \frac{\partial v_0}{\partial n}|_\gamma$, which is unknown, and $q=-v_0|\gamma=u_0|\gamma$, which is known through (12). When (ξ,η) tends to a point $M=(x,1-h(x))$ of γ, and utilizing the properties of single and double layer potentials (see [9]), we obtain the integral equation of the first kind

$$S\psi = \frac{1}{2}q + Dq \quad \text{on } \gamma, \tag{21}$$

where S and D are respectively the single- and double-layer potentials defined for $M=(x,y)\in\gamma$ by

$$S\psi(M) = \int_\gamma G(M,M')\psi(M')\,ds(M'), \qquad Dq(M) = \int_\gamma \frac{\partial G}{\partial n}(M,M')q(M')\,ds(M').$$

The solvability of (21) follows from the potential theory developed in the framework of Sobolev spaces $H\sim s(\gamma)$, $s>1/2$ (see [9, 10, 12, 14]). Once equation (21) is solved, we have the solution

$$v_0(M) = \int_\gamma \left(q(M') \frac{\partial G}{\partial n}(M,M') - G(M,M')\psi(M') \right) ds(M'), \quad M \in \Omega. \tag{22}$$

For the numerical solution of the integral equation (21), a parametrization of γ is required.

Parametrization of the Integral Equation

In order to use the results of integral equation on a contour, we consider the extension $\tilde{\gamma}$ of γ defined by $\gamma\sim=\gamma\cup\gamma+$, where $\gamma+=\{(x,1+h(x)),-a<x<a\}$. Extending the data q on $\gamma+$ by symmetry $q(x,1+h(x))=-q(x,1-h(x))$ and using the parametrization of $\tilde{\gamma}$, $z(t)=(x(t),y(t))$, $0\leq t\leq 2\pi$, we transform (21) as follows:

$$\frac{1}{2\pi}\int_0^{2\pi} L(t,s)\varphi(s)\,ds = \frac{1}{2}q(t) + \int_0^{2\pi} M(t,s)q(s)\,ds, \quad 0 \leq t \leq 2\pi, \tag{23}$$

where

$$\varphi(t) = \psi(z(t))|z'(t)|, \qquad q(t) = q(z(t)), \tag{24}$$

and

$$L(t,s) = \log\left[\frac{\cosh(x(t)-x(s)) - \cos(y(t)-y(s))}{\cosh(x(t)-x(s)) - \cos(y(t)+y(s))}\right], \tag{25}$$

$$M(t,s) = \frac{\sinh(x(t)-x(s))y'(s) + \sin(y(t)+y(s))x'(s)}{\cosh(x(t)-x(s)) - \cos(y(t)+y(s))} + \frac{-\sinh(x(t)-x(s))y'(s) + \sin(y(t)-y(s))x'(s)}{\cosh(x(t)-x(s)) - \cos(y(t)-y(s))}. \tag{26}$$

For the discretization of the integral operators, we note that kernel L(t,s) can be decomposed in the form

$$L(t,s) = \log\left(4\sin^2\left(\frac{t-s}{2}\right)\right) + k(t,s), \tag{27}$$

where *k* and *M* are smooth with diagonal values

$$k(t,t) = \log\left(\frac{16\sin^2 y(t)}{|z'(t)|^2}\right) \tag{28}$$

and

$$M(t,t) = \frac{-y''(t)}{|z'(t)|^2} + \pi\cot(\pi y(t)). \tag{29}$$

Using the decomposition (27), we write equation (23) in operational form

$$(A+K)\varphi = \frac{1}{2}q + Bq, \tag{30}$$

where

$$A\varphi(t) = \frac{1}{2\pi}\int_0^{2\pi}\left[\log\left(4\sin^2\left(\frac{t-s}{2}\right)\right) - 2\right]\varphi(s)\,ds, \quad 0 \le t \le 2\pi \tag{31}$$

and

$$K\varphi(t) = \frac{1}{2\pi}\int_0^{2\pi}[k(t,s)+2]\varphi(s)\,ds, \qquad Bq(t) = \int_0^{2\pi} M(t,s)q(s)\,ds, \quad 0 \le t \le 2\pi. \tag{32}$$

The properties of the operators *A*, *K* and *B* are given in the following lemma [9].

Lemma 1

For r≥0r≥0, *we have*:

- *The operator* A:Hr[0,2π]→Hr+1[0,2π] *is bounded and bijective.*
- *The operator K is compact from* Hr[0,2π] *to* Hr+1[0,2π].
- *B is bounded in* Hr+1[0,2π].

The potentials *S* and *D* defined in (22) have the following properties (see [9]).

Lemma 2

The single-layer potential defines a bounded linear operator from H~−1/2(γ) *into* Hloc1(Ω). *The double-layer potential defines a bounded linear operator from* H~1/2(γ) *into* Hloc1(Ω).

Theorem 4

The integral equation (30) *has a unique solution, i.e.,* N(A+K)={0}.

Proof

Assume that (A+K)φ=0. We associate to *φ* the potential $v_0(M)$ defined by the integral (22) (with q=0). Using Lemma 2 and the asymptotic behavior of G(M,M′) when x|→+∞ (see (16)–(17)), we can prove that v_0 satisfies the uniqueness Theorem 3. Then $v_0=0$ in *Ω*, it follows that $\psi = \frac{\partial v_0}{\partial n}|_\gamma = 0$. Therefore φ(t)=ψ(z(t))|z′(t)|=0.

Nyström's Method

In this section we use Nyström's method for the numerical approximation of the integral equation (30) of the first kind with weakly singular kernels.

Following Kress [9], we construct numerical quadratures for the improper integral

$$(Q\varphi)(t) = \frac{1}{2\pi}\int_0^{2\pi} \log\left[4\sin^2\left(\frac{t-s}{2}\right)\right]\varphi(s)\,ds \tag{33}$$

by replacing the continuous periodic function *φ* by its trigonometric interpolation polynomial described in [9, Sect. 11.3]. Using the Lagrange basis (Lj), we obtain

$$(Q_n\varphi)(t) \simeq \sum_{j=0}^{2n-1} R_j^n(t)\varphi(t_j), \quad t_j = \frac{j\pi}{n}, \tag{34}$$

with the quadrature weights

$$R_j^n(t) = \frac{1}{2\pi}\int_0^{2\pi} \log\left[4\sin^2\left(\frac{t-s}{2}\right)\right] L_j(s)\,ds, \quad j = 0,\dots,2n-1. \tag{35}$$

More precisely, we have $R_j^n(t_k) = R_{|j-k|}^n$ such that

$$R_j^n = -\frac{1}{n}\left\{\sum_{m=1}^{n-1}\frac{1}{m}\cos\left(\frac{mj}{\pi}\right) + \frac{(-1)^j}{2n}, j = 0,\dots,2n-1\right\}. \tag{36}$$

Thus, we have the algebraic system

$$\sum_{j=0}^{2n-1}\left(R_{|i-j|}^n + \frac{1}{n}k(t_i,t_j)\right)\varphi(t_j) = \frac{1}{2}g(t_i) + \frac{1}{n}\sum_{j=0}^{2n-1} M(t_i,t_j)g(t_j), \tag{37}$$

Remark 1

In the numerical computation, we approach the Fourier integral (12) by the trigonometric polynomial

$$u_{0n}(x,y) = \pi\sum_{j=-n}^{n} a(\xi_j,y)\hat{f}(\xi_j)e^{ix\xi_j}, \quad \xi_j = j\pi. \tag{38}$$

In particular, if f(x)=qa(x−x0), where qa is an even function (wavelet), then $f\hat{}(\xi)=e^{i\xi x0}q\hat{}_a(\xi)$ f, and we have

$$u_{0n}(x,y) = \pi\sum_{j=1}^{n+1} H_j(y)\cos\left(\pi j(x-x_0)\right), \tag{39}$$

where

$$H_j(y) = a(\xi_j,y)\hat{q}_a(\xi_j). \tag{40}$$

Numerical Implementation and Examples

To illustrate our method, we show a numerical test with the boundary condition f(t) and the parametrization of the arc h(t) such that

$$f(t) = a_1\exp\left(-c_1(t-t_1)^2\right) + a_2\exp\left(-c_2(t-t_2)^2\right),$$

with $a_1=3$, $c_1=2$, $a_2=1$, $c_2=1$, $t_1=-1$, $t_2=1$, and

$$h(t) = H\left(a^2 - t^2\right)\times\left[(t-t_0)^2 + b\right],$$

with a=2, b=1, H=0.03, t0=1.

In the discretization we use the parameters a=2, n=120, which is the number of points in [0,π], and $x_j = a\cos(\frac{j\pi}{n}), j = 1,\ldots,n$, the points of [−a,a].

In the approximation of the Fourier integral (38), we use the interval $\xi\in[0,\xi_{max}]$ with ξ_{max}=Mπ, M=n/2 and the points $\xi_j = \frac{j\pi}{2}, j = 1,\ldots,n$. Figures 2 and 3 show the results of the simulation.

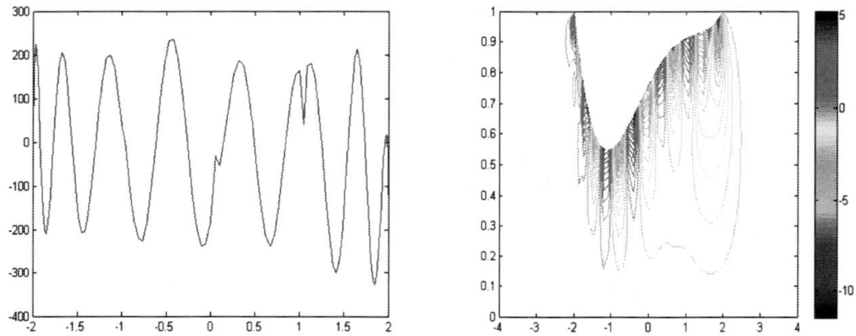

Figure 2: (left) Solution ψ of the integral equation (30); (right) Solution v_0 given by (22) in the domain ΩR, R=4

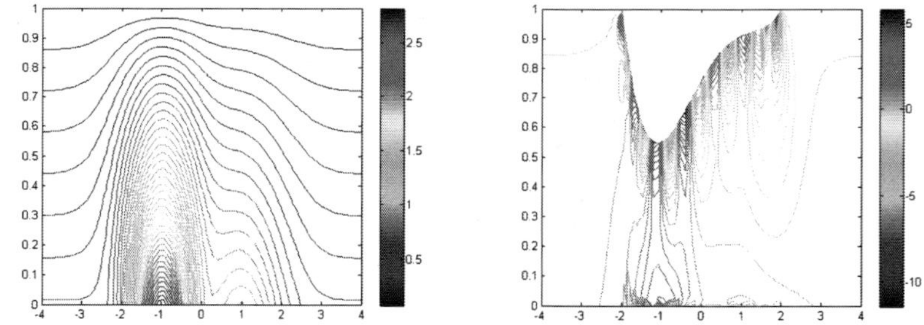

Figure 3: (left) Solution u_0 of the unperturbed problem; (right) Solution $u=v_0+u_0$ of the direct problem (2)–(3)

THE INVERSE PROBLEM

It is shown (in Sect. 2) that, for $f\in H^{3/2}(R)$, there exists a unique solution $u\in H^1(\Omega)\cap H^2_{loc}(\Omega)$ of (2)–(3). Recall that our inverse problem can be formulated as follows: given ff:=u|Γ_0 and $g := \frac{\partial u}{\partial y}|_{\Gamma_0}, g \in L^2(\mathbb{R})$, determine h(x) the parametrization of γ, i.e.,

$$\gamma : y = 1 - h(x), \quad -a \le x \le a.$$

We proceed with giving a uniqueness theorem.

Uniqueness for the Inverse Problem

Theorem 5

Let Ω_1 and Ω_2 be two perturbed strips with two boundary arcs γ_1 and γ_2 of class C^2, respectively. Denote by u_1 and u_2 the solutions to the problem (2)–(3) *for the domains Ω_1 and Ω_2, respectively, and assume that*

$$u_1 = u_2 \quad \text{and} \quad \frac{\partial u_1}{\partial y} = \frac{\partial u_2}{\partial y} \quad \text{on } \Gamma_0. \tag{41}$$

Then $\gamma_1=\gamma_2$.

Proof

(1) Suppose that $\Omega_1 \neq \Omega_2$ and put $u=u_1-u_2$, then $\Delta u=0$ in $\Omega_{12}=\Omega_1 \cap \Omega_2$ and u|y=0=∂u∂y|y=0=0. Holmgren's uniqueness theorem (see [16]) implies that u=0 in a neighbourhood of the axis y=0, and by analyticity $u_1=u_2$ in Ω_{12}.

(2) Without loss of generality, we may assume that the open set $W=\Omega_2 \setminus \overline{\Omega}_1$ is connected and not empty (see Fig. 4). Then from the boundary conditions we can deduce that $u_2=0$ on the boundary of *W*. Now by the maximum–minimum principle for harmonic functions, we can conclude that $u_2=0$ in *W*, and consequently, by analyticity it follows that $u_2=0$ in Ω_2. However, this contradicts the fact that *f* is not identically zero, and the proof is complete.

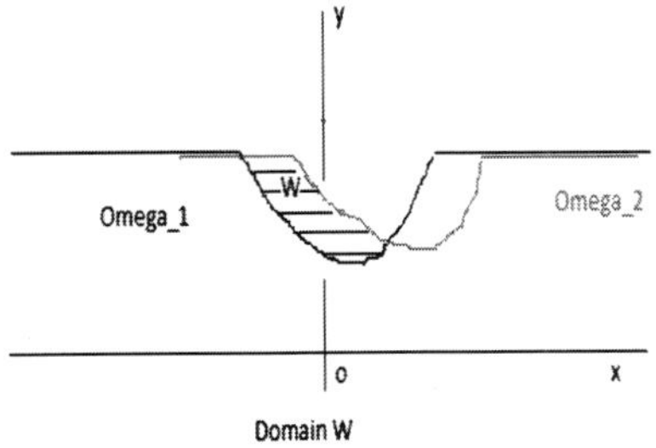

Figure 4: Domain W

Nonlinear Integral Equations

Our approach for solving the inverse problem is based on a system of nonlinear and ill-posed integral equations. We shall use identity (22) to derive two nonlinear integral equations for the unknowns (ψ,h), hence we obtain

$$H'q-H\psi=g, \tag{42}$$

where

$$\mathcal{H}\psi(x) = \int_{-a}^{a} \psi(x')H(x,x',1-h(x'))\omega(x')\,dx', \quad \omega(x') = \sqrt{1+h'^2(x')}, \tag{43}$$

with

$$H(x,x',y') = \frac{\partial G}{\partial y}((x,y),(x',y'))\Big|_{y=0} \tag{44}$$

and

$$\mathcal{H}'q(x) = \int_{-a}^{a} q(x')H'(x,x',h(x'))\,dx', \tag{45}$$

with

$$H'(x,x',h(x')) = \frac{\partial H}{\partial x'}(x,x',1-h(x'))h'(x') + \frac{\partial H}{\partial y'}(x,x',1-h(x')). \tag{46}$$

Then we obtain the system:

$$S(h)\psi = \frac{1}{2}q + D(h)q, \qquad q(x) = u_0(x,1-h(x)), \tag{47}$$

$$\mathcal{H}'(h)q - \mathcal{H}(h)\psi = g. \tag{48}$$

The kernels of the integral operators are given in Appendix 2.

System (47)–(48) is linear in ψ and nonlinear in h. We remark that H and H′ are C∞ in x and decay faster than e−π|x| (we assume that $h\in C^1[-a,a]$). Hence g∈C∞(R) and |g(x)|=O(e−π|x|).

The Iterative Procedure

We suggested the following iterative method (Newton-type method) for approximately solving system (47)–(48). It involves partial linearization of the system with respect to the variable h, the boundary parametrization.

Given an approximation h_0, we first solve the linear equation

$$S(h_0)\psi = \frac{1}{2}q_0 + D(h_0)q_0, \qquad q_0(x) = u_0\big(x, 1 - h_0(x)\big) \tag{49}$$

for ψ. Then, keeping ψ fixed, we replace (48) by the linearized equation

$$D\mathcal{H}'(h_0; wq_0) - D\mathcal{H}(h_0, w\psi) = \mathcal{H}(h_0)\psi - \mathcal{H}'(h_0)q_0 + g, \tag{50}$$

which we have to solve for w in order to improve an approximate boundary given by the parametrization h into the new approximation given by $h=h_0+w$. The method consists in iterating this procedure. For a theoretical foundation of such a regularized Newton-type method for nonlinear ill-posed problems, in general, we refer to [4].

The Fréchet derivatives DH(h;k) and DH'(h;k) can be found by differentiating their kernels with respect to *h*, their representations are given in Appendix 2.

Equation (47) is solved by Nyström's method. Since the integral operators DH′ and DH have kernels in $L^2(R\times]-a,a[)$, they are compact from $L^2(]-a,a[)$ to $L^2(R)$, and therefore the integral equation (50) is ill-posed and requires a regularization. We use a regularization based on the least squares method [8]. All the computational codes are written in MATLAB.11 and in particular the MATLAB codes developed by Hansen [7] for solving the discrete ill-posed equation (50) has been adopted in our computations.

Our algorithm consists in the following steps:

- **Step 1**: Given exact *h* and *f*, we compute Cauchy data *g* by solving integral equation (23) and using equation (42).
- **Step 2**: Perturb the data $g\delta=g+\delta\varepsilon(t)$, where $\varepsilon(t)\varepsilon(t)$ is a Gaussian noise.
- **Step 3**: Initialize $h=h_0$ and fix the parameters *m*, *n* of the discrete problem.
- **Step 4**: For it=1 to *itermax* do
  - compute $q_0(x)=u_0(x,1-h(x))$ and solve equation (49) by Nyström's method
  - solve the linearized equation (50) for *w* using a least squares algorithm.
  - put $h=h_0+w$

end do

Numerical Examples and Discussion

In this final section we present some numerical results to illustrate the accuracy and effectiveness of the reconstruction method as described in the previous section. We choose three profiles of γ to test our recovery algorithm.

Example 1 $h_1(x) = 0.2(1 - \frac{x^2}{4}), -2 \leq x \leq 2.$

Example 2 $h_2(x) = 0.03(4 - x^2)(1 + x^2), -2 \leq x \leq 2.$

Example 3 $h_3(x) = 0.01(4 - x^2)(2 - 2x + x^2), -2 \leq x \leq 2.$

The initial guess was taken to be h_0(x)=0.1(1−x24) for all examples.

The synthetic Cauchy data ((f,g) were obtained by solving the Dirichlet problem in Ω (see Fig. 5), with boundary condition u=f with

$$f(t) = 3e^{-2t^2}.$$

In the examples we approximate function *h* for the unknown boundary curve γ(t)=(acost,1−h(t)), t∈[0,π], by a trigonometric polynomial

$$h(t) \approx \sin t\left(a_0 + \sum_{k=1}^{m} a_k \cos kt + b_k \sin kt\right).$$

In all the tests we used n=50 grid points yj=acos(jπn), j=0,…,n, for the arc γ(y)=(y,1−h(y)) (n=150 for the simulated data) and xi=−2a+i2an, i=0,…,2n, points of collocation. Linear equation (50) is reduced to a ((2n+1)×(2m+1)) system. Then the problem becomes determining (2m+1) coefficients ($_{a0}$,a_k,b_k) of the approximation. In view of the ill-posedness, this approximating linear system is solved via the least squares algorithm *lsqr_b.m* from Hansen's package "Regularization tools" [7].

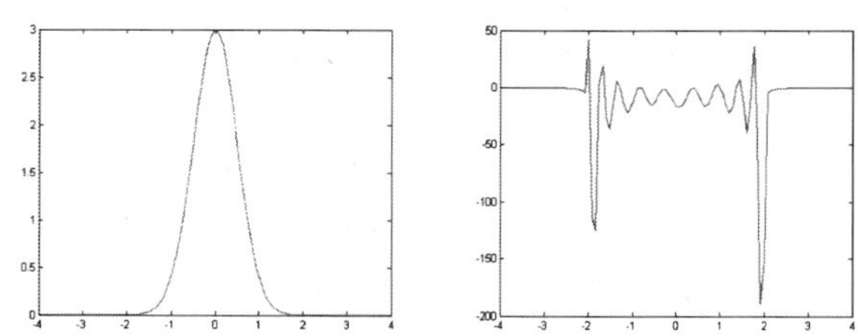

Figure 5: Cauchy data calculated by the direct algorithm, function *f* (left), function *g* for Example 1 (right)

To obtain data with noise, we add random noise of given level δ (relative to l2-discreet norm||g||2) to the simulated data *g* as

$$g_\delta(t) = g(t) + \delta \|g\|_2 \sigma(t),$$

where σ(t) represents random numbers uniformly distributed on the interval (−1;1).

In Figs. 6–9, we present the numerical results for Examples 1, 2 and 3, respectively, involving both exact and noisy data. The reconstructions are obtained after one iteration for all examples. The numerical experiments show rather satisfying reconstructions under the restrictions:

- $0 \le h(x) \le 0.2$,
- with adequate initial guess h_0,
- m=p the degree of the polynomial which interpolates h (p=2 for h_1 and p=4 for h_2 and h_3).
- $0 \le \delta \le 0.1$.

We remark that with arbitrary initial guess (in particular, for $h_0=0$) the algorithm fails to recover the boundary. In general, the quality of the reconstruction is affected by the shape of γ (see Figs. 8–9). As expected, the first iteration is stable with respect to the addition of random noise.

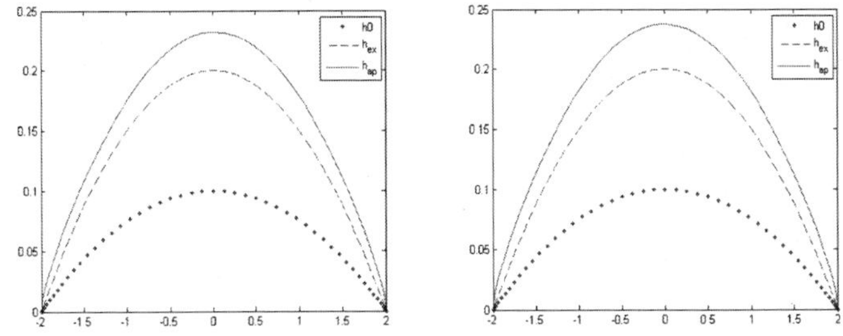

Figure 6: Reconstruction of h_1 after one iteration for exact data (left), and noisy data with δ=0.05 (right)

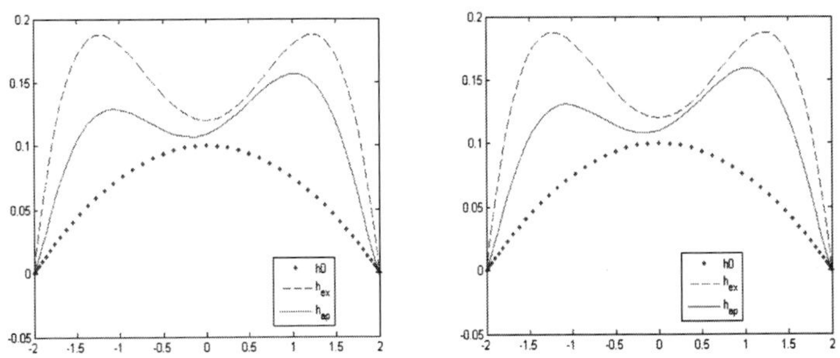

Figure 7: Reconstruction of h_2 after one iteration for exact data (left), and noisy data with δ=0.01 (right)

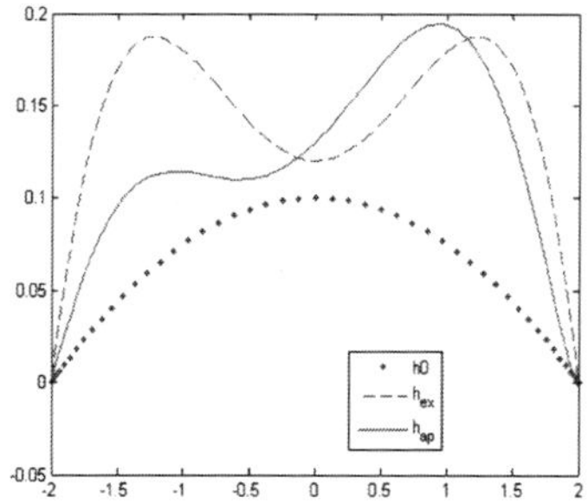

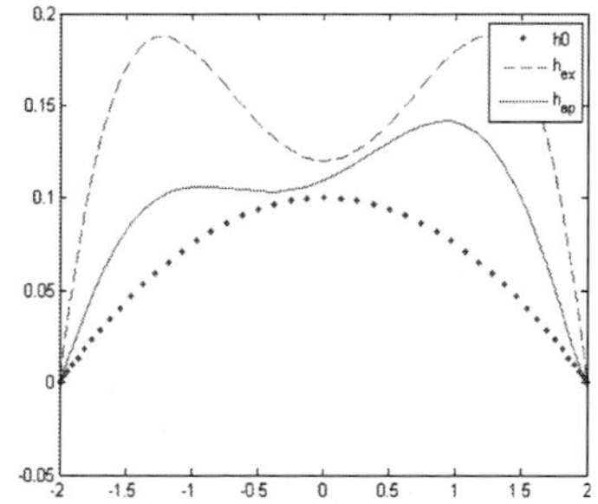

Figure 8: Reconstruction of h_2 for δ=0.05 (left), for δ=0.1 (right)

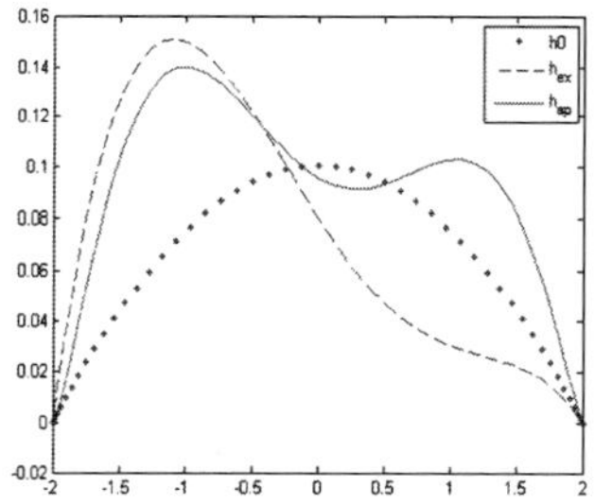

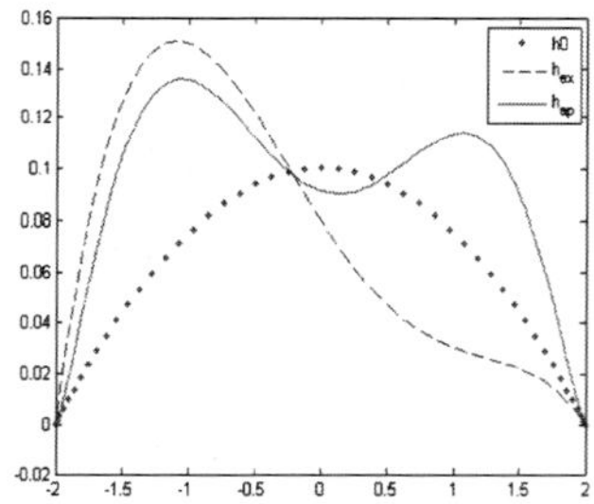

Figure 9: Reconstruction of h_3 after one iteration for exact data (left), and noisy data with δ=0.01 (right)

CONCLUSION

In this paper, direct and inverse boundary value problems associated with the two-dimensional harmonic equation have been investigated. We obtained the following results:

- The direct problem was studied systematically by employing the boundary integral method.
- The inverse problem was reduced to a nonlinear system which was approximated by a Gauss–Newton method (partially).
- The unknown arc of the boundary was approximated by a trigonometric polynomial. The resulting ill-conditioned system of linear algebraic equations has been regularized by using a least squares method.

- The numerical results obtained showed satisfying reconstructions for exact and noisy data with suitable initial guess.

Further research is required in order to improve the performance of the algorithm for the reconstruction of the boundary. It is necessary to introduce additional regularization treatments such as the choice of the parametrization of *h*, initial guess h0h0 and a stopping rule for terminating the iterations.

APPENDICES

Appendix 1: Derivative of nonlinear integral operator

In the following we state a useful formula. Let f(x,y,z) be a real function defined and of class $C^1(D)$ in an open set *D* of R^3. To a function g∈C1([a,b]) we associated a function *F* defined by an integral

$$F(x) = \int_a^b f(x, y, g(y))\, dy, \quad x \in [c, d].$$

We consider the nonlinear operator

F:V→W

defined by the mapping g→F between the Banach spaces V=C01[a,b] and W=C[c,d]. We can prove the theorem.

Theorem 6

FF *is Fréchet differentiable. The derivative, in the direction w,* Aw=DF(g;w) *is given by*

$$(Aw)(x) = \int_a^b \frac{\partial f}{\partial z}(x, y, g(y))\, w(y)\, dy, \quad x \in [c, d]. \tag{51}$$

Now we consider the case of the function F_1 defined by

$$F_1(x) = \int_a^b f_1(x, y, g(y))\, g'(y)\, dy, \quad x \in [c, d],$$

then $A_1w=DF_1(g;w)$ is given by

$$(A_1 w)(x) = \int_a^b \left[\frac{\partial f_1}{\partial z}(x, y, g(y))(1 - g'(y)) - \frac{\partial f_1}{\partial y}(x, y, g(y)) \right] w(y)\, dy, \quad x \in [c, d]. \tag{52}$$

Appendix 2: Kernels of integral operators

We proceed with giving the expressions of the kernels of the nonlinear integral operators H and H′ and their derivatives. Following the definitions (44)–(46), we deduce the formulas

$$H(x,x',y') = \frac{1}{2}\left[\frac{\sin \pi y'}{\cosh \pi (x-x') - \cos \pi y'}\right], \tag{53}$$

$$\left(\frac{\partial H}{\partial x'}, \frac{\partial H}{\partial y'}\right)(x,x',y') = \frac{\pi}{2}\left(\frac{\sin \pi y' \sinh \pi (x-x')}{(\cosh \pi (x-x') - \cos \pi y')^2}, \frac{\cos \pi y' \cosh \pi (x-x') - 1}{(\cosh \pi (x-x') - \cos \pi y')^2}\right), \tag{54}$$

and

$$H'(x,x',h(x')) = \frac{\partial H}{\partial x'}(x,x',1-h(x'))h'(x') + \frac{\partial H}{\partial y'}(x,x',1-h(x')). \tag{55}$$

The Fréchet derivatives of the integral operators H and H′ with respect to h can be obtained by differentiating their kernels with respect to h (see (51)–(52)). More precisely, we have

$$D\mathcal{H}(h;k)(x) = \frac{\pi}{2}\int_{-a}^{a}\left[\frac{\cos \pi h(x') \cosh \pi (x-x') - 1}{(\cosh \pi (x-x') + \cos \pi h(x'))^2}\right]k(x')\omega(x')\,dx',$$
$$\omega(x') = \sqrt{1 + h'^2(x')}, \tag{56}$$

and

$$D\mathcal{H}'(h;k)(x) = \int_{-a}^{a}\left[\left(\frac{\partial^2 H}{\partial y'^2}(x,x',1-h(x')) - \frac{\partial^2 H}{\partial x'^2}(x,x',1-h(x'))\right)q(x') - \frac{\partial H}{\partial x'}(x,x',1-h(x'))q'(x')\right]k(x')\,dx'$$
$$= \int_{-a}^{a}\left[K_1(x,x')q(x') - K_2(x,x')q'(x')\right]k(x')\,dx', \tag{57}$$

with

$$K_1(x,x') = -2\pi^2 \times \frac{\sin \pi h(x') \sinh^2 \pi (x-x')}{(\cosh \pi (x-x') + \cos \pi h(x'))^3} \tag{58}$$

and

$$K_2(x,x') = \frac{\pi}{2} \times \frac{\sin \pi h(x') \sinh \pi (x-x')}{(\cosh \pi (x-x') + \cos \pi h(x'))^2}. \tag{59}$$

In the previous equations we assume that $h \in C^1[-a,a]$ and $h(\pm a)=0$, hence $q \in C^1[-a,a]$. For simplicity, we suppose in the differentiation that $\omega(x')$ is independent of *h*, i.e., depends on h_0 of the previous step.

ACKNOWLEDGEMENTS

The authors would like to thank the anonymous reviewers for their careful work and comments that helped to improve the clarity of the paper.

CONTRIBUTIONS

All of the authors contributed equally in writing this paper. All authors read and approved the final manuscript.

REFERENCES

1. Alessandrini, G., Beretta, E., Rosset, E., Vessella, S.: Inverse boundary value problems with unknown boundaries: optimal stability. C. R. Acad. Sci. Paris **328**, 607–611 (2000)
2. Cakoni, F., Kress, R.: Integral equations for inverse problems in corrosion detection from partial Cauchy data. Inverse Probl. Imaging **1**(2), 229–245 (2007)
3. Chipot, M.: Elliptic Equations: An Introductory Course. Birkhäuser, Basel (2009)
4. Engl, H.W., Hanke, M., Neubauer, A.: Regularization of Inverse Problems. Kluwer Academic, Dordrecht (1996)
5. Fang, W., Yan, L., Zen, S.: Numerical recovery of Robin boundary from boundary measurements for the Laplace equation. J. Comput. Appl. Math. **224**, 573–580 (2009)
6. Grisvard, P.: Elliptic Problem in Nonsmooth Domains. Classics in Applied Mathematics, vol. 69. SIAM, Philadelphia (1985)
7. Hansen, P.C.: Regularization tools version 4.0 for Matlab 7.3. Numer. Algorithms **46**, 189–194 (2007)
8. Kirsch, A.: An Introduction to the Mathematical Theory of Inverse Problems. Springer, New York (2011)
9. Kress, R.: Linear Integral Equations. Applied Mathematical Sciences, vol. 82. Springer, New York (2014).
10. Kress, R., Rundel, W.: Nonlinear integral equations and the iterative solution for an inverse boundary value problem. Inverse Probl. **21**, 1207 (2005)
11. Kurt, B., Caudill, L.: An inverse problem in thermal imaging. SIAM J. Appl. Math. **56**(3), 715–735 (1996)
12. McLean, W.: Strongly Elliptic Systems and Boundary Integral Equations. Cambridge University Press, Cambridge (2000)
13. Melnikov, M.Y.: Green's Functions: Construction and Application. de Gruyter, Berlin (2012)
14. Monch, L.: On the inverse acoustic scattering problem by an open arc: the sound-hard case. Inverse Probl. **13**, 1379 (1997)
15. Nachaoui, A.: An improved implementation of an iterative method in boundary identification problems. Numer. Algorithms **33**, 381–398 (2003)

16. Smoller, J.: Shock Waves and Reaction–Diffusion Equations. Springer, Berlin (1983)
17. Yang, F., Yan, L., Wei, T.: Reconstruction of the corrosion boundary for the Laplace equation¦a¯fÿÿ by using a boundary collocation method. Math. Comput. Simul. **79**, 2148–2156 (2009)

CHAPTER
10

Solving Large-Scale Inverse Magnetostatic Problems Using the Adjoint Method

Florian Bruckner[1], ClaasAbert[1], GregorWautischer[1], Christian Huber[1], ChristophVogler[2], Michael Hinze[3] & Dieter Suess[1]

[1]Christian Doppler Laboratory of Advanced Magnetic Sensing and Materials, Institute of Solid State Physics, Vienna University of Technology, Austria

[2] Institute of Solid State Physics, Vienna University of Technology, Austria

[3] Department of Mathematics, University of Hamburg, Germany.

ABSTRACT

An efficient algorithm for the reconstruction of the magnetization state within magnetic components is presented. The occurring inverse magnetostatic problem is solved by means of an adjoint approach, based on the Fredkin-Koehler method for the solution of the forward problem. Due to the use of

Citation: Bruckner, F., Abert, C., Wautischer, G., Huber, C., Vogler, C., Hinze, M., & Suess, D. (2017). Solving large-scale inverse magnetostatic problems using the adjoint method. Scientific reports, 7, 40816. (7 pages) DOI: https://doi.org/10.1038/srep40816

hybrid FEM-BEM coupling combined with matrix compression techniques the resulting algorithm is well suited for large-scale problems. Furthermore the reconstruction of the magnetization state within a permanent magnet as well as an optimal design application are demonstrated.

INTRODUCTION

Magnetic materials are used in a wide range of applications, ranging from permanent magnets, magnetic machines, up to magnetic sensors and magnetic recording devices. Solving the Inverse Magnetostatic Problem allows to reconstruct the internal magnetization state of a magnetic component, by means of magnetic field measurements outside of the magnetic part, which is of importance for quality control. Compared with the forward problem, where the magnetic state is known and the strayfield is calculated, inverse problems are much harder to solve, since they typically are much worse conditioned and often not uniquely solveable. Inverse problem solvers are based on stable and reliable solvers for the forward problem. In the case of magnetostatic Maxwell equations, Finite Element (FEM) formulations, combined with methods to handle the open-boundary, have proven to be the methods of choice for many efficient and accurate methods[1,2,3].

The applications of inverse problems can be coarsely divided into optimal design and source identification. Optimal design problems define a desired strayfield and try to calculate optimal material distributions or geometries to reach these requirements as accurate as possible[4,5]. In contrast to this, source identification problems try to reconstruct the state of existing magnetic components.

The identification of (permanent) magnetic materials has e.g. been successfully applied for reconstructing the state of magnetic rollers used in copy machines[6], magnetically biased chokes[7], or even for the magnetic anomaly created by ferromagnetic ships[8].

The presented solver for the inverse 3D magnetostatic Maxwell equations, is based on the well-established Fredkin-Koehler-Method[3], which uses a hybrid FEM-BEM coupling for efficient handling of the open-boundary conditions. Combined with a hierarchical matrix compression technique for the dense boundary integral matrices, the algorithm is able to handle large-scale problems. Additionally, the use of a general Tikhonov regularization (see e.g. ref. 9), provides a very flexible means to define application specific regularizations.

ADJOINT METHOD

Forward Problem

The demagnetization field of a magnetic body is defined as $h_d = -\nabla u$, where the magnetic scalar potential *u* is given by

$$\Delta u = \nabla \cdot \boldsymbol{m} \text{ in } \Omega \tag{1}$$

$$\Delta u = 0 \text{ in } \mathbb{R}^3 \backslash \Omega \tag{2}$$

with jump and boundary conditions

$$[u]_{\partial\Omega} = 0 \tag{3}$$

$$\left[\frac{\partial u}{\partial \boldsymbol{n}}\right]_{\partial\Omega} = -\boldsymbol{m} \cdot \boldsymbol{n} \tag{4}$$

$$u(\boldsymbol{r}) \rightarrow \mathcal{O}(1/|\boldsymbol{r}|) \quad \text{for} \quad |\boldsymbol{r}| \rightarrow \infty \tag{5}$$

where ***m*** is the magnetization and Ω is the magnetic region.

The forward problem requires the solution of the potential u on the region Ωh generated by the magnetization in a magnetic region Ω_m (see Fig. 1). This problem is solved by considering a single region $\Omega = \Omega_m \cup \Omega_h$ with $m(x) = 0$ for $x \in \Omega_h$. The hybrid FEM/BEM method introduced by Fredkin and Koehler [3] is one of the most accurate methods for the solution of this problem and will be used in the following. Consider the following splitting of the solution u:

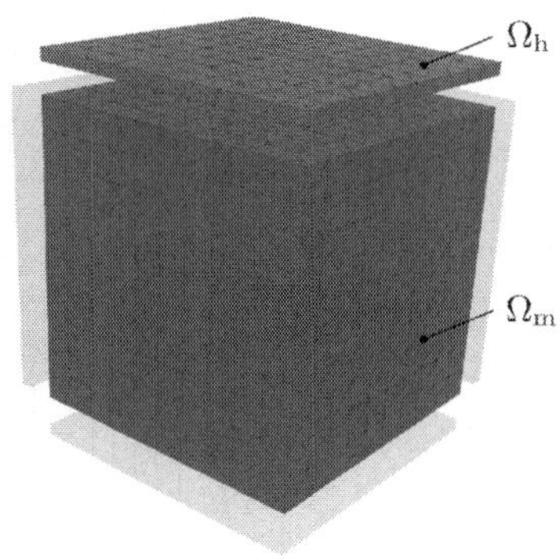

Figure 1: Discretized magnetization region Ω_m **(blue) and measurement region** Ω_h **(brown).** Since the strayfield problem is scale-invariant, length units are omitted. The magnetization is defined on a unit cube. Measurement boxes of

thickness 0.04 are located next to each side of the cube, using an airgap of 0.1 (the 2 boxes in front are not shown in the figure).

$$u = u_1 + u_2 \tag{6}$$

Here u_1 is defined by

$$\Delta u_1 = \nabla \cdot \boldsymbol{m} \quad \text{in} \quad \Omega \tag{7}$$

$$\frac{\partial u_1}{\partial \boldsymbol{n}} = -\boldsymbol{m} \cdot \boldsymbol{n} \quad \text{on} \quad \partial\Omega \tag{8}$$

$$u_1 = 0 \qquad \text{in} \quad \mathbb{R}^3 \setminus \Omega. \tag{9}$$

This Neumann problem is solved with the finite-element method. While u_1 solves for the right-hand side $\boldsymbol{m}$ and fulfills the jump condition of the normal derivative $-\boldsymbol{m} \cdot \boldsymbol{n}$, it is not continuous across $\partial\Omega$ as required. This jump is compensated by u_2 which is defined as

$$\Delta u_2 = 0 \qquad \text{in} \quad \Omega \tag{10}$$

$$[u_2] = -[u_1] \qquad \text{on} \quad \partial\Omega \tag{11}$$

$$\left[\frac{\partial u_2}{\partial \boldsymbol{n}}\right] = 0 \qquad \text{on} \quad \partial\Omega \tag{12}$$

$$u_2(\boldsymbol{x}) = \mathcal{O}(1/|\boldsymbol{x}|) \quad \text{if} \quad |\boldsymbol{x}| \to \infty \tag{13}$$

This system is solved by the following double-layer potential

$$u_2 = \int_{\partial\Omega} u_1 \frac{\partial}{\partial \boldsymbol{n}} \frac{1}{|\boldsymbol{x} - \boldsymbol{x}'|} \, \mathrm{d}\boldsymbol{x} \tag{14}$$

For efficiency reasons the double-layer potential is only computed on the boundary $\partial\Omega$ using a Galerkin boundary-element method. Subsequently these values are used as Dirichlet boundary conditions to solve u_2 within Ω using the finite-element method.

All potentials are calculated using piecewise linear basis function (P_1) and the derived strayfield would be constant within each element. Thus, a mass lumping procedure needs to be used to project the field onto piecewise linear basis functions which are defined on each vertex of the mesh.

Inverse Problem

The inverse problem can be understood as a PDE constrained optimization problem. Due to the ill-posedness of the inverse problem, some additional information needs to be provided to allow reasonable results. This can be achieved by using Tikhonov regularization, which uses an additional penalty term which should be considered for the optimization. A possible candidate for the objective function is

$$J = \frac{1}{2}\int_{\Omega_{\mathrm{h}}} \|-\boldsymbol{\nabla} u - \boldsymbol{h}_{\mathrm{m}}\|^2 \,\mathrm{d}\boldsymbol{x} + \alpha T(\boldsymbol{m}) \tag{15}$$

where $\boldsymbol{h}_{\mathrm{m}}$ is the prescribed (measured) field in Ω_{h} and α is the Tikhonov constant corresponding to the regularization functional $T(\boldsymbol{m})$. This functional should be minimized, constrained by

$$F = \Delta u - \boldsymbol{\nabla}\cdot\boldsymbol{m} = 0 \quad \text{on} \quad \Omega \tag{16}$$

with boundary conditions as given above. We aim to solve this problem using a gradient based iterative minimizer. The constraint F gives an implicit expression for $u(\boldsymbol{m})$ which allows to directly calculate the desired gradient by

$$\frac{\mathrm{d}J}{\mathrm{d}\boldsymbol{m}} = \frac{\partial J}{\partial u}\frac{\partial u}{\partial \boldsymbol{m}} + \frac{\partial J}{\partial \boldsymbol{m}} \tag{17}$$

The inefficient direct calculation of the term $\frac{\partial u}{\partial \boldsymbol{m}}$ can be avoided using the adjoint approach, which makes use of the derivative of the constraint equation to eliminate the problematic term. Compared with a naive implementation using a dense system matrix $\frac{\partial u}{\partial \boldsymbol{m}}$, the computational as well as storage costs can be reduced from $O(N^2)$ to $O(N\log N)$:

$$\frac{\mathrm{d}J}{\mathrm{d}\boldsymbol{m}} = \lambda^T\frac{\partial F}{\partial \boldsymbol{m}} + \frac{\partial J}{\partial \boldsymbol{m}} \tag{18}$$

where λ is given by the adjoint equation

$$\left(\frac{\partial F}{\partial u}\right)^T \lambda = -\left(\frac{\partial J}{\partial u}\right)^T. \tag{19}$$

Since the Poisson problem is self-adjoint, the adjoint system (19) can be solved along the lines of the forward problem. Computing the variational derivative on the RHS yields

$$\Delta\lambda = \boldsymbol{\nabla}\cdot(\boldsymbol{\nabla} u + \boldsymbol{h}_{\mathrm{m}}) \quad \text{on} \quad \Omega. \tag{20}$$

where the sources (RHS) live on Ω_h and the solution is only computed on Ω_m. The same boundary conditions as for the forward problem hold. Thus, the above described hybrid method is applied. The gradient of J is then finally given by

$$\frac{\mathrm{d}J}{\mathrm{d}\boldsymbol{m}} = \boldsymbol{\nabla}\lambda + \alpha \frac{\partial T}{\partial \boldsymbol{m}} \tag{21}$$

Note that ∇u and $\nabla\lambda$ are projected onto P_1 before computing (20) and (21) respectively. The algorithm is implemented using Magnum.Fe [10], which is based on the finite element library FEniCS [11]. This allows a very comfortable definition of the regularization functional. Furthermore, automatic differentiation can be used for the calculation of the corresponding partial derivatives. The algorithm was verified by comparison with a FEMonly implementation, using the dolfin-adjoint library [12], which allows to automatically derive the adjoint equation for a given forward problem

NUMERICAL EXPERIMENTS

The presented algorithm is validated by the reconstruction of the flower-state within a magnetic unit cube. The strayfield is calculated within measurement planes next to each side of the cube (see Fig. 1). The magnetic state of the cube is parametrized by

$$\boldsymbol{m} = \begin{pmatrix} \sin(c_{\mathrm{tilt}}\theta)\,\cos(\phi) \\ \sin(c_{\mathrm{tilt}}\theta)\,\sin(\phi) \\ \cos(c_{\mathrm{tilt}}\theta) \end{pmatrix} \qquad \begin{matrix} \theta = z\,\sqrt{x^2+y^2} \\ \phi = \tan^{-1}(y/x) \end{matrix} \tag{22}$$

where c_{tilt} is an open parameter that allows to change the strength of the flower state. For the proper reconstruction of the magnetic state additional knowledge about the solution needs to be provided. For all presented results a smooth reconstructed magnetization is desired which suggests using the following default regularization functional

$$T(\boldsymbol{m}) = \int_{\Omega_{\mathrm{m}}} (\nabla \boldsymbol{m})^2 \,\mathrm{d}\boldsymbol{x} \tag{23}$$

For this specific flower-state the absolute value of the magnetization is known to be constant. Thus, the solution of the inverse problem could be simplified by using the following penaltization functional

$$T^*(m) = \int_{\Omega_m} (m^2 - 1)^2 \, dx \quad (24)$$

The assumption of a constant magnetization may be a good approximation for (isotropic) permanent magnetic materials. Due to the large magnetic remanence and the relatively small susceptibility the induced magnetization may be negligible (see ref. 13 for a simple model of isotropic permanent magnetic materials).

A Gaussian noise with zero mean and a standard deviation $\sigma=10^{-4}$ has been added to the field, calculated by the forward problem, which should simulate unavoidable measurement errors. The minimization problem is solved by a gradient descent method combined with a line-search strategy. As expected, reconstruction without using a proper regularization leads to large magnetization vectors near to the edges of the unit cube. Increasing the regularization parameter α, first avoids the over-fitting of the noisy measurement data, but finally leads to blurring of the reconstruction results if α gets too large. Determining the optimal alpha is a crucial step for the solution of an inverse problem. The results for the reconstruction of a flower state with $c_{tilt}=2$ using $\alpha=10^{-3}$ is visualized in Fig. 2. Although there are some deviations of the reconstructed magnetization from the reference state. It can be seen that the created strayfield is nearly identical. As stated above this is a clear indication of the bad condition of the inverse problem.

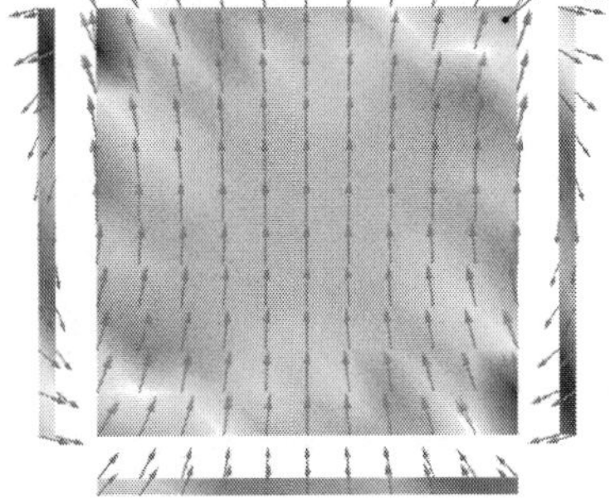

Figure 2: Reconstruction of a flower-state within a unit cube according to Eqn. (22) using $c_{tilt}=2$. A cut through the $y=0$ plane is visualized. Starting from the initial parametrized flower state in Ω_m the magnetic strayfield is calculated within the fieldboxes Ω_h (green arrows). In order to simulate measurement errors a Gaussian noise with $\sigma=10^{-4}$ has been added to the forward strayfield. The reconstructed magnetization as well as strayfield are computed using $\alpha=10^{-3}$ (red

arrows). The relative differences of initial and reconstructed states are indicated by the gray-scale. Maximal relative errors of the x-components amount to 0.25 for the magnetization, and $5 \cdot 10^{-3}$ for the induced magnetic field, respectively.

Using the so-called L-curve method [14], allows to visualize the trade-off between the reconstruction of the strayfield and the fulfillment of the regularization constraint. Plotting the regularization norm (also called solution norm) $\|T(m)\|$ over the residual norm $\|-\nabla u - hm\|$ for different regularization parameters α, shows an L-shaped curve. The optimal α_{opt} can be selected at the corner of the L-curve which means that α is large enough to reduce the regularization norm significantly, but it does not change the residual norm too much. The resulting L-curves for the reconstruction of the flower state for different noise levels are summarized in Fig. 3. An optimal value $\alpha_{opt} \approx 5 \cdot 10^{-5}$ can be found.

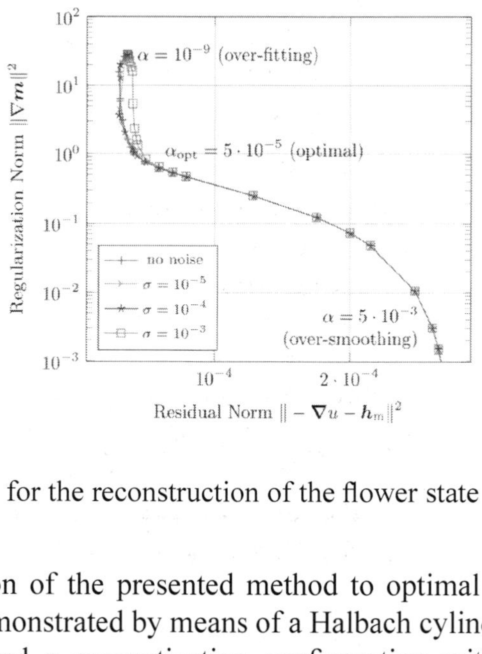

Figure 3: L-curves for the reconstruction of the flower state for different noise levels σ.

The application of the presented method to optimal design problems should now be demonstrated by means of a Halbach cylinder configuration. The goal is to find a magnetization configuration within a cylindrical domain, which creates a homogeneous strayfield inside of the cylinder. The magnetization domain Ω_m has an outer radius $r_o = 1.0$, an inner radius $r_i = 0.6$, and a height of $h = 2.0$, while a cylindrical measurement domain Ω_h with radius $r_m = 0.5$ and the same height is used. The magnetization vectors should have constant norm, which suggests using the constant-norm penaltization functional (24). The analytic solution for a cylindical Halbach array15 can be expressed in cylindrical coordinates as

$$\mathbf{m}(\rho, \phi) = \cos(\phi)\, \mathbf{e}_\rho - \sin(\phi)\, \mathbf{e}_\phi \tag{25}$$

where ρ, ϕ are the cylindrical coordinates, with the corresponding unit vectors $\mathbf{e}_\rho$, $\mathbf{e}_\phi$.

As visualized in Fig. 4 there is a nearly perfect match of the reconstructed and the analytical Halbach configuration.

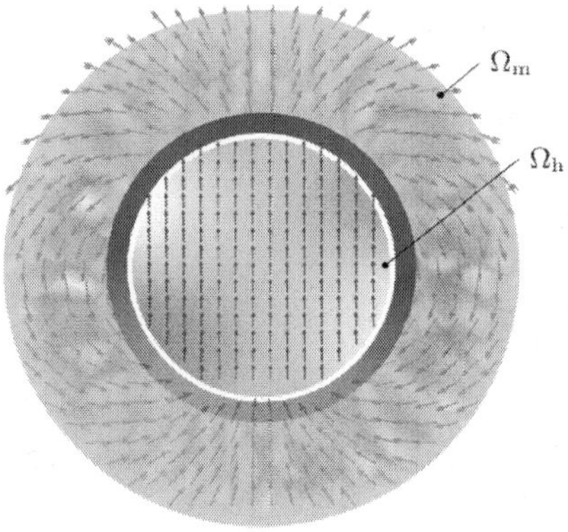

Figure 4: Optimal design problem of a Halbach cylinder creating a homogeneous strayfield inside of the cylinder. Starting from a homogeneous strayfield the presented algorithm reproduces a Halbach like magnetization configuration within Ω_m (red arrows). A constant-norm regularization with $\alpha = 10^4$ is used and shows a nearly perfect match with the analytical solution (green arrows). The resulting strayfield is calculated inside Ω_h and shows a nearly homogeneous distribution (red arrows). The relative errors of the magnetization magnitude and the reconstruced strayfield are indicated by the gray-scale, and their maximum amount to 2% and 6%, respectively.

CONCLUSION

An efficient algorithm for the solution of inverse problems has been introduced. The use of the Finite Element library FEniCS allows to easily define application specific regularization functionals in a very flexible way. Thus, the implemented algorithm is suitable for a wide range of applications including reverse engineering of magnetic components, design and optimization of magnetic circuits and topology optimization, respectively. Using the hybrid FEM-BEM method proposed by Fredkin-Koehler allows to handle the open-boundary problem accurately and without the need for global mesh including a large airbox. Source identification has been validated by the successful reconstruction of the magnetic flower-state within a unit cube by means of Tikhonov regularization. The selection of a suitable regularization parameter has been demonstrated using the L-curve method. Finally, the application of the method to an optimal design problem has been demonstrated by means of an Halbach cylinder, which is nearly perfectly reproduced.

ACKNOWLEDGEMENTS

The authors acknowledge the CD-Laboratory AMSEN (financed by the Austrian Federal Ministry of Economy, Family and Youth, the National Foundation for Research, Technology and Development), the Vienna Science and Technology Fund (WWTF) under Grant No. MA14-044 and the Austrian Science Fund (FWF) under grant No. I2214-N20 for financial support.

CONTRIBUTIONS

F.B. and C.A. developed and implemented the algorithm. M.H. provided the mathematical background. G.W., C.H., C.V. and D.S. developed the numerical experiments and helped developing and extending the inverse problem code. All authors contributed to the paper writing.

REFERENCES

1. Brunotte, X., Meunier, G. & Imhoff, J. Finite element modeling of unbounded problems using transformations: A rigorous, powerful and easy solution. *IEEE Trans. Magn.* **28**, 1663–1666 (1992).
2. Khebir, A., Kouki, A. & Mittra, R. Asymptotic Boundary Conditions for Finite Element Analysis of Three-Dimensional Transmission Line Discontinuities. *IEEE Trans. Microwave Theory Tech.* **38**, 1427–1432 (1990).
3. Fredkin, D. & Koehler, T. Hybrid method for computing demagnetizing fields. *IEEE Trans. Magn.* **26**, 415–417 (1990).
4. Neittaanmki, P., Rudnicki, M. & Savini, A. *Inverse Problems and Optimal Design in Electricity and Magnetism* (Clarendon Press, 1996).
5. Yoo, J. & Kikuchi, N. Topology optimization in magnetic fields using the homogenization design method. *Int. J. Numer. Methods Eng.* **48**, 1463–1479 (2000).
6. Igarashi, H., Honma, T. & Kost, A. Inverse inference of magnetization distribution in cylindrical permanent magnets. *IEEE Trans. Magn.* **36**, 1168–1171 (2000).
7. Husstedt, H. & Kaltenbacher, M. Detailed analysis of permanent magnets by means of free field measurements. *J. Appl. Phys.* **115**, 17E302 (2014).
8. Chadebec, O., Coulomb, J. L., Bongiraud, J. P., Cauffet, G. & Thiec, P. L. Recent improvements for solving inverse magnetostatic problem applied to thin shells. *IEEE Trans. Magn.* **38**, 1005–1008 (2002).
9. Calvetti, D., Morigi, S., Reichel, L. & Sgallari, F. Tikhonov regularization and the L-curve for large discrete ill-posed problems. *J. Comput. Appl. Math.* **123**, 423–446 (2000).
10. Abert, C., Exl, L., Bruckner, F., Drews, A. & Suess, D. magnum.fe: A micromagnetic finite-element simulation code based on FEniCS. *J. Magn. Magn. Mater.* **345**, 29–35 (2013).
11. Logg, A., Mardal, K.-A. & Wells, G. *Automated Solution of Differential Equations by the Finite Element Method: The FEniCS Book* (Springer Science & Business Media, 2012).
12. Farrell, P. E., Ham, D. A., Funke, S. W. & Rognes, M. E. Automated Derivation of the Adjoint of High-Level Transient Finite Element Programs. *SIAM J. Sci. Comput.* **35**, C369–C393 (2013).

13. Bruckner, F. et al. Macroscopic simulation of isotropic permanent magnets. *J. Magn. Magn. Mater.* **401**, 875–879 (2016).
14. Hansen, P. C. & OLeary, D. P. The Use of the L-Curve in the Regularization of Discrete Ill-Posed Problems. *SIAM J. Sci. Comput.* **14**, 1487–1503 (1993).
15. Halbach, K. Design of permanent multipole magnets with oriented rare earth cobalt material. *Nucl. Instrum. Methods* **169**, 1–10 (1980).

CHAPTER

11

The Problem of the Inverse Lyapunov Exponent and its Applications

Marcin Lawnik

Faculty of Applied Mathematics, Silesian University of Technology, Kaszubska 23, 44-100 Gliwice, Poland

ABSTRACT.

The problem of the inverse Lyapunov exponent was formulated and solved, involving to find such chaotic transformation for which the value of the Lyapunov exponent is given in advance. The solution procedure was presented by a numerical example. Furthermore, applications of the discussed model in chaos based cryptography were discussed.

Keywords: Lyapunov exponent, piecewise linear map, inverse problem.

Citation: Lawnik, M. (2018). The problem of the inverse Lyapunov exponent and its applications. Nonlinear Anal., Model. Control, 23, 951-960. (10 pages) DOI: https://doi.org/10.15388/NA.2018.6.9

INTRODUCTION

Chaotic functions play a very important role in many fields of science and technology. For example, they are used in cryptography as generators of pseudo-random numbers [1, 18] for advanced technical models of chemical reactors [5, 6, 20] or description of physical phenomena such as weather forecasts [24]. They are also applied outside the range of technical sciences, for example, in economics – for modelling certain processes [12, 15] or even medicine [28, 30].

One of the basic measures showing if a given function

$$x_{k+1} = f(x_k) \tag{1}$$

could generate chaotic solutions is the Lyapunov exponent. For maps in the form of (1), it may be expressed as

$$\lambda = \lim_{N \to \infty} \frac{1}{N} \sum_{i=0}^{N-1} \ln |f'(x_i)| \tag{2}$$

and measures the rate of the propagation of infinitely approximate trajectories of the system (1). Positive value of (2) is a necessary condition for function (1) to generate a chaotic solution; otherwise, the solution is stable.

The Lyapunov exponent for discrete dynamical systems may be easily determined in a numerical manner. Accordingly, for a logistics map expressed as

$$x_{k+1} = a x_k (1 - x_k), \tag{3}$$

where a ∈ [0, 4] and x ∈ [0, 1], the Lyapunov exponent is presented in the graph in Fig. 1. As shown in the graph, its structure is fairly complicated and has a fractal character. Moreover, its maximum value is ln 2 for the parameter value a = 4.

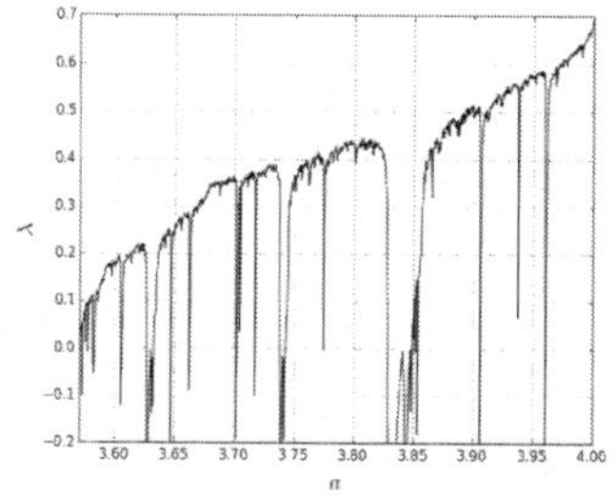

Figure 1. Lyapunov exponent λ of Eq. (3) in the interval [3.57, 4]. Outside this interval the value of λ is non-positive.

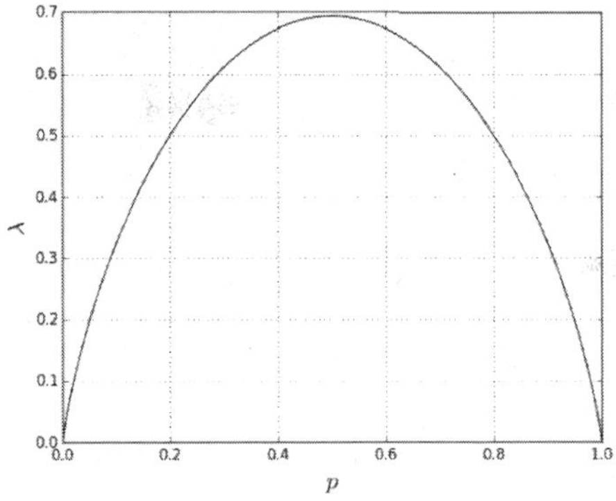

Figure 2. Lyapunov exponent λ given by Eq. (5).

Another example of a transformation for which the Lyapunov exponent may be designated in an easy manner is the asymmetric (skew) tent map given by the equation

$$x_{k+1} = \begin{cases} \frac{x_k}{p}, & 0 < x_k \leqslant p, \\ \frac{1-x_k}{1-p}, & p < x_k < 1, \end{cases} \tag{4}$$

where p ∈ (0, 1). Its Lyapunov exponent λ is expressed as [2]

$$\lambda = -p \ln p - (1-p)\ln(1-p). \tag{5}$$

For each value of parameter p, the value of (5) is positive and its graph is plotted in Fig. 2. Likewise, in the case of (3) its maximum is ln 2 for p = 1/2.

Let us now consider another example of piecewise linear map [25]:

$$x_{k+1} = \begin{cases} \frac{x_k}{p_1}, & x_k \in I_1, \\ \frac{x_k - p_1}{p_2}, & x_k \in I_2, \\ \vdots & \vdots \\ \frac{x_k - \sum_{i=1}^{n-1} p_i}{p_n}, & x_k \in I_{n-2}, \end{cases} \tag{6}$$

where $I_1 = [0, p_1]$, $I_i = [\sum_{j=1}^{i-1} p_j, \sum_{j=1}^{i} p_j]$ and $\sum_{i=1}^{n} p_i = 1$.

The Lyapunov exponent for (6) is expressed by the following equation [2]:

$$\lambda = -\sum_{i=1}^{n} p_i \ln p_i.$$

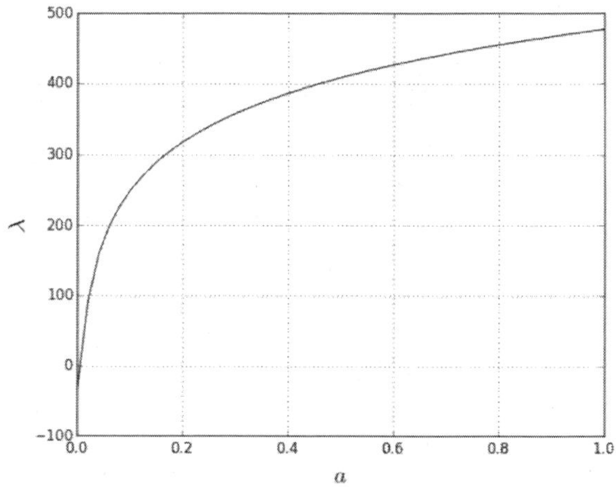

Figure 3. Lyapunov exponent for transformation (7) with b = 123 and n = 100.

Similarly to (5), for each set of pi it has a positive value. Its maximum is equal to ln n for pi = 1/n (i= 1, 2, . . . , n). Furthermore, its value increases together with the increase of n.

An interesting function for which the Lyapunov exponent also increases indefinitely, is the so called Weierstrass recurrence given as [7, 17]

$$x_{k+1} = \sum_{i=0}^{n} a^i \cos\left(b^i \pi x_k\right). \tag{7}$$

Its Lyapunov exponent is shown in Fig. 3.

As inferred from the examples discussed above, despite the fact that the Lyapunov exponent may be expressed by means of complicated dependencies, it may also reach very high values. Nevertheless, while analysing the discussed cases, a question may arise whether it is possible to construct such chaotic function that would have a pre-determined value of the Lyapunov exponent. Such formulated problem may be labelled as the inverse designation of the Lyapunov exponent and will be discussed in the next part of the paper.

Such types of tasks are often analyzed in technical sciences, e.g. the inverse Stefan problem [13,29], which has wide applications in metals solidification. Moreover, it should also be emphasised that other inverse problems are considered in the theory of dynamical systems, for example, the inverse Frobenius–Perron problem [10, 11, 16, 27], involving the designation of a map with the assumed invariant density.

The posed problem can be used in cryptography based on the theory of chaos. In this case, the security of the algorithm depends on the used map and, consequently, from its properties. Mappings with a small and variable value of Lyapunov exponent are not the best solutions. More on this subject is given in Section 5

FORMULATION OF THE PROBLEM

Let us assume a given λ value. The task is to find such chaotic function for which the value of the Lyapunov exponent is exactly equal to λ. Such task may be described as the inverse Lyapunov exponent problem.

RESULTS

The problem of the inverse Lyapunov exponent will be solved by means of the following expression:

$$x_{k+1} = \begin{cases} nx_k, & x_k \in [0, \frac{1}{n}), \\ nx_k - 1, & x_k \in [\frac{1}{n}, \frac{2}{n}), \\ \vdots & \vdots \\ nx_k - n + 3, & x_k \in [\frac{n-3}{n}, \frac{n-2}{n}), \\ \frac{nx_k - n + 2}{np}, & x_k \in [\frac{n-2}{n}, \frac{n-2}{n} + p), \\ \frac{nx_k - n + 2 - np}{2 - np}, & x_k \in [\frac{n-2}{n} + p, 1], \end{cases} \tag{8}$$

where n > 3 and p ∈ (0, 2/n). Its exemplary graph for settled values of n = 4 and p = 0.1 is plotted in Fig. 4

Let us now assume some basic facts concerning Eq. (8).

Theorem 1. The invariant density ρ(x) of (8) is 1.

Proof. The invariant density ρ(x) for mapping (1) may be designated by solving the Frobenius–Perron equation in the following form [9]:

$$\rho(x) = \sum_{i=1}^{m} \frac{\rho(y_i)}{|f'(y_i)|}, \tag{9}$$

where yi is the ith inverse image of point x (f(yi) = x, i = 1, 2, . . . , m). By transcribing Eq. (9) with (8) the following equation is derived:

$$\rho(x) = \frac{1}{n}\rho\left(\frac{x}{n}\right) + \frac{1}{n}\rho\left(\frac{x+1}{n}\right) + \cdots + \frac{1}{n}\rho\left(\frac{x+n-3}{n}\right) \\ + p\rho\left(\frac{npx+n-2}{n}\right) + \left(\frac{2}{n} - p\right)\rho\left(\frac{x(2-np)+n-2+np}{n}\right). \tag{10}$$

By substituting ρ(x) = 1 to (10), the equality is designated.

Theorem 2. The Lyapunov exponent for Eq. (8) is given by the following formula:

$$\lambda = \frac{n-2}{n}\ln n - p\ln p - \left(\frac{2}{n} - p\right)\ln\left(\frac{2}{n} - p\right). \tag{11}$$

Proof. Let us take advantage of dependence [8]

$$\lambda = \int_0^1 \rho(x)\ln\left|f'(x)\right| \mathrm{d}x,$$

which bids the Lyapunov exponent with the invariant density.

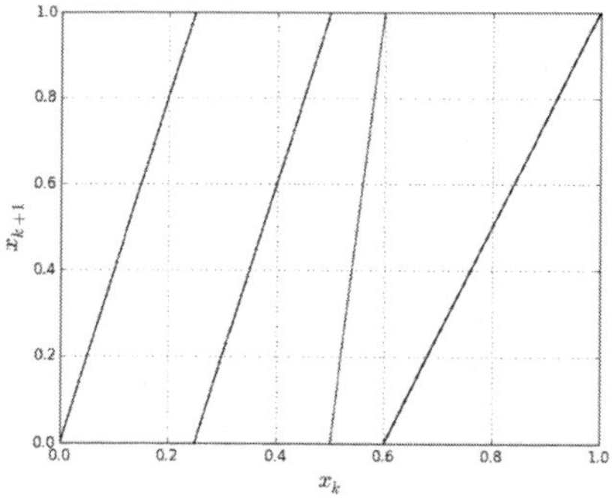

Figure 4. Exemplary transformation of (8) with n = 4 and p = 0.1.

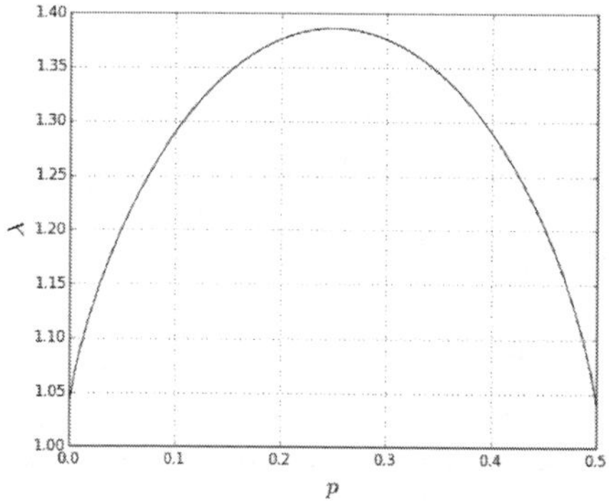

Figure 5. Lyapunov exponent (11) with n = 4.

On the grounds of the fact proven in Theorem 1 that ρ(x) = 1, the following equation is derived:

$$\lambda = \int_0^{1/n} \ln n \, dx + \int_{1/n}^{2/n} \ln n \, dx + \cdots + \int_{(n-3)/n}^{(n-2)/n} \ln n \, dx$$

$$+ \int_{(n-2)/n}^{(n-2)/n+p} \ln \frac{1}{p} \, dx + \int_{(n-2)/n+p}^{1} \ln \frac{1}{\frac{2}{n} - p} \, dx$$

$$= \frac{1}{n} \ln n + \frac{1}{n} \ln n + \cdots + \frac{1}{n} \ln n - p \ln p - \left(\frac{2}{n} - p\right) \ln\left(\frac{2}{n} - p\right)$$

$$= \frac{n-2}{n} \ln n - p \ln p - \left(\frac{2}{n} - p\right) \ln\left(\frac{2}{n} - p\right).$$

Lemma 1. The Lyapunov exponent given by Eq. (11) fulfils the dependence

$$\frac{n-2}{n} \ln n - \frac{2}{n} \ln \frac{2}{n} \leqslant \lambda \leqslant \ln n, \quad n \geqslant 2.$$

Proof. Function (11) reaches its maximum value for p = 1/n, which proves the right side of the inequality. In turn, the lowest value of function (11) is obtained for p = 0 or p = 2/n, directly leading to the left side of the inequality.

Furthermore, we shall prove a supporting lemma enabling the solution of the discussed problem.

Lemma 2. The following inequality holds:

$$\frac{n-2}{n} \ln n - \frac{2}{n} \ln \frac{2}{n} \leqslant \ln(n-1), \quad n \geqslant 2.$$

Proof. We shall prove that

$$\ln(n-1) - \frac{n-2}{n} \ln n + \frac{2}{n} \ln \frac{2}{n} \geqslant 0. \tag{12}$$

By multiplying both sides of (12) by n we derive

$$n \ln(n-1) - (n-2) \ln n + 2 \ln \frac{2}{n} \geqslant 0.$$

Next, by performing elementary procedures on the logarithms, we obtain

$$\ln \frac{4(n-1)^n}{n^n} \geqslant 0. \tag{13}$$

In consequence, inequality (13) leads to

$$\frac{4(n-1)^n}{n^n} \geqslant 1. \tag{14}$$

Dependence (14) is true for n ⩾ 2 as the sequence $a_n = (n-1)^n/n^n$ is incremental and $\lim_{n\to\infty} an = e^{-1}$.

Conclusion 1. It is possible to find such representation of (8), for which the Lyapunov exponent (11) fulfils the following dependence:

$$\ln(n-1) \leqslant \lambda \leqslant \ln n. \tag{15}$$

Proof. The proof directly follows from Lemmas 1 and 2.

On the grounds of Conclusion 1, it is possible to solve the inverse Lyapunov exponent by means of the method described below:

Method 1. Let λ denote the assumed value of the Lyapunov exponent. Then:

- If λ 6 ln 2, the searched equation has the form of (4). The value of parameter p is designated by solving Eq. (5).
- If λ > ln 2, the searched dynamical system has the form of (8). The first step involves the determination of the value of n in accordance with dependence (15). The value of parameter p is designated by solving Eq. (11).

EXAMPLE

By applying Method 1 we shall find the chaotic map in the form of (8), for which the Lyapunov exponent λ = 1.1. n = 4 because

$$\ln 3 < 1.1 < \ln 4.$$

By solving numerically Eq. (11) with λ = 1.1 and n = 4, p ~ 0.079787 is derived. The obtained equation is presented in Fig. 6.

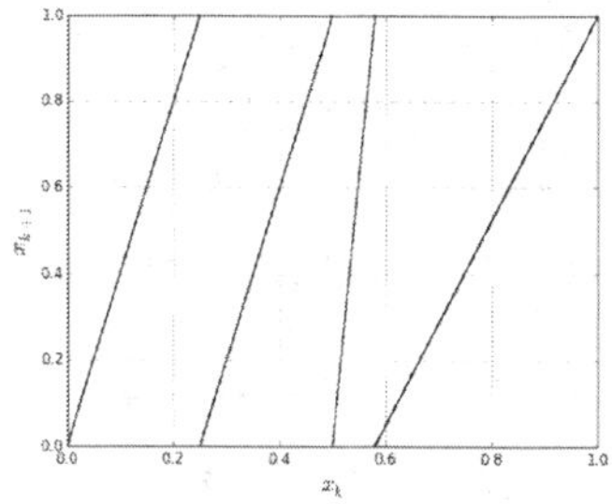

Figure 6. Transformation of the form of (8) with λ = 1.1.

APPLICATIONS

Chaotic maps are used, for example, in chaotic cryptography, enabling, on the grounds of the values derived from the functions, construction of various algorithms for data protection.

The secret keys for such algorithms are the values of parameters and initial conditions. Very frequently the logistics map (3) (among others, in [4, 23, 26, 31]) and the asymmetric tent map (4) (for example, in [14, 21, 22]) are used in such an encryption process. However, as far as cryptography is concerned, both types of these functions are often the weakest points of algorithms referred to in publications. This fact results, among other factors, from:

- i nonuniform distribution of the iterated variable;
- ii too narrow range of parameters values within which the chaotic solution is generated;
- iii unstable value of the Lyapunov exponent, i.e. its value changes in the range [0, ln 2] (as in the case of logistic (3) and tent map (4)) and its very fragile, i.e. a small change in the parameter value eliminates the phenomenon of chaos (logistic map (3), see Fig. 1).

The above mentioned drawbacks have been noticed by some researchers and attempts made at determining other functions that could successfully replace logistic map (3) or skew tent map (4) [3, 19].

The presented map (8) is devoid of the above-mentioned disadvantages. It may be used as a function suitable for encryption, due to its features such as: uniform distribution of the iterated variable or the length of intervals of the parameter values where the function is chaotic. Moreover, Eq. (8) is a representant of the family of functions for which the Lyapunov exponent λ is designated by expression (11). Other equations that belong to this family are determined in such a manner that the direction coefficients of the straight lines in (8) are positive or negative with the same modulus value. Likewise, it is possible to re-permute the sub-intervals on which particular straight lines are designated. The number of such functions is determined by the lemma given below.

Lemma 3. The number P_n of a piecewie linear chaotic functions in the form of (8), for which the Lyapunov exponent λ is equal to (11) and the invariant density is ρ(x) = 1, is

$$P_n = 2^n n(n-1).$$

Proof. By changing the signs of the direction components of the straight lines in (8) we derive 2^n maps. In turn, by changing the position of the sub-intervals in which parameter p is contained we obtain n(n−1) new maps. The procedures described above do not alter the values of the Lyapunov exponent and invariant density.

As it follows from Lemma 3, it is not only possible to hide the value of parameter p, but also the form of the function, rendering more opportunities for the selection of secret keys and making it more resistant, for example, to brutal attacks.

CONCLUSIONS

The problem of the inverse Lyapunov exponent was formulated and solved. The solution involved finding a chaotic function in the form of (1), for which the value of the Lyapunov exponent is assumed in advance. The solution was based on the construction of a piecewise linear model of (8), for which the Lyapunov exponent was given by dependence (11). The solution procedure was illustrated by a numerical example. Furthermore, it was indicated that the discussed model may be used in chaotic cryptography.

REFERENCES

1. T. Addabbo, A. Fort, S. Rocchi, V. Vignoli, Digitized chaos for pseudo-random number generation in cryptography, in Kocarev L., Lian S. (Eds.), Chaos-Based Cryptography, Stud. Comput. Intell., Vol. 354, Springer, Berlin, Heidelberg, 2011, pp. 67–97.
2. V.M. Anikin, S.S. Arkadaksky, S.N. Kuptsov, A.S. Remizov, L.P. Vasilenko, Lyapunov exponent for chaotic 1D maps with uniform invariant distribution, Bull. Russ. Acad. Sci., Phys., 72(12):1684–1688, 2008.
3. D. Arroyo, G. Alvarez, V. Fernandez, On the inadequacy of the logistic map for cryptographic applications, in A.M. Hernandez (Ed.), X Reunin Espanola sobre Criptologıa y Seguridad de la Informacion (X RECSI), Universidad de Salamanca, Salamanca, Spain, 2008, pp. 77–82.
4. M.S. Baptista, Cryptography with chaos, Phys. Lett. A, 240(1–2):50–54, 1998.
5. M. Berezowski, Spatio-temporal chaos in tubular chemical reactors with the recycle of mass, Chaos Solitons Fractals, 11:1197–1204, 2000.
6. M. Berezowski, D. Dubaj, Chaotic oscillations of coupled chemical reactors, Chaos Solitons Fractals, 78:22–25, 2015.
7. M. Berezowski, M. Lawnik, Identification of fast-changing signals by means of adaptive chaotic transformations, Nonlinear Anal. Model. Control, 19(2):172–177, 2014.
8. P. Biswas, H. Shimoyama, L.R. Mead, Lyapunov exponents and the natural invariant density determination of chaotic maps: An iterative maximum entropy ansatz, J. Phys. A, Math. Theor., 43:125103, 2010.
9. T. Bohr, T. Tél, The thermodynamics of fractals, in B.-L. Hao (Ed.), Directions in Chaos, Vol. 2, Ser. Dir. Condens. Matter Phys., Vol. 4, World Scientific, Singapore, 1988, pp. 194– 237.
10. A. Boyarsky, P. Göra, Laws of Chaos Invariant Measures and Dynamical Systems in One Dimension, Springer, New York, 1997.
11. S. Grossmann, S. Thomae, Invariant distributions and stationary correlation functions of onedimensional discrete processes, Z. Naturforsch., 32:1353–1363, 1977.
12. D. Guégan, Chaos in economics and finance, Annu. Rev. Control, 33(1):89–93, 2009.
13. E. Hetmaniok, D. Słota, R. Wituła, A. Zielonka, Solution of the one-phase inverse Stefan problem by using the homotopy analysis method, Appl. Math. Modelling, 39(22):6793–6805, 2015.
14. A. Kadir, A. Hamdulla, W.-Q. Guo, Color image encryption using skew tent map and hyper chaotic system of 6th-order CNN, Optik, 125(5):1671–1675, 2014.
15. J. Kemp, New methods and understanding in economic dynamics: An introductory guide to chaos and economics, Economic Issues, 2(Part 1):1–26, 1997.

16. S. Koga, The inverse problem of Flobenius–Perron equations in 1D difference systems: 1D map idealization, Prog. Theor. Phys., 86(5):991–1002, 1991.
17. M. Lawnik, The approximation of the normal distribution by means of chaotic expression, J. Phys., Conf. Ser., 490:012072, 2014.
18. M. Lawnik, Generation of numbers with the distribution close to uniform with the use of chaotic maps, in M.S. Obaidat, J. Kacprzyk, T. Ören (Eds.), Proceedings of the 4th International Conference on Simulation and Modeling Methodologies, Technologies and Applications, Vienna, August 28–30, 2014, Scitepress, Setúbal, 2014, pp. 451–455.
19. M. Lawnik, Generalized logistic map and its application in chaos based cryptography, J. Phys., Conf. Ser., 936:012017, 2017.
20. M. Lawnik, M. Berezowski, Identification of the oscillation period of chemical reactors by chaotic sampling of the conversion degree, Chem. Process Eng., 35(3):387–393, 2014.
21. C. Li, G. Luo, K. Qin, C. Li, An image encryption scheme based on chaotic tent map, Nonlinear Dyn., 87(1):127–133, 2017.
22. X. Liao, Improved tent map and its applications in image encryption, International Journal of Security and Its Applications, 9(1):25–34, 2015.
23. L. Liu, S. Miao, A new image encryption algorithm based on logistic chaotic map with varying parameter, SprigerPlus, 5:289, 2016.
24. E.N. Lorenz, Deterministic nonperiodic flow, J. Atmos. Sci., 20(2):130–141, 1963.
25. N. Nagaraj, P.G. Vaidya, K.G. Bhat, Arithmetic coding as a non-linear dynamical system, Commun. Nonlinear Sci. Numer. Simul., 14:1013–1020, 2009.
26. N.K. Pareek, V. Patidar, K.K. Sud, Image encryption using chaotic logistic map, Image Vision Comput., 24(9):926–934, 2006.
27. D. Pingel, P. Schmelcher, F.K. Diakonos, Theory and examples of the inverse Frobenius–Perron problem for complete chaotic maps, Chaos, 9(2):357–366, 1999.
28. J.E. Skinner, M. Molnar, T. Vybiral, M. Mitra, Application of chaos theory to biology and medicine, Integr. Physiol. Behav. Sci., 27(1):39–53, 1992.
29. D. Słota, Direct and inverse one-phase Stefan problem solved by the variational iteration method, Comput. Math. Appl., 54(7–8):1139–1146, 2007.
30. B.J. West, Fractal Physiology and Chaos in Medicine, 2nd ed., Stud. Nonlinear Phenom. Life Sci., Vol. 16, World Scientific, Singapore, 2012.
31. Y. Wu, J.P. Noonan, G. Yang, H. Jin, Image encryption using the two-dimensional logistic chaotic map, J. Electron. Imaging, 21(1):1–16, 2012.

CHAPTER 12

Nonlinear Damping Identification in Nonlinear Dynamic System Based on Stochastic Inverse Approach

S. L. Han[1] and Takeshi Kinoshita[1]

[1]Institute of Industrial Science, The University of Tokyo, 4-6-1 Komaba, Meguro-ku, Tokyo 153-8505, Japan

ABSTRACT

The nonlinear model is crucial to prepare, supervise, and analyze mechanical system. In this paper, a new nonparametric and output-only identification procedure for nonlinear damping is studied. By introducing the concept of the stochastic state space, we formulate a stochastic inverse problem for a nonlinear damping. The solution of the stochastic inverse problem is designed as probabilistic expression via the hierarchical Bayesian formulation by considering various uncertainties such as the information insufficiency in

Citation: Han, S. L., & Kinoshita, T. (2012). Nonlinear damping identification in nonlinear dynamic system based on stochastic inverse approach. Mathematical Problems in Engineering, 2012. (20 pages) DOI: https://doi.org/10.1155/2012/574291

parameter of interests or errors in measurement. The probability space is estimated using Markov chain Monte Carlo (MCMC). The applicability of the proposed method is demonstrated through numerical experiment and particular application to a realistic problem related to ship roll motion.

INTRODUCTION

The energy dissipation in dynamic system is often neglected or overly simplified in engineering design. However, there are many practical cases where an appropriate nonlinear model of damping is essential and an effective identification is needed. One example could be a material with high damping which is widely used for vibration and noise reduction. In addition, the increasing demands for enhanced and reliable performance of vibrating structures are requiring appropriate modeling of nonlinear damping for mechanical system. This is because of the fact that different types of nonlinear model could yield different responses of a system.

The identification problems of nonlinear system are very important in engineering in order to avoid unwanted instabilities or failure of the system. There have been considerable numbers of studies on the identification of nonlinear damping. For example, Iourtchenko et al. [1] successfully proposed an identification method. Iourtchenko and Dimentberg [2] further described a procedure based on the stochastic averaging method for in-service identification of the damping characteristic from a measured stationary response. Mohammad et al. [3] introduced a method to estimate nonlinear damping for linear and nonlinear structures. Tomme [4] introduced a method, based on modal analysis, to evaluate damping from measurements in materials and structures. The related studies mentioned previously are parametric identification methods, which focus on direct estimations of coefficients of assumed form for nonlinear damping in the nonlinear system.

Recently, Jang et al. [5] proposed a nonparametric identification method based on deterministic inverse approach, Landweber regularization method. Jang et al. [6] proved the applicability of the method through experimental work. The results presented in the literatures [5, 6] are very encouraging because this method does not require any assumption of the form of nonlinear damping unlike parametric identification method. However, this method has some limitations because it is dependent on the deterministic inverse approach. This kind of technique is content with singling out one solution without quantifying the related uncertainties or rigorously considering the stochastic nature of data noise driven by approximation errors, rounding errors, and measurement errors. Furthermore, it is not possible to explain the variability of the possible solutions because there exists an ensemble of inverse solutions consistent with the data due to unstable nature.

In this study, we proposed a new method based on a stochastic inverse approach to identify a nonlinear damping of nonlinear oscillatory system. The unique features of this paper are as follows. Firstly, an original method for the identification of nonlinear damping is proposed. The method can also be classified as nonparametric method which does not require any assumption of the form of nonlinear damping. That is, this method can be applicable to a system when no or little information is available about the actual form of nonlinear damping unlike the parametric method [1–4] which is limited to be used when there is enough knowledge about the nonlinear damping. Secondly, we formulate a stochastic inverse problem of identifying nonlinear damping by introducing a stochastic state space. The stochastic inverse problem studied here can naturally resolve the computational difficulty of the deterministic inverse model [5–10], that is, lack of stability. In addition, it can quantify various uncertainties arising from an insufficiency of information on parameter of interests or measurement errors. Thirdly, we develop methodologies enabling probabilistic modeling for the solution based on hierarchical Bayesian formulation [11–14]. There have been lots of studies on the use of the hierarchical models applied to very challenging problems which arise in atmospheric chemistry, environmental sciences, or metabolic models [15–17]. Although this study considers a relatively simple mathematical model which is expressed as nonlinear single degree of freedom system, the present work may provide some insights into the work involved in nonlinear system identification. The main novelty is that we do not focus directly on the hierarchical model, but rather, on the identification procedure. Lastly, the method presented in this paper only requires system output, that is, motion responses. If the input, the prescribed external force, is acted on the system, then the nonlinearity of the system is easily found through the "black-box" model.

The outline of this paper is as follows. Section 2 introduces the mathematical description for the problem. In Section 3, nonlinear damping identification problem is discussed with its instable characteristics. Section 4 illustrates the formulation of stochastic inverse problem. In addition, probabilistic modeling of the stochastic inverse solution with the explanation of Markov random field [18, 19] is also described. Hierarchical Bayesian formulation and exploration of the posterior state space via Markov chain Monte Carlo [20] are introduced in Sections 5 and 6, respectively. Section

7 illustrates the identification procedure with numerical examples. The particular application to a realistic problem is demonstrated in Section 8. The problem is concerned with identifying nonlinear roll damping of a ship from free-roll decay experiment. Finally, we summarize the result and make conclusion in Section 9.

MATHEMATICAL DESCRIPTION

Consider nonlinear single degree of freedom system as

$$m\ddot{y} + B(y, \dot{y}) + M(y) = 0, \tag{2.1}$$

with initial conditions

$$y(0) = \alpha, \dot{y}(0) = \beta. \tag{2.2}$$

In (2.1), B and M are nonlinear damping and restoring force, respectively. In this study, the nonlinear restoring function is considered to be known a priori and assumed to have the form

$$M(y) = M_L(y) + M_{NL}(y) = ky + M_{NL}(y), \tag{2.3}$$

where $M_L(y)$ and $M_{NL}(y)$ are linear and nonlinear restoring components. Then, (2.1) can be rewritten as

$$m\ddot{y} + B(y, \dot{y}) + k_1 y + M_{NL}(y) = 0. \tag{2.4}$$

From the concept of variation of parameters, nonlinear motion equation (2.1) can be transformed to following nonlinear Volterra integral equation of the second kind [21]:

$$y(t) = \frac{\alpha}{\mu}\xi_1(t) + \frac{\beta}{\nu}\xi_2(t) - \int_0^t \frac{\xi_1(\tau)\xi_2(t) - \xi_1(t)\xi_2(\tau)}{mW(\tau)}\{B(y, \dot{y}) + M_{NL}(y)\}d\tau, \tag{2.5}$$

where ξ_1 and ξ_2 satisfy

$$m\ddot{\xi}_1 + k_1\xi_1 = 0, \qquad \xi_1(0) = \mu, \dot{\xi}_1(0) = 0,$$
$$m\ddot{\xi}_2 + k_1\xi_2 = 0, \qquad \xi_2(0) = 0, \dot{\xi}_2(0) = \nu, \tag{2.6}$$

and Wronskian W is

$$W \equiv \begin{vmatrix} \xi_1 & \xi_2 \\ \dot{\xi}_1 & \dot{\xi}_2 \end{vmatrix}. \tag{2.7}$$

NONLINEAR DAMPING IDENTIFICATION

The focus of this study is to identify nonlinear damping of the nonlinear dynamic system when dynamic responses are observed. The identification of this unknown nonlinear damping becomes feasible with measurement of responses during $0 < t < T$. Let y denote the measured response data, that is, $y = [y_1, y_2, \ldots, y_m]$; then we can obtain the following system for unknown nonlinear damping $\theta\tau j = B[y(\tau j), y^{\cdot}(\tau j)]$:

$$g_i = H_{ij}\theta_j, \tag{3.1}$$

where

$$H_{ij} = \Delta\tau \frac{\xi_1(t_i)\xi_2(\tau_j) - \xi_1(\tau_j)\xi_2(t_i)}{mW(\tau_j)} \quad \text{for } t_i < \tau_j, \tag{3.2}$$

$$g(t_i) = y(t_i) - \frac{\alpha}{\mu}\xi_1(t_i) - \frac{\beta}{\nu}\xi_2(t_i) + \Delta\tau \sum_{t_i<\tau_j} \frac{\xi_1(t_i)\xi_2(\tau_j) - \xi_1(\tau_j)\xi_2(t_i)}{mW(\tau_j)} M_{NL}[y;\tau_j]. \tag{3.3}$$

The identification of the unknown nonlinear damping B can be achieved by solving (3.1) regardless of any assumption on the form of nonlinear damping. Therefore, this method can be classified as the nonparametric method. The merit of nonparametric method is that this method does not require the specific form of a system explicitly.

However, unfortunately, (3.1) becomes ill-conditioned system causing computational difficulty to obtain inverse solution. This is mainly because the system (3.1) is obtained by approximating the first-kind integral operator. According to inverse problem theory [22, 23], the approximation of the first-kind integral operator yields ill-conditioned system regardless of the choice of approximate methods such as the quadrature method and the Galerkin method [22] with orthonormal basis functions. Thus, the identification of nonlinear damping is very difficult to achieve both mathematically and numerically because even very small amount of noisy data in the measured displacement can be amplified and can give an effect on an inverse solution.

STOCHASTIC INVERSE FORMALISM

To address the difficulty discussed in the previous section, we newly formulate a stochastic inverse problem for the identification of the nonlinear damping by introducing a stochastic state space. A stochastic inverse problem can be formulated by considering variables g and θ in (3.1) as random variables. Let us define a multivariate random variables $\Theta : \Omega \rightarrow R^m$; then a multivariate random variable $G : \Omega \rightarrow R^n$ can be defined by

$$\mathbf{G} = \mathbf{H}\boldsymbol{\Theta}. \tag{4.1}$$

For each realization $g = \{g_1, g_2, ..., g_n\}$ of $g = G$, the following relationship is given:

$$\mathbf{g} = \mathbf{H}\boldsymbol{\theta}, \tag{4.2}$$

where $\theta = \{\theta_1, \theta_2, ..., \theta_m\}$ is a realization of the random variable Θ. Equation (4.2) can be interpreted as a stochastic function corresponding to (3.1) because it is a function of random variables. Therefore, the problem of finding θ given a realization g is the stochastic inverse problem. Unlike the ill-conditioned system in (3.1), (4.2) becomes wellposed in an expanded stochastic space and gives a solution with distribution of random unknowns.

Assume that the dynamic response y is observed, then a single realization of directly observable parameter g can be obtained from (3.3), and then the solution to the stochastic inverse problem can be expressed by Bayes' rule [11–14]:

$$p(\boldsymbol{\theta} \mid \mathbf{g}) = \frac{p(\mathbf{g} \mid \boldsymbol{\theta})p(\boldsymbol{\theta})}{p(\mathbf{g})}, \tag{4.3}$$

where $p(g \mid \theta)$, $p(\theta)$, and $p(g)$ are known as the likelihood, the prior probability density function, and the normalizing constant, respectively. It is noted that the normalizing constant is not necessary to be computed for sampling procedure [11–14]. Posterior probability density function $p(\theta \mid g)$ in (4.3) can then be evaluated as follows:

$$p(\boldsymbol{\theta} \mid \mathbf{g}) \propto p(\mathbf{g} \mid \boldsymbol{\theta})p(\boldsymbol{\theta}). \tag{4.4}$$

It is usually easy to obtain the likelihood $p(g \mid \theta)$ provided that the measurement data is contaminated by Gaussian random noise ε with zero-mean and standard deviation σ. The likelihood then depends on the distribution of random noise. Consequently, the likelihood is given by

$$p(\boldsymbol{\theta} \mid \mathbf{g}) \propto \left(\sigma^2\right)^{-m/2} \exp\left(-\frac{\|\mathbf{H}\boldsymbol{\theta} - \mathbf{g}\|_2^2}{2\sigma^2}\right), \tag{4.5}$$

where $\|\cdot\|_2$ and m refer to Euclidean norm and the number of measurements, respectively.

Next, we should consider the prior probability density function $p(\theta)$ which reflects the information of the unknown θ prior to collecting the data. In this paper, we adopt Gaussian model with Markov random field [18, 19], which is the most popular model for prior. A Gaussian Markov random field for the unknown θ is of the following form:

$$p(\boldsymbol{\theta}) \propto \lambda^{n/2} \exp\left(-\frac{1}{2}\lambda \boldsymbol{\theta}^{\mathrm{T}} \mathbf{W} \boldsymbol{\theta}\right). \tag{4.6}$$

Here, n is dimension of θ, λ is the scaling parameter, and the matrix W ∈ $R^{n \times n}$ is determined as

$$W_{ij} = \begin{cases} n_i, & i = j, \\ -1, & i \sim j, \\ 0, & \text{otherwise}, \end{cases} \tag{4.7}$$

where n_i is the number of neighbors for the point i. Then, with the likelihood (4.5) and the prior distribution (4.6), posterior probability density function (4.3) can be formulated as

$$p(\boldsymbol{\theta} \mid \mathbf{g}) \propto \left(\sigma^2\right)^{-m/2} \exp\left(-\frac{\|\mathbf{H}\boldsymbol{\theta} - \mathbf{g}\|_2^2}{2\sigma^2}\right) \lambda^{n/2} \exp\left(-\frac{1}{2}\lambda \boldsymbol{\theta}^{\mathrm{T}} \mathbf{W} \boldsymbol{\theta}\right). \tag{4.8}$$

HIERARCHICAL BAYESIAN FORMULATION

The maximum a posteriori (MAP) estimate [11–14] for posterior probability density function (4.8) can be derived as

$$\boldsymbol{\theta}_{\mathrm{map}} = \arg\min_{\theta}\left\{ \|\mathbf{H}\boldsymbol{\theta} - \mathbf{g}\|_2^2 + \lambda\sigma^2 \boldsymbol{\theta}^T \mathbf{W} \boldsymbol{\theta} \right\}, \tag{5.1}$$

and the MAP estimation is similar to Tikhonov regularization method [22–24], the representative deterministic inverse approach, which minimizes the functional

$$\mathcal{A}(\boldsymbol{\theta}) = \|\mathbf{H}\boldsymbol{\theta} - \mathbf{g}\|_2^2 + \eta \boldsymbol{\theta}^T \mathbf{W} \boldsymbol{\theta} \tag{5.2}$$

with η λσ2. Therefore, it can be easily found that the scaling parameter λ and the standard deviation σ play an important role like the regularization parameter in the deterministic approaches. Furthermore, the standard deviation σ describes the noise level of the measurement data, and it is hard to quantify directly, except when the experiment for acquisition of data is not carried out repeatedly. In almost all deterministic approaches, the performance of methods highly depends on the regularization parameter and this parameter needs to be chosen a priori [5, 6, 22–24]. However, the choice of the optimal regularization parameter has never been trivial.

Unlike the deterministic inverse approach, a hierarchical Bayesian model, which is known as modern Bayesian analysis, provides a flexible way to choose hyperparameters σ and λ by treating these parameters as random variables [11–14]. As a result, the hyperparameters are automatically chosen through the hierarchical Bayesian formulation and there is no need to worry about the choice of optimal regularization parameter unlike deterministic inverse approach. This is one of distinct features of stochastic inverse problem. The hierarchical Bayesian model [11–14] for the present identification problem can then be formulated as

$$p(\boldsymbol{\theta}, \lambda, \sigma \mid \mathbf{g}) \propto \left(\sigma^2\right)^{-m/2} \exp\left(-\frac{\|\mathbf{H}\boldsymbol{\theta} - \mathbf{g}\|_2^2}{2\sigma^2}\right) \lambda^{n/2} \exp\left(-\frac{1}{2}\lambda \boldsymbol{\theta}^T \mathbf{W} \boldsymbol{\theta}\right)$$

$$\cdot \lambda^{\alpha_1 - 1} \exp(-\beta_1 \lambda) \frac{1}{(\sigma^2)^{\alpha_2 + 1}} \exp\left(-\frac{\beta_2}{\sigma^2}\right), \quad 5.3)$$

where (α_1, β_1) and (α_2, β_2) are pairs for gamma distribution for λ and inverse gamma distribution for σ:

$$p(\lambda) \propto \lambda^{\alpha_1 - 1} \exp(-\beta_1 \lambda), \qquad p\left(\sigma^2\right) \propto \frac{1}{(\sigma^2)^{\alpha_2 + 1}} \exp\left(-\frac{\beta_2}{\sigma^2}\right). \quad (5.2)$$

POSTERIOR STATE EXPLORATION

The posterior probability density function (5.3) corresponding to the solution of inverse problem in (4.2) was formulated. This probability function quantifies associated uncertainties and includes a stochastic nature of data driven by various errors. The posterior probability density function (5.3) needs to be estimated for the purpose of identification of nonlinear damping which is desired solution of the stochastic inverse problem.

The estimation can be achieved through Markov chain Monte Carlo combined with Metropolis-Hastings and Gibbs algorithms [25, 26].

In Algorithm 1, N_{MCMC} is the total number of samples, $\theta_{-j}^{(i+1)}$ refers to $\{\theta_1^{(i+1)}, \ldots, \theta_{j-1}^{(i+1)} \theta_{j+1}^{(i)}, \ldots, \theta_m^{(i)}\}$, $q(\sigma^{(*)} \mid \sigma^{(i)})$, is an easy-to-sample proposal distribution, and the conditional $p(\theta_i \mid \theta_{-i}, \lambda, \sigma)$ which follows a multivariate Gaussian distribution can be expressed by

$$p(\theta_i \mid \theta_{-i}, \lambda, \sigma) \propto \exp\left(-\frac{(\theta_i - \mu_i)^2}{2\sigma_i^2}\right),$$

$$\mu_i = \frac{b_i}{2a_i}, \qquad \sigma_i = \sqrt{\frac{1}{a_i}},$$

$$a_i = \sum_{s=1}^{n} \frac{H_{si}}{\sigma^2} + \lambda W_{ii}, \qquad b_i = 2\sum_{s=1}^{n} \frac{\mu_s H_{si}}{\sigma^2} - \lambda \mu_p,$$

$$\mu_s = \sum_{t \neq i} H_{st} \theta_t, \qquad \mu_p = \sum_{j \neq i} W_{ji} \theta_j + \sum_{k \neq i} W_{ik} \theta_k. \quad (6.1)$$

In a similar manner, the full conditional p(λ | θ, σ) and p(σ | θ, λ) can also be derived as

(i) initialize $\boldsymbol{\theta}^{(0)}$, $\lambda^{(0)}$ and $\sigma^{(0)}$
(ii) For $i = 0 : N_{\text{MCMC}} - 1$
Sample $\theta_1^{(i+1)} \sim p(\theta_1 \mid \theta_{-1}^{(i+1)}, \lambda^{(i)}, \sigma^{(i)})$
Sample $\theta_2^{(i+1)} \sim p(\theta_2 \mid \theta_{-2}^{(i+1)}, \lambda^{(i)}, \sigma^{(i)})$
...
Sample $\theta_n^{(i+1)} \sim p(\theta_n \mid \theta_{-n}^{(i+1)}, \lambda^{(i)}, \sigma^{(i)})$
Sample $\lambda^{(i+1)} \sim p(\lambda \mid \theta^{(i+1)}, \sigma^{(i)})$
Sample $u_1 \sim U(0,1)$
Sample $\lambda^{(*)} \sim q(\lambda^{(*)} \mid \lambda^{(i)})$
if $u_1 < A(\lambda^{(i)}, \lambda^{(*)}) = \min\left\{1, \dfrac{p(\lambda^{(*)} \mid \boldsymbol{\theta}^{(i+1)}, \sigma^{(i)})q(\lambda^{(i)} \mid \lambda^{(*)})}{p(\lambda^{(i)} \mid \boldsymbol{\theta}^{(i+1)}, \sigma^{(i)})q(\lambda^{(*)} \mid \lambda^{(i)})}\right\}$
$\lambda^{(i+1)} = \lambda^{(*)}$
else
$\lambda^{(i+1)} = \lambda^{(i)}$
Sample $u_2 \sim U(0,1)$
Sample $\sigma^{(*)} \sim q(\sigma^{(*)} \mid \sigma^{(i)})$
if $u_2 < A(\sigma^{(i)}, \sigma^{(*)}) = \min\left\{1, \dfrac{p(\sigma^{(*)} \mid \boldsymbol{\theta}^{(i+1)}, \lambda^{(i+1)})q(\sigma^{(i)} \mid \sigma^{(*)})}{p(\sigma^{(i)} \mid \boldsymbol{\theta}^{(i+1)}, \lambda^{(i+1)})q(\sigma^{(*)} \mid \sigma^{(i)})}\right\}$
$\sigma^{(i+1)} = \sigma^{(*)}$
else
$\sigma^{(i+1)} = \sigma^{(i)}$

Algorithm 1

NUMERICAL EXPERIMENTS

In order to demonstrate the workability of the method proposed to identify nonlinear damping, the numerical experiments are carried out. As a numerical example, Van der Pol Equation [27] is considered as

$$\ddot{y} - \mu\left(1 - y^2\right)\dot{y} + y = 0, \tag{7.1}$$

where μ =1. Comparison of (2.1) and (7.1) shows that $B(y, \dot{y}) = \mu(1 - y^2)\dot{y}$ and $M(y) = y$.

As first step toward the identification of nonlinear damping, the nonlinear model (7.1) is simulated using numerical integration scheme such as the Runge-Kutta method with the initial conditions y(0)= 1 and ˙y(0)= 0. Figure 1 plots the simulated responses and phase diagram with respect to time. In this study, the simulated displacement is considered as the measured data for the inverse identification of nonlinear damping. If the displacement is measured, the pseudo-displacement g which is necessary data for the identification of the nonlinear damping can be calculated through (3.3) and is plotted in Figure 2. In practice, the measured data is contaminated with the various noises such as approximation errors and rounding errors. In order to examine this kind of uncertainties, the synthetic noisy data is generated by adding Gaussian random noise with zero mean and unit standard deviation. The noisy data $g^\delta = g + e$ with 10.1% noise is also plotted in Figure 2.

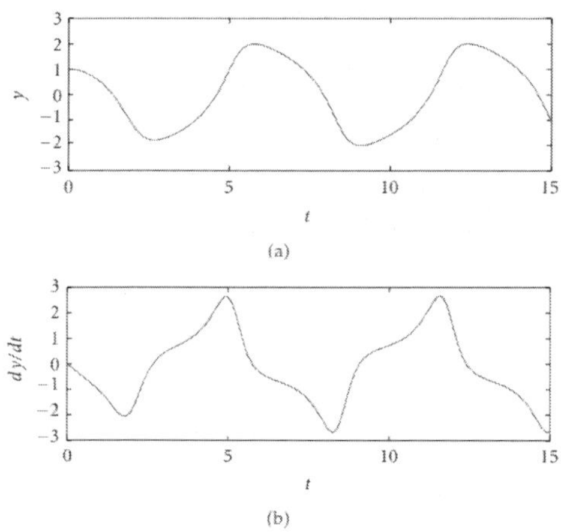

Figure 1: The simulated responses of the numerical example.

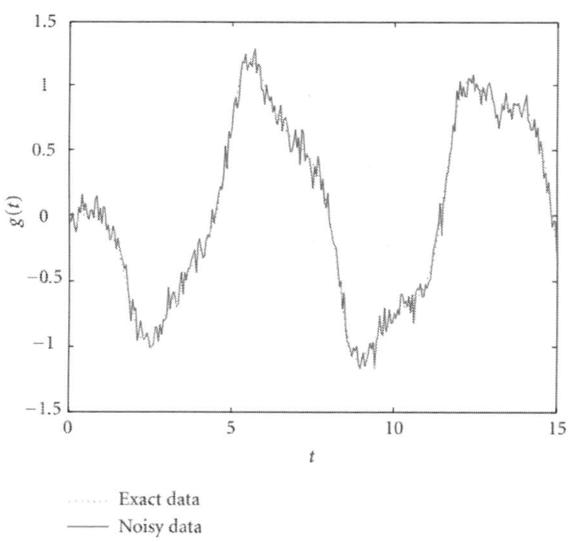

Figure 2: The calculated pseudo-displacement g and noisy data g^δ for the first example.

To show the instability of the original mathematical model in (3.1), the numerical solution without any numerical treatments is first shown in Figure 3. That is, after decomposing (3.1) into singular systems [28] with singular values and vectors, the pseudoinverse [22, 23] is conducted. For the numerical experiments, the number of measurement and the dimension of unknown θ are considered as (m, n) = (301, 301) and the elapsed time T is 15 s. The magnitude of the solution in Figure 3 is unrealistically high and thus this solution is completely useless. Figure 4 shows the distribution of the singular values from the singular value decomposition [28] of the approximated system. Some of singular values become very small, and this makes the original mathematical model very unstable. In order to address this difficulty, we follow the solution procedure by formulating the stochastic inverse problem.

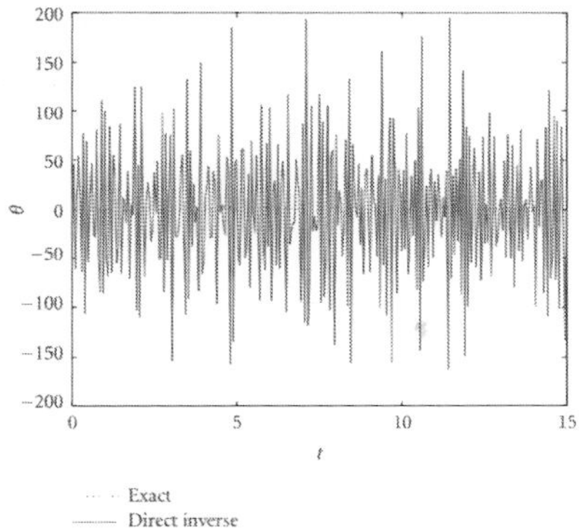

Figure 3: The numerical solution of the numerical example through pseudoinversion.

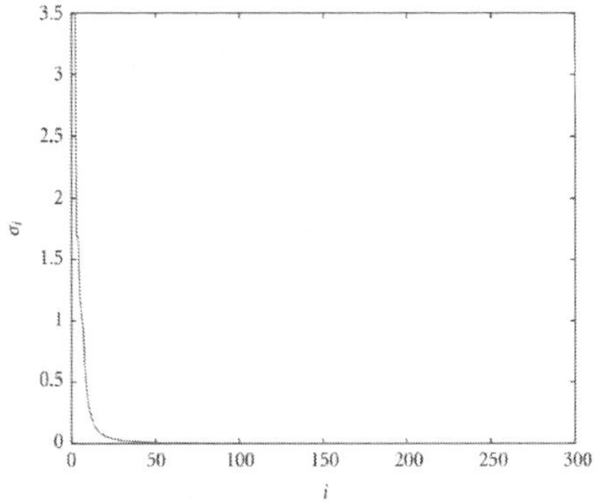

Figure 4: The distribution of the singular values.

The hierarchical Bayesian formulation (5.3) is adopted as a probabilistic model for the stochastic inverse solution. In order to explore the posterior state, Markov chain Monte Carlo is used as described in Section 6. For the algorithm, the initial guesses (λ_0) and (σ_0) for hyperparameters are taken to be 1 and 0.1 and $\theta^{(0)}$ is considered to be zero vectors. In addition, the pairs of parameters (α_1, β_1) and (α_2, β_2) are taken to be (1, 0.1) and (1, 0.1), respectively. The total number of samples N_{MCMC} for Algorithm 1 is taken to be 50,000, and the last 25,000 realizations are used to estimate statistics such as mean and standard deviation for the nonlinear damping θ.

The numerical results are shown in Figure 5. The upper and lower dotted lines denote the 95% credible interval. It is easily found that the posterior mean $E\{\theta\}$ is fairly accurate compared with exact solution, and the credible interval shows the quantification of its associated uncertainty corresponding to the measured data. Note here that the number of samples means the number of realization for the unknown θ $\{\theta_1, \theta_2,...,\theta_m\}$, and for each element θ_i of the unknown θ has 25,000 realizations. Examples of trace plots are illustrated in Figure 6 for θ_1 and θ_{100} : θ_1 and θ_{100} correspond to the samples for t =0 s and 5 =s, respectively. The quality of the MCMC sample may be estimated by inspecting the autocorrelation of a solution sequence [13]. The scale parameter σ_q for a proposal distribution $q(\sigma \mid \sigma^*) \propto \exp(-(\sigma - \sigma^*)^2/(2\sigma_q^2))$ can affect the efficiency of the Markov chain. The stronger correlation may result in poor sampling efficiency. Here, we used σ_q =0.05 to produce a sample with reasonably low autocorrelation.

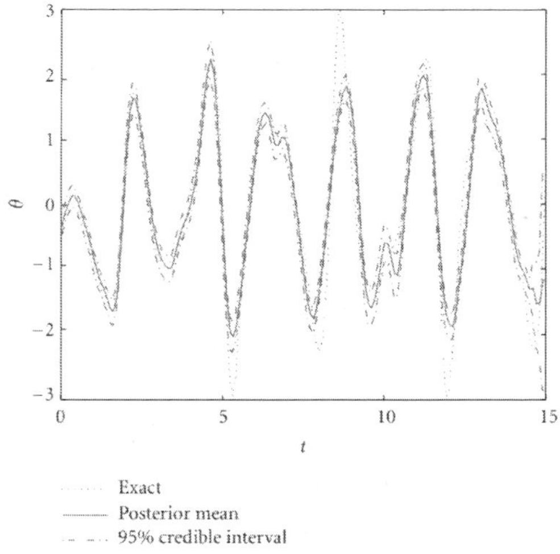

Figure 5: The numerical results from MCMC computations.

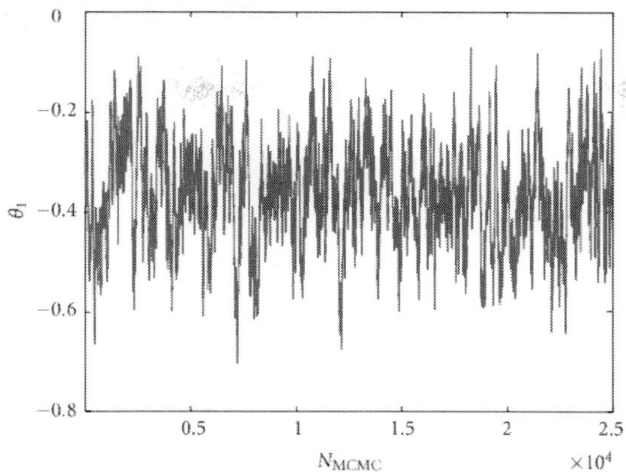
0
−0.2
−0.4
−0.6
−0.8
θ_1
0.5
1
1.5
2
2.5
N_{MCMC}
$\times 10^4$

(a)

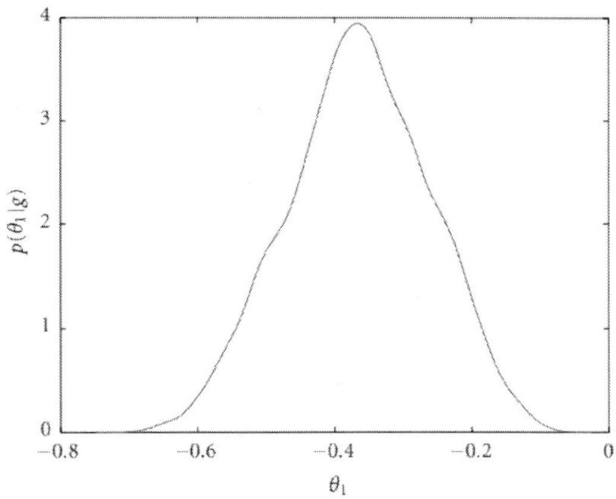
4
3
2
1
0
$p(\theta_1|g)$
−0.8
−0.6
−0.4
−0.2
0
θ_1

(b)

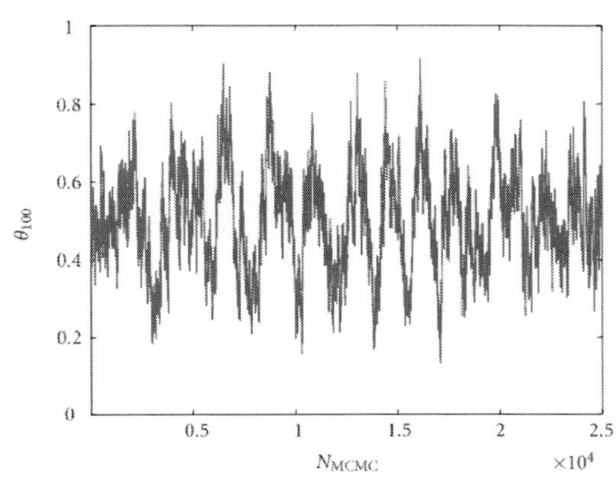
1
0.8
0.6
0.4
0.2
0
θ_{100}
0.5
1
1.5
2
2.5
N_{MCMC}
$\times 10^4$

(c)

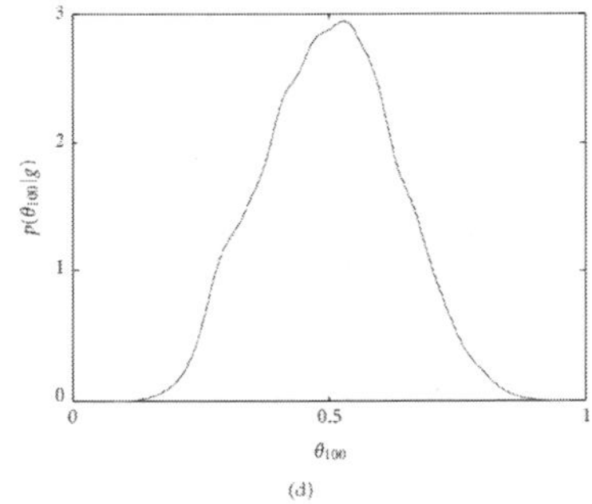

Figure 6: The trace plots for θ_1 and θ_{100} and their posterior distributions.

Now it is time to determine the nonlinear damping. Once the solution is obtained, the nonlinear damping can be determined from the relationship $\theta(t) = -B[y(t), \dot{y}(t)]$. In this study, the posterior mean E{θ} is used and the result is depicted in Figure 7. Finally, the displacement and velocity are resimulated by using the identified nonlinear damping in Figure 8. It is confirmed that the results are in well coincidence with exact motion response.

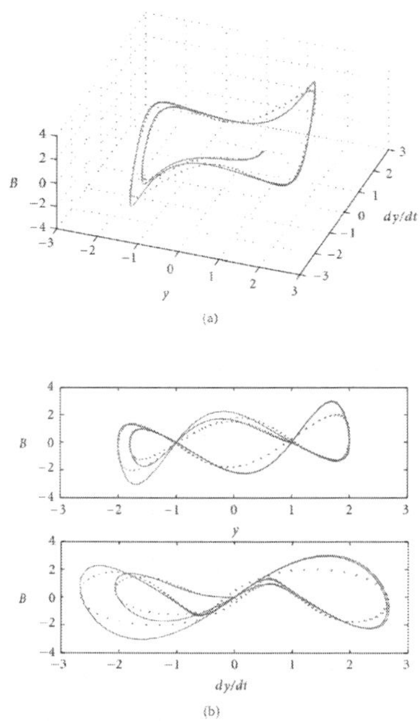

Figure 7: (a) The identified nonlinear damping 3D plots and (b) its 2D projection on the planes (y, B) and (dy/dt, B).

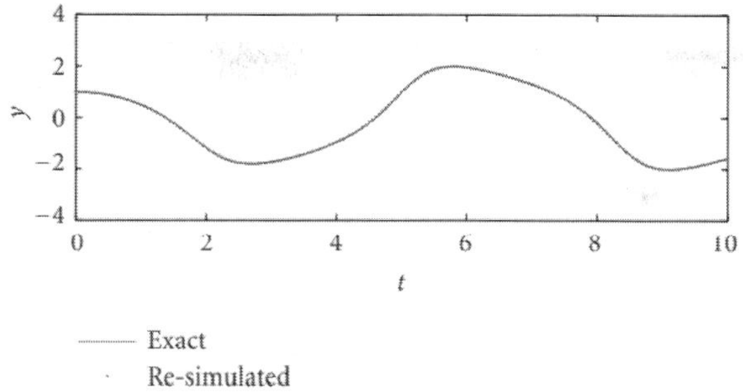

(a)

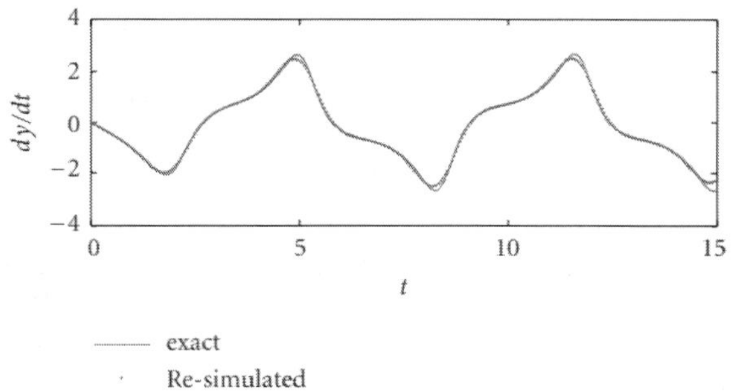

(b)

Figure 8: The resimulated responses using the identified nonlinear damping in Figure 7.

A PARTICULAR APPLICATION TO REALISTIC PROBLEM: SHIP ROLL MOTION

The results in the numerical example illustrated in Section 7 are fairly good. However, the results have a limitation that the order of damping magnitude is same order with the restoring. In practice, the order of damping magnitude is so low in relative to restoring force; the identification of damping is thus always more problematic. Considering this point, we also applied the present identification procedure to a realistic problem related to ship roll motion by conducting some experiments.

The roll motion of ships, barges, and similar floating structures are highly involved in strong nonlinearity. This is mainly because of the complex interaction between floating body and surrounding fluid; see Figure 9. The more important thing is that the dynamic stability of floating body in realistic sea is dependent on its rolling motion. For this purpose, accurate prediction of motion response to various loading conditions is necessary and it requires the precise information on roll damping.

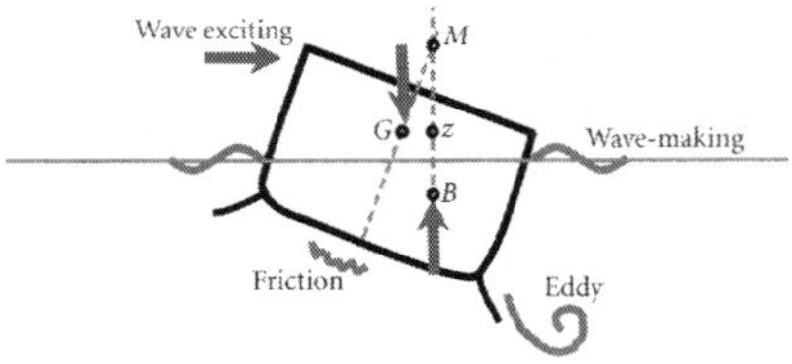

Figure 9: Interaction between ship hull and surrounding water.

Experimental Set-Up

All experiments were conducted in Ocean Engineering Basin at the University of Tokyo. The basin, which is also called as towing tank, is designed to perform various tests related to various kinds of floating structures. A model tested is shown in Figure 10 and its particulars are summarized in Table 1. For experiments, the model ship was restrained in all degrees of freedom during experiments, excepting roll motion. The model ship is placed at the center of the basin and roll angle is measured with potentiometer attached to the center of gravity of the model ship.

Table 1: Particulars of the model ship.

Length Lpp (m)	2500
Breadth B (m)	0.387
Draft D (m)	0.132
Displacement volume (m^3)	0.110
GM (m)	0.074

Figure 10: Test model of vessel.

Analysis of Model Test Data

The mathematical model of ship rolling in calm water is generally written as a second-order differential equation:

$$(I + A_{44})\ddot{\phi} + B(\dot{\phi}) + C_{44}\phi = 0, \quad (8.1)$$

where I represents the actual mass moment of inertia; A_{44} is the added mass moment of inertia; B represents the nonlinear roll damping; C_{44} represents the coefficient of restoring moment that is hydrostatic force induced by hydrostatic pressure (the static buoyancy force) exerted by a fluid at equilibrium due to the force of gravity. The coefficient C_{44} is simply expressed as the multiplication of the displacement of ship and the distance between metacentric height and center of gravity GM: $C_{44} = W \cdot GM$. By eliminating the inertia coefficient, (8.1) can be rewritten as

$$\ddot{\phi} + \widetilde{B}(\dot{\phi}) + \omega_n{}^2\phi = 0, \quad (8.2)$$

where ω_n is the resonance frequency of a ship and it can be obtained by analyzing the free-roll decay curve in Figure 11. For the test model, ω_n 6.905 rad/s.

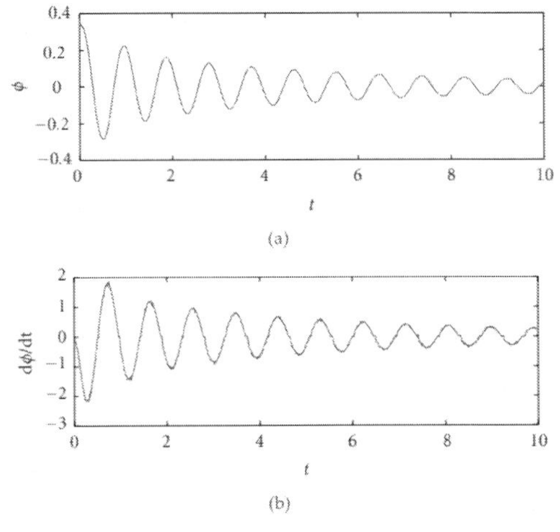

Figure 11: Roll responses of free-decay test.

In a similar way described in previous section, we first start with the roll response data in Figure 11. The calculated pseudo-displacement g is shown in Figure 12. Now, we can construct the approximate system as in (3.1). Figure 13 shows the solution through pseudo inverse and it turns out to be very illconditioned. To address this difficulty, we also try to consider the original ill-conditioned system in the stochastic space by following the procedures described in Section 4. By adopting the hierarchical Bayesian formulation (5.3), the original system can be treated as well-posed system in an expanded stochastic space. The solution θ can be obtained through Markov chain Monte Carlo and illustrated in Figure 14. The nonlinear damping in Figure 15 can be identified by the relationship $\theta(t) = -\widetilde{B}[\dot{\phi}(t)]$.

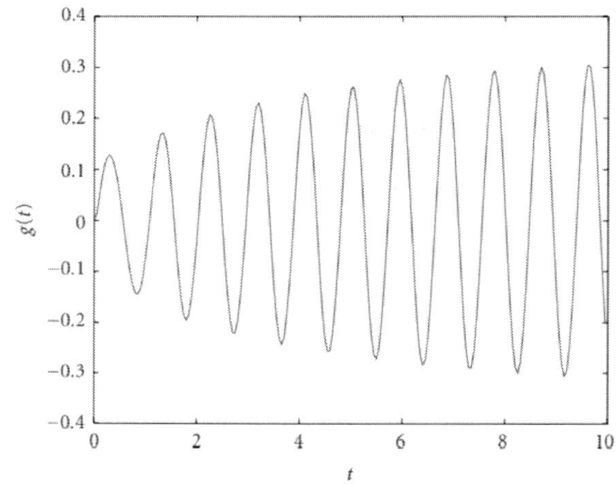

Figure 12: The calculated pseudo-displacement g for the identification of roll damping of a ship.

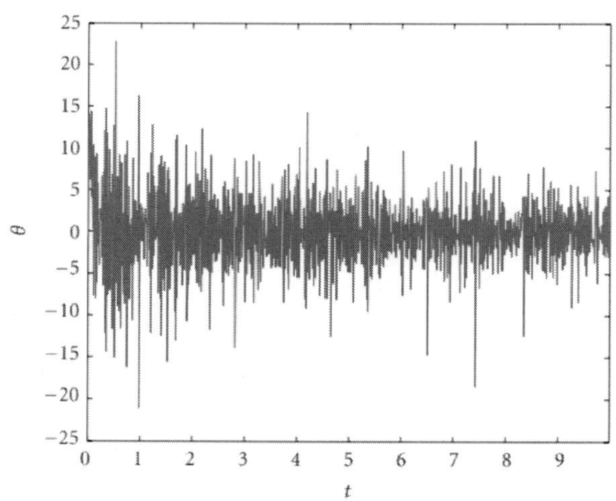

Figure 13: The numerical solution through pseudoinversion for the identification of roll damping of a ship.

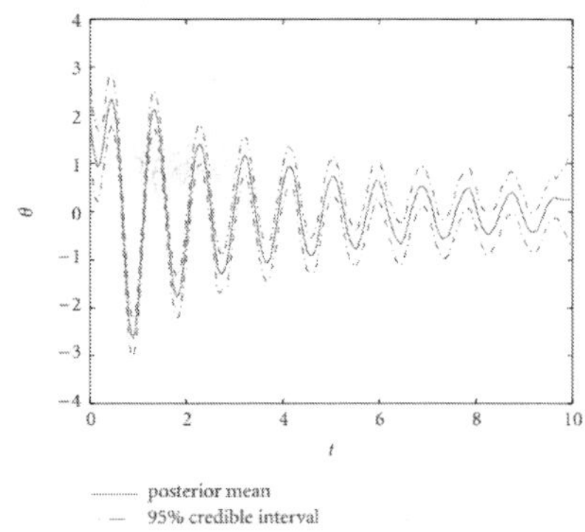

Figure 14: The numerical results from MCMC computations for the identification of roll damping of a ship.

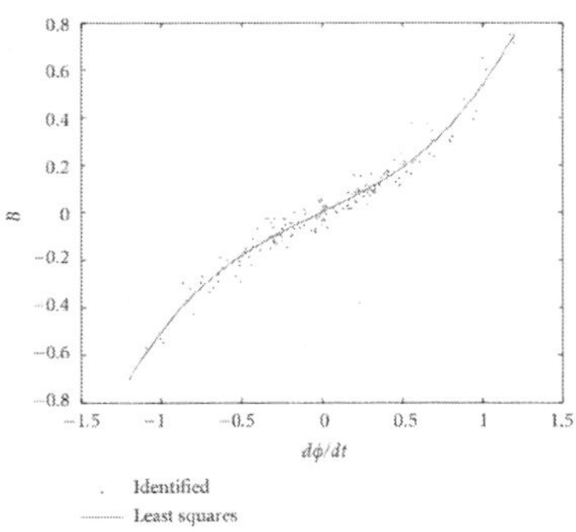

Figure 15: Identified nonlinear roll damping and its polynomial approximation.

Sometimes, it may be convenient to express nonlinear damping as a class of functions rather than a set of data. We caution that it is sometimes extremely difficult to find an analytic function for nonlinear damping as illustrated in the numerical example in Section 7. For such cases, we have to pay attention to choose a class of functions for the identified damping. However, for the present example, we can find that the damping is mainly dependent on the roll angular velocity. In this case, we can approximate the nonlinear damping as a polynomial of the roll angular velocity depicted in Figure 15 through the least squares method which minimizes the residual

$$R = \sum_{n=1}^{N} \left(B(\dot{\phi}) - p_n(\dot{\phi})\right)^2. \tag{8.3}$$

The solid-line in Figure 15 shows a polynomial approximation of the third order for the identified nonlinear damping: $p(\dot{\phi}) = 0.2144\dot{\phi}^3 + 0.0264\dot{\phi}^2 + 0.3113\dot{\phi}$

. If we know the analytical solution of the nonlinear damping, the order of polynomial can then be decided by comparing the norm between the analytic solution and the polynomial approximation $p_n(\dot{\phi})$ with respect to the order n. However, the analytical solution to the nonlinear damping B is not available for this application. To ensure the validity of the identified solution, we compare the resimulated roll response using identified nonlinear roll damping with the measured roll response data. For the simulation, the Runge-Kutta method is used. The result is shown in Figure 16. The result is well coincident with the measured roll response. This proves the validity of the present identification procedure and thus the polynomial approximation.

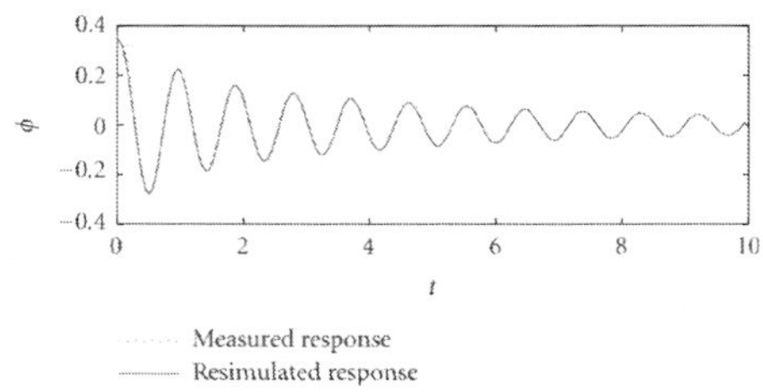

Figure 16: The resimulated responses using the identified nonlinear damping.

CONCLUSIONS

An original output-only and nonparametric procedure for the identification of nonlinear damping in nonlinear oscillatory motion was proposed. The nonlinear damping identification was formulated as a stochastic inverse problem defined on the state space. Probabilistic modeling for the stochastic inverse problem was developed. The way to design computational tools for stochastic inverse solutions with full-probabilistic specification was also illustrated.

Numerical experiments were made to show the workability and applicability of the proposed method. In addition, the present procedure was also applied to a realistic problem related to ship roll motion. From the results, it was concluded that the proposed method is well justified for detecting the nonlinear damping in the nonlinear oscillatory system. Through the results, it was also shown that the stochastic inverse formalism has lots of distinct features over deterministic inversion. Although this study has some limitations that the only nonlinear single degree of system is considered, the results of the present work may give an insight applicable to many other nonlinear system identification procedures.

REFERENCES

1. D. V. Iourtchenko, L. Duval, and M. F. Dimentberg, "The damping identification for certain SDOF systems: developments in theoretical and applied mechanics," in *Proceedings of the SECT AM XX Conference*, 2000.
2. D. V. Iourtchenko and M. F. Dimentberg, "In-service identification of non-linear damping from measured random vibration," *Journal of Sound and Vibration*, vol. 255, no. 3, pp. 549–554, 2002.
3. K. S. Mohammad, K. Worden, and G. R. Tomlinson, "Direct parameter estimation for linear and non-linear structures," *Journal of Sound and Vibration*, vol. 152, no. 3, pp. 471–499, 1992.
4. J. V. Tomme, "Evaluation of damping measurements in materials and structures," in *Proceedings of the 13th International Seminar on Modal Analysis*, 1988.
5. T. S. Jang, H. S. Choi, and S. L. Han, "A new method for detecting non-linear damping and restoring forces in non-linear oscillation systems from transient data," *International Journal of Non-Linear Mechanics*, vol. 44, no. 7, pp. 801–808, 2009.
6. T.S. Jang, J. W. Son, S. L. Han, H. G. Sung, S. K. Lee, and S. C. Shin, "A numerical investigation on nonparametric identification of nonlinear roll damping moment of a ship from transient response," *The Open Ocean Engineering Journal*, vol. 3, pp. 100–107, 2010.
7. F. Yang, H. Guo, and X. Li, "The simplified Tikhonov regularization method for identifying the unknown source for the modified Helmholtz equation," *Mathematical Problems in Engineering*, vol. 2011, Article ID 953492, 14 pages, 2011.
8. M. Dehghan, "New schemes for a two-dimensional inverse problem with temperature overspecification," *Mathematical Problems in Engineering*, vol. 7, no. 3, pp. 283–297, 2001.
9. T. S. Jang, H. G. Sung, S. L. Han, and S. H. Kwon, "Inverse determination of the loading source of the infinite beam on elastic foundation," *Journal of Mechanical Science and Technology*, vol. 22, no. 12, pp. 2350–2356, 2008.
10. T. S. Jang and S. L. Han, "Numerical experiments on determination of spatially concentrated time-varying loads on a beam: an iterative regularization method," *Journal of Mechanical Science and Technology*, vol. 23, no. 10, pp. 2722–2729, 2009.

11. X. Ma and N. Zabaras, "An efficient Bayesian inference approach to inverse problems based on an adaptive sparse grid collocation method," *Inverse Problems*, vol. 25, no. 3, Article ID 035013, 27 pages, 2009.
12. J. Wang and N. Zabaras, "Hierarchical Bayesian models for inverse problems in heat conduction," *Inverse Problems*, vol. 21, no. 1, pp. 183–206, 2005.
13. B. Jin and J. Zou, "Hierarchical Bayesian inference for ill-posed problems via variational method," *Journal of Computational Physics*, vol. 229, no. 19, pp. 7317–7343, 2010.
14. J. Kaipio and E. Somersalo, *Statistical and Computational Inverse Problems*, vol. 160 of *Applied Mathematical Sciences*, Springer, New York, NY, USA, 2005.
15. A. Riccio, G. Barone, E. Chianese, and G. Giunta, "A hierarchical Bayesian approach to the spatio-temporal modeling of air quality data," *Atmospheric Environment*, vol. 40, no. 3, pp. 554–566, 2006.
16. A. Riccio, "A Bayesian approach for the spatiotemporal interpolation of environmental data," *Monthly Weather Review*, vol. 133, no. 2, pp. 430–440, 2005.
17. G. Barone, E. Chianese, G. Giunta, and A. Riccio, "A hierarchical Bayesian approach to the spatio-temporal assessment of the CAMx model," in *Proceedings of the 2nd GLOREAM/EURASAP Workshop*, Copenhagen, Denmark, September 2004.
18. H. K. H. Lee, D. M. Higdon, Z. Bi, M. A. R. Ferreira, and M. West, "Markov random field models for high-dimensional parameters in simulations of fluid flow in porous media," *Technometrics*, vol. 44, no. 3, pp. 230–241, 2002.
19. J. Besag, P. Green, D. Higdon, and K. Mengersen, "Bayesian computation and stochastic systems," *Statistical Science*, vol. 10, no. 1, pp. 3–41, 1995.
20. W. R. Gilks, S. Richardson, and D. J. Spiegelhalter, *Markov Chain Monte Carlo in Practice*, Interdisciplinary Statistics, Chapman & Hall, London, UK, 1996.
21. J. A. Murdock, *Perturbations: Theory and Methods*, A Wiley-Interscience Publication, John Wiley & Sons, New York, NY, USA, 1991.
22. A. Kirsch, *An Introduction to the Mathematical Theory of Inverse Problems*, vol. 120 of *Applied Mathematical Sciences*, Springer, New

York, NY, USA, 1996.

23. C. W. Groetsch, *Inverse Problems in the Mathematical Sciences*, Vieweg Mathematics for Scientists and Engineers, Friedr. Vieweg & Sohn, Braunschweig, Germany, 1993.
24. A. N. Tikhonov, A. Goncharsky, V. V. Stepanov, and A. G. Yagola, *Numerical Methods for the Solution of Ill-Posed Problems*, Springer, Berlin, Germany, 2010.
25. S. Chib and E. Greenberg, "Understanding the Metropolis-Hastings algorithm," *The American Statistician*, vol. 49, pp. 327–335, 1995.
26. G. Casella and E. I. George, "Explaining the Gibbs sampler," *The American Statistician*, vol. 46, no. 3, pp. 167–174, 1992.
27. A. H. Nayfeh and D. T. Mook, *Nonlinear Oscillations*, Pure and Applied Mathematics, John Wiley & Sons, New York, NY, USA, 1979.
28. G. Strang, *Linear Algebra and Its Applications*, Academic Press, New York, NY, USA, Second edition, 1980.

CHAPTER
13

Model Order Reduction for Bayesian Approach to Inverse Problems

Ngoc-Hien Nguyen[1], Boo Cheong Khoo[2] and Karen Willcox[3]

[1]Singapore-MIT Alliance, National University of Singapore, Singapore 117576, Singapore

[2]Department of Mechanical Engineering, National University of Singapore, Singapore 117576, Singapore

[3]Massachusetts Institute of Technology, Cambridge, MA 02139-4307, USA.

ABSTRACT

This work presents an approach to solve inverse problems in the application of water quality management in reservoir systems. One such application is contaminant cleanup, which is challenging because tasks such as inferring

Citation: Nguyen, N. H., Khoo, B. C., & Willcox, K. (2014). Model order reduction for Bayesian approach to inverse problems. Asia Pacific Journal on Computational Engineering, 1(1), 2. (17 pages) DOI: https://doi.org/10.1186/2196-1166-1-2

the contaminant location and its distribution require large computational efforts and data storage requirements. In addition, real systems contain uncertain parameters such as wind velocity; these uncertainties must be accounted for in the inference problem. The approach developed here uses the combination of a reduced-order model and a Bayesian inference formulation to rapidly determine contaminant locations given sparse measurements of contaminant concentration. The system is modelled by the coupled Navier-Stokes equations and convection-diffusion transport equations. The Galerkin finite element method provides an approximate numerical solution-the 'full model', which cannot be solved in real-time. The proper orthogonal decomposition and Galerkin projection technique are applied to obtain a reduced-order model that approximates the full model. The Bayesian formulation of the inverse problem is solved using a Markov chain Monte Carlo method for a variety of source locations in the domain. Numerical results show that applying the reduced-order model to the source inversion problem yields a speed-up in computational time by a factor of approximately 32 with acceptable accuracy in comparison with the full model. Application of the inference strategy shows the potential effectiveness of this computational modeling approach for managing water quality.

BACKGROUND

Hydrodynamic processes such as contaminant transport in lakes and reservoirs have a direct impact on water quality. The contaminants will appear, spread out, and decrease in concentration, etc. because of some processes such as convection, diffusion, time rate release of contaminants, and distance of travel. To simulate such processes, a coupled system of partial differential equations (PDEs) including the Navier-Stokes equations (NSEs) and contaminant transport equations needs to be solved. A better understanding of these processes is important in managing water resources effectively.

The direct or forward problems compute the distribution of contaminant directly from given input information such as contaminant location, contaminant properties, fluid flow properties, boundary conditions, initial conditions, etc. On the contrary, the inverse problems infer the unknown physical parameters, boundary conditions, initial conditions, or geometry given a set of measured data. These known data can be obtained experimental or computational. In realistic applications, data are not perfect because of error due to sensor noise. In addition, the model may contain some uncertain parameters such as wind velocity. Thus, inverse problems are generally stochastic problems.

A Bayesian inference approach to inverse problems is one way to deal with the stochastic problem. This approach takes into account both the information of the model parameters with uncertainty and the inaccuracy of data in terms of a probability distribution [1, 2]. The Bayesian approach provides a general framework for the formulation of wide variety of problems such as climate modeling [3], contaminant transport modeling [4–6], and heat transfer [7]. Under the Bayesian framework, simulated solutions need to be evaluated repeatedly over different samples of the input parameters. There are available sampling strategies associated with Bayesian computation such as Markov chain Monte Carlo (MCMC) methods [8–11]. However, if we use traditional PDE discretization methods, such as finite element or finite volume methods, the resulting numerical models describing the system will be very large and expensive to solve.

This paper presents an efficient computational approach to solve the statistical inverse problem. The approach uses the combination of a reduced-order model and a Bayesian inference formulation. The Galerkin finite element method [12] provides an approximate numerical solution - the 'full model' or 'forward model'. The MCMC method is applied to solve the Bayesian inverse problems. In particular, we are interested in inferring an arbitrary source location in a time-dependent convection-diffusion transport equation, given a velocity field and a set of measured data of the concentration field at sparse spatial and temporal locations. We obtain the velocity field by solving the Navier-Stokes equations.

Solution of large-scale inverse problems can be accelerated by applying model order reduction [13–15]. The idea is to project the large-scale governing equations onto the subspace spanned by a reduced-space basis, yielding a low-order dynamical system. Proper orthogonal decomposition (POD) is the most popular method to find the reduced basis for a given set of data. The snapshot POD method, which was proposed by Sirovich [16], provide an efficient way for determining the POD basis vectors. Snapshots are solutions of a numerical simulation at selected times or choices of the parameters. The choice of snapshots should ensure that the resulting POD basis captures the most important characteristics of the system.

Since the convection term in contaminant transport equations is a velocity-dependent term, we cannot apply the standard POD reduction framework. The evaluation of the velocity-dependent convection term in the reduced system still depends on the full finite element dimension and has the same complexity as the full-order system. Thus, a system with velocity-dependent convection term needs an additional treatment to obtain an efficient reduced-order model.

This paper is outlined as follows. In the section 'Methods', we introduce the mathematical model, the finite element approximate technique, and the Bayesian inference formulation and its solution using an MCMC method. In section 'Model order reduction', we describe the model order reduction technique and the treatment of the velocity-dependent convection term. In the section 'Results and discussion', we use the numerical example to demonstrate the solution of the statistical inverse problem and the reduced-order model performance. We provide some concluding remarks in the final section.

METHODS

In this section, the mathematical model, finite element approximation technique, Bayesian inference formulation, and MCMC method are briefly introduced.

Transport model

Consider the fluid flow through a physical domain $D \subset R^d, d=1,2,3$ with boundary Γ. The contaminant concentration $c :\equiv c(\mathbf{x},t;\theta)$ satisfies the dimensionless parabolic PDEs, boundary conditions, and initial conditions as follows:

$$\frac{\partial c}{\partial t} + \mathbf{u} \cdot \nabla c - \frac{1}{Pe} \Delta c = f \quad \text{in } \mathcal{D} \times [t_0, t_f], \tag{1}$$

$$c = g \quad \text{on } \Gamma_D \times [t_0, t_f], \tag{2}$$

$$\frac{\partial c}{\partial \mathbf{n}} = 0 \quad \text{on } \Gamma_N \times [t_0, t_f], \tag{3}$$

$$c(\mathbf{x}, t_0) = c_0(\mathbf{x}) \quad \text{in } \mathcal{D}, \tag{4}$$

where $x \in D$ denotes the spatial coordinates, $t \in [t_0, t_f]$ denotes time, and c_0 is the given initial condition. The inlet boundary Γ_D is subjected to a Dirichlet condition, while the remainder of the boundary $\Gamma_N = \Gamma \setminus \Gamma_D$ satisfies Neumann conditions. The external source $f(\mathbf{x},t;\theta)$ in Equation 1 is used here as a input parameter, where $\theta \in D$ is the source location. In this study, we suppose that the contaminant source drives the transport process, given a velocity field u∈Rd. Thus, the Péclet number Pe=‖u‖Lκ, where κ is the diffusivity and L is the characteristic length. The velocity field is a function of $\mathbf{x}$ and t, $\mathbf{u}(\mathbf{x},t)$, which can be obtained by solving the Navier-Stokes equations, as follows:

$$\frac{\partial \mathbf{u}}{\partial t} + (\mathbf{u} \cdot \nabla)\mathbf{u} - \frac{1}{\mathrm{Re}}\nabla^2\mathbf{u} + \nabla p = \mathbf{f}_c \quad \text{in } \mathcal{D}\times [t_0, t_f], \tag{5}$$

$$\nabla \cdot \mathbf{u} = 0 \quad \text{in } \mathcal{D}\times [t_0, t_f], \tag{6}$$

$$\mathbf{u} = \mathbf{u}_D \text{ on } \Gamma\times [t_0, t_f], \tag{7}$$

$$\mathbf{u}(\mathbf{x}, t_0) = \mathbf{u}_0(\mathbf{x}) \quad \text{in } \mathcal{D}, \tag{8}$$

where Re is the Reynolds number, p is the pressure field, and $\mathbf{f}_c$ is the body force. So, for any given f and $\mathbf{u}$, one can solve for the solution $c(\mathbf{x},t;\theta)$.

Finite Element Approximations

The finite element method [12] associated with the stabilized second-order fractional-step method is applied to solve the incompressible Navier-Stokes equations (5 to 8). The detailed discussion and derivation of this method is beyond the scope of this paper. For more details, please refer to [17, 18].

We also use the finite element method for spatial discretization of the contaminant transport equations in Equations 1 to 4. This leads to the semi-discrete equations

$$\mathbf{M}\dot{\mathbf{c}} + \left(\mathbf{C}(\mathbf{u}) + \mathbf{K}\right)\mathbf{c} = \mathbf{F}(t;\theta), \tag{9}$$

$$\mathbf{c}(t_0) = \mathbf{c}_0, \tag{10}$$

with the output of contaminant solution at sensor locations in the domain

y=Bc. (11)

Here, c(t)∈RN is the discretized approximation of $c(\mathbf{x},t)$ and contains N state unknowns where N is the number of grid points in the spatial discretization. ċ is the derivative of **c** with respect to time. The vector y∈RNo contains the N_o outputs of the system related to the state through the matrix B∈RNo×N,No≪N. The matrices M,K,C(u)∈RN×N and the vector F(t;θ)∈RN are defined in the finite element space as follows:

$$\mathbf{M}_{i,j} = \int_{\mathcal{D}} \phi_i(\mathbf{x})\phi_j(\mathbf{x})\, d\mathbf{x}, \tag{12}$$

$$\mathbf{K}_{i,j} = \int_{\mathcal{D}} \frac{1}{Pe} \nabla\phi_i(\mathbf{x}) \cdot \nabla\phi_j(\mathbf{x})\, d\mathbf{x}, \tag{13}$$

$$\left(\mathbf{C}(\mathbf{u})\right)_{i,j} = \int_{\mathcal{D}} \mathbf{u} \cdot \nabla\phi_i(\mathbf{x})\phi_j(\mathbf{x})\, d\mathbf{x}, \tag{14}$$

$$\left(\mathbf{F}(t;\theta)\right)_j = \int_{\mathcal{D}} f(t;\theta)\phi_j(\mathbf{x})\, d\mathbf{x}. \tag{15}$$

Here ϕ_i with $i=1,\cdots,N$ is the finite element basis and θ∈Rdθ∈Rd is the coordinate vector of the source location.

Bayesian Inference to Inverse Problems

The Bayesian formulation of an inference problem is derived from Bayes' theorem [1]:

$$p(\theta|\mathbf{y}) = \frac{p(\mathbf{y}|\theta)p(\theta)}{p(\mathbf{y})}, \tag{16}$$

where y∈RNo is the observed data and θ∈Rd⊆D is the vector of input parameters. $p(\theta)$ denotes the prior distribution of the parameters, $p(\mathbf{y}|\theta)$ is the likelihood function, and $p(\theta|\mathbf{y})$ is the posterior distribution of parameters given the observed data. $p(\mathbf{y})$ is the prior predictive distribution of **y** and may be set to an unknown constant that is not needed in the MCMC solution method. In this case, we write Bayes's formula as

$$p(\theta|y) \propto p(y|\theta)p(\theta). \tag{17}$$

A simple model for the likelihood function can be obtained from the following relationship,

$$y=G(\theta)+\eta, \tag{18}$$

where η is the noise which we assume to be white Gaussian noise $\eta\sim N(0,\sigma^2 I)$. The input-output relative in Equations 1 to 4 is denoted as the forward model $G(\theta)$. Subsequently, the likelihood can be written as

$$p(\mathbf{y}|\theta) = \frac{1}{\sigma\sqrt{2\pi}} \exp\left(-\frac{1}{2\sigma^2}\|\mathbf{y} - G(\theta)\|^2\right). \tag{19}$$

There are many approaches to define the prior information, such as Gaussian Markov random fields or uniform distribution models, etc. In this work, we assume that our only prior information on the source location is given by the bounds on the domain. Thus, using the principle of maximum entropy [19], we take our prior to be a uniform distribution. Therefore, Equation 17 becomes

$$p(\theta|\mathbf{y}) \propto \prod_{i=1}^{K} \exp\left[-\frac{1}{2\sigma^2}(\mathbf{y}_i - G(\theta)_i)^T(\mathbf{y}_i - G(\theta)_i)\right]. \tag{20}$$

Here, K is the number of time steps over which we collect the output data.

Markov chain Monte Carlo for Posterior Sampling

MCMC simulation provides a sampling strategy from a proposal distribution to the target distribution using the Markov chain mechanism. Hastings introduced the Metropolis-Hastings (MH) algorithm [10], which is a generalized version of the Metropolis algorithm [8, 9], in which we can apply both symmetric and asymmetric proposal distributions. In this work, the MH algorithm is used to solve the Bayesian inverse problem. The MH algorithm is summarized as follows:

1. Initialize the chain θ^0 and set n=0
2. Repeat

- n=n+1
- Generate a proposal point $\theta^* \sim q(\theta^*|\theta)$
- Generate U from a uniform U(0,1) distribution
- Update the state to θ^{n+1} as

$$\theta^{n+1} = \begin{cases} \theta^*, \text{ if } \beta < \text{U} \\ \theta^n, \text{ otherwise} \end{cases}$$

3. Until n=N_{mcmc} → stop.

Here, β is the acceptance-rejection ratio, given by

$$\beta = \min\left(1, \frac{\pi(\theta^*)\, q(\theta^{n-1}|\theta^*)}{\pi(\theta^{n-1})\, q(\theta^*)\,|\theta^{n-1}}\right). \tag{21}$$

N_{mcmc} is the total number of samples, $\pi(\theta)$ is the target distribution, $q(\theta^*|\theta)$ is a proposal distribution, and U is a random number.

The discretized model in the form of ordinary differential equations (ODEs) (Equations 9 to 11) may have very large dimensions and be expensive to solve. The MCMC method requires evaluating repeatedly the solution of the forward model (many thousands or even millions of times). Hence, these simulations can be a computationally expensive undertaking. In such situations, the reduced-order model is needed to approximate the large-scale model, which allows efficient simulations.

Model Order Reduction

This section presents the model order reduction framework. This includes the reduction via projection, the proper orthogonal decomposition method, the additional treatment for the velocity-dependent convection term, and the error estimation.

Reduction Via Projection

A reduced-order system approximating the ODEs (9 to 11) can be obtained by approximating the full state vector **c** as a linear combination of *m* basis vectors as follows:

$$c \approx V c_r, \tag{22}$$

where $c_r \in R^m$ is the reduced order state and $V=[v_1 v_2 \cdots v_m] \in R^{N\times m}$ is an orthonormal basis, i.e., $V^T V = I$. The low-order model can be derived by projecting the governing system (9 to 11) onto the reduced space formed by the column span of basis V. This yields

$$\mathbf{M}_r \dot{\mathbf{c}}_r + \left(\mathbf{C}_r(\mathbf{u}) + \mathbf{K}_r\right)\mathbf{c}_r = \mathbf{F}_r(t;\theta), \tag{23}$$

$$c_r(t_0) = c_{0r}, \tag{24}$$

$$y_r = B_r c_r, \tag{25}$$

where

$$\mathbf{M}_r = V^T \mathbf{M} V, \tag{26}$$

$$\mathbf{K}_r = V^T \mathbf{K} V, \tag{27}$$

$$\mathbf{C}_r(\mathbf{u}) = V^T \mathbf{C}(\mathbf{u}) V, \tag{28}$$

$$\mathbf{F}_r(t;\theta) = V^T \mathbf{F}(t;\theta), \tag{29}$$

$$B_r = BV, \tag{30}$$

$$\mathbf{c}_{0r} = V^T \mathbf{c}_0. \tag{31}$$

The model reduction task is to find a suitable basis V so that $m \ll N$ and the reduced-order model yields accurate results. This study will consider POD as the method to compute the basis.

Proper Orthogonal Decomposition

POD provides a method to compute the reduced-order basis V and construct the low-order system. Here, we briefly describe the general POD method (more details may be found in [16]). Supposed that we have the snapshot matrix X∈RN×Q, where Q is the number of snapshot solutions, which are built from the instantaneous solution $\mathbf{c}^s(t_k)$ of the forward model, corresponding to $s=1,\cdots,S$ random input source locations, picked up at $k=1,\cdots,T$ different time steps (so that $Q=S\ T$). The POD basis vectors are the m left singular vectors of X corresponding to the largest singular values ($m\leq Q$). Let $\sigma_i, i=1,2,\cdots,Q$ be the singular values of X in decreasing order. We choose the number of POD basis vectors to retain in the reduced-order model as $m\leq Q$ so that

$$\sum_{i=1}^{m} \sigma_i^2 / \sum_{j=1}^{Q} \sigma_j^2 \geq \epsilon_E, \tag{32}$$

where $\varepsilon_E(\%)$ is the required amount of energy, typically taken to be 99% or higher.

After obtaining the POD basis vectors for the contaminant, we have defined our reduced-order system. However, as mentioned earlier, the dimension of the system of Equations 24 to 26 is reduced only in state (concentration). The reduced convection matrix ($\mathbf{C}_r(\mathbf{u})$) in Equation 29 still depends on the full dimension of the velocity field as in Equation 14. This means that we have to re-compute the full convection matrix ($\mathbf{C}(\mathbf{u})$) at each time-step before projecting it onto the reduced-space basis to obtain the reduced convection matrix ($\mathbf{C}_r(\mathbf{u})$). Hence, we need an additional treatment to avoid this computational cost.

Linear Expansion Techniques For Velocity-Dependent Term

We proceed by representing the velocity field as a linear combination of POD velocity basis vectors. The velocity-dependent convection matrix can then be expanded based on this velocity expansion. In particular, let $\mathbf{u}(\mathbf{x},t)$ be a given velocity field and u(x,tk)k=1T be a set of 'snapshots' of velocities which are obtained from Equations 5 to 8 and taken over T time steps. The velocity field is decomposed as

$$\mathbf{u}(\mathbf{x}, t) = \mathbf{u}_m(\mathbf{x}) + \mathbf{u}'(\mathbf{x}, t), \tag{33}$$

where $\mathbf{u}_m(\mathbf{x}) = \frac{1}{T}\sum_{k=1}^{T} \mathbf{u}(\mathbf{x}, t^k)$ is the mean velocity field and $\mathbf{u}'(\mathbf{x},t)$ is the fluctuating velocity field. The fluctuating velocity field is then represented by

$$\mathbf{u}'(\mathbf{x}, t) = \sum_{k=1}^{T} \alpha_k(t)\Psi_k(\mathbf{x}), \tag{34}$$

where $\Psi_k(\mathbf{x})$ is the kth POD velocity basis and $\alpha_k(t)$ is the corresponding time dependent amplitude. We now consider the expansion of the velocity field as

$$\mathbf{u}(\mathbf{x}, t) = \mathbf{u}_m(\mathbf{x}) + \sum_{k=1}^{N_u} \alpha_k(t)\Psi_k(\mathbf{x}), \tag{35}$$

where $N_u \ll T$ is the number of POD basis vectors use to represent the velocity.

Galerkin projection of the time-dependent velocity term in Equation 1 has the form as

$$\begin{aligned}(\mathbf{u} \cdot \nabla c, V_i) &= \left(\left[\mathbf{u}_m + \sum_{k=1}^{N_u} \alpha_k \Psi_k\right] \cdot \nabla c, V_i\right) \\ &= [\mathbf{u}_m \cdot \nabla c, V_i] + \left(\sum_{k=1}^{N_u} \alpha_k \Psi_k \cdot \nabla c, V_i\right).\end{aligned} \tag{36}$$

The discrete form of Equation 37 then becomes

$$\begin{aligned}\mathbf{C}_r(\mathbf{u}) &= [V_i, \mathbf{u}_m \cdot \nabla V_j] + \sum_{k=1}^{N_u} \alpha_k [V_i, \Psi_k \cdot \nabla V_j] \\ &= \mathbf{C}_{rm}(\mathbf{u}_m) + \sum_{k=1}^{N_u} \alpha_k \mathbf{C}_{rk}(\Psi_k).\end{aligned} \tag{37}$$

Here, $\mathbf{C}_{rm}(\mathbf{u}_m(\mathbf{x}))$ and $\mathbf{C}_{rk}(\Psi_k(\mathbf{x})), k=1,\cdots,N_u$ are the reduced-order forms of the full-order convection matrices $\mathbf{C}(\mathbf{u}_m(\mathbf{x}))$ and $\mathbf{C}(\Psi_k(\mathbf{x}))$, respectively. They are computed only once in the offline step. Thus, the first case of the reduced-order model (with Equation 29) is only reduced in state (concentration), but the second case (with Equation 38) is now reduced in both state (concentration) and parameter (velocity).

Error Estimation

In order to estimate the efficiency of the reduced model relative to the full model, we use the relative errors of state solutions and relative errors of outputs. These errors are defined as follows:

$$\varepsilon_F\left(t^k, \theta\right) = \frac{\|\mathbf{c}\left(t^k; \theta\right) - V\mathbf{c}_r\left(t^k; \theta\right)\|_{L_2(\mathcal{D})}}{\|\mathbf{c}\left(t^k; \theta\right)\|_{L_2(\mathcal{D})}}, \tag{38}$$

$$\varepsilon_S\left(t^k,\theta\right) = \frac{\|\mathbf{y}\left(t^k;\theta\right) - \mathbf{y}_r\left(t^k;\theta\right)\|_{L_2(\mathcal{D})}}{\|\mathbf{y}\left(t^k;\theta\right)\|_{L_2(\mathcal{D})}}. \quad (39)$$

Here, $\mathbf{c}(t^k;\theta),\mathbf{c}_r(t^k;\theta),1\leq k\leq T$ are the full and reduced solutions, while $\mathbf{y}(t^k;\theta)$ and $\mathbf{y}_r(t^k;\theta),1\leq k\leq T$ are the full and reduced outputs of interest, respectively.

RESULTS AND DISCUSSION

We consider a numerical example based on a 2D mathematical model. The velocity field in the reservoir is obtained by solving a 2D laterally averaged Navier-Stokes model. The Bayesian formulation of the inverse problem is then solved to determine an uncertain contaminant source location. We compare the effectiveness of solving the inverse problem using both the full model and reduced model techniques.

Model Setup

The physical domain is illustrated in Figure 1a, which represents a simplified model of a 2D laterally averaged reservoir system. The reservoir system includes a main reservoir section and the river connections. We assume that the contaminant is present within the main reservoir section and that the contaminant transport processes are mainly affected by the inflow and wind velocity. We assume that other factors (such as heat exchange, etc.) have little influence on the processes. The spatial domain is discretized with a finite element mesh, as shown in Figure 1b. The total number of grid points is N=1,377 and the total number of elements is N_E=2,560. The simulation is run from initial time t_0=0 to final time t_f=40, with time-step size Δt=0.08. Thus, the number of time steps is T=500.

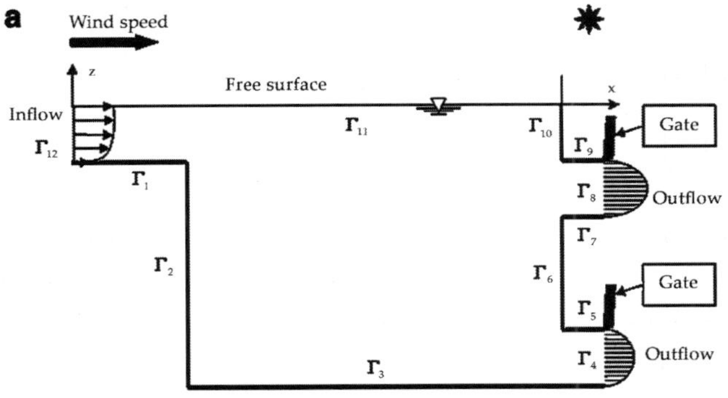

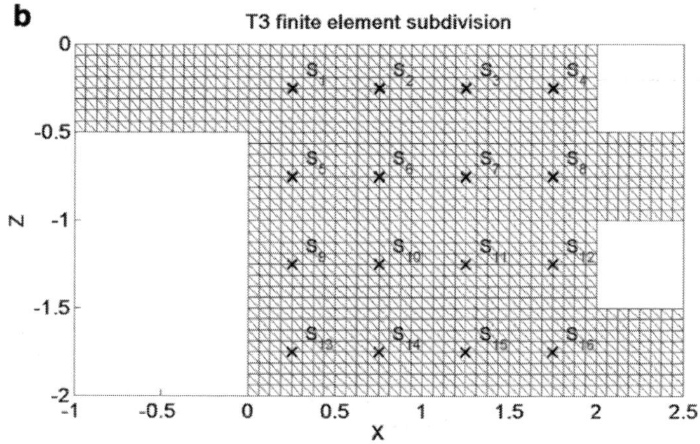

Figure 1: The domain of interest. (a) Physical domain. (b) Computational domain.

A time-dependent velocity field is obtained from the 2D laterally averaged system, which is given in Equations 5 to 8, where the body force $\mathbf{f}_c = [f_{cx}; f_{cz}]^T = [0; g]^T$ with g the gravitational acceleration.

The boundary conditions are set up as follows (refer to Figure 1a for the boundary definitions):

$$(u,w)=(1,0) \text{ on } \Gamma_{12}, \quad (40)$$

$$(u,w)=(-16*(2.0+z)*(1.5+z),0) \text{ on } \Gamma_{4}, \quad (41)$$

$$(u,w)=(-16*(1.0+z)*(0.5+z),0) \text{ on } \Gamma_{8}, \quad (42)$$

$$(u,w)=(0.03V_a,0) \text{ on } \Gamma 11, \quad (43)$$

$$p=0 \text{ on } \Gamma_{11}. \quad (44)$$

The velocity on remaining boundaries are set to zero. Here, V_a is the wind speed at 10 m above the water surface. In this example, we assumed that V_a=2 m/s for the entire simulation time. The Reynolds number is set to Re=5.0e^5, and a mixing length turbulence model is used [20]. The full model system is given in Equations 9 to 10. The Crank-Nicolson method [21] is used to discretize the system in time.

The setup for the contaminant transport part of the problem is as follows. We use a contaminant source that is defined as the superposition of Gaussian sources, each one active on the time interval $t_{0k} \in [t_0, t_{off}]$ and centered at θk∈Dθk∈D, with strength h_k and width σ_{sk}:

$$f(\mathbf{x}, t; \theta) = \sum_{k=1}^{n_s} \frac{h_k}{2\pi\sigma_{sk}^2} \exp\left(-\frac{|\theta_k - \mathbf{x}|^2}{2\sigma_{sk}^2}\right) \delta(t - t_{0k}). \tag{45}$$

Here, to simplify the problem, we choose the number of sources to be $n_s=1$, located at $\theta_1=(x_c, z_c)$, with strength $h_1=0.2$ and width $\sigma_{s1}=0.05$. The active time of the source is $t_{01}\in[0, t_{\text{off}}]$ with $t_{\text{off}}=10$. The inflow boundary and other solid boundaries satisfy a homogeneous Dirichlet condition on the contaminant concentration; the outflow boundaries and free surface boundary satisfy a homogeneous Neumann condition. The diffusivity coefficient is assumed to be constant, $\kappa=0.005$. Thus, the Péclet number Pe=||u||Lκ=100, where the length of the inflow section is used as the characteristic length $L=0.5$. The contaminant concentration is assumed to be zero at initial time $t_0=0$.

With this setup, we have specified our forward problem with the input parameter θ∈R2 in the range $\Theta:=[0, 2]\times[-2, 0]$. Figure 2 shows the contaminant solution $c(\mathbf{x},t;\theta)$ of the forward model with $\theta_1=(0.5,-0.5)$ at specific times. The contaminant field grows while the source is active. After the shutoff time of the source, the contaminant moves away, spreads out, and decreases in concentration due to convection and diffusion, until it flows out of the domain.

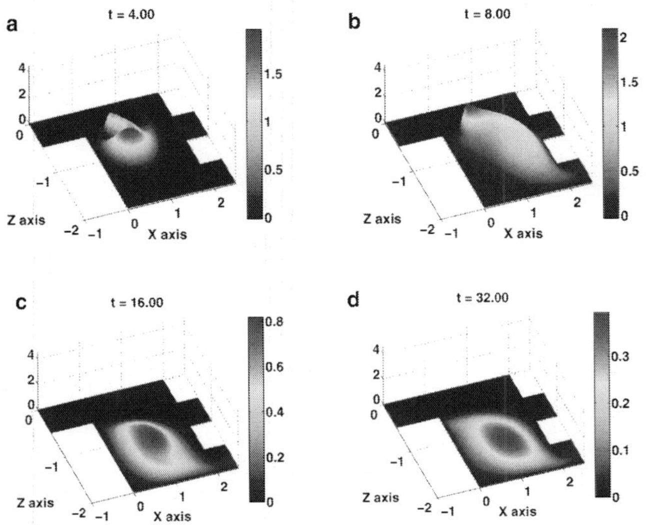

Figure 2: Contaminant field of forward model at specific times. Figures are taken at (a) t=4.0s, (b) t=8.0s, (c) t=16.0s and (d) t=32.0s.

The outputs of interest are the values of contaminant solution $c(\mathbf{x},t;\theta)$ at selected sensor locations in the computational domain. These sensors are located on a 4×4 uniform grid covering the reservoir domain as shown in Figure 1b.

Model Order Reduction

To perform the model order reduction, we need to obtain the POD expansion of velocity field.

Figure 3a shows all eigenvalues of the velocity field. As the number of eigenvalues increases, the significant of eigenvalues decreases, respectively. Figure 3b shows the relative errors of representing the velocity field with various numbers of POD velocity modes. The goal of this step is to find a suitable N_u that can provide a fast computation of the velocity-dependent convection term and to ensure that the relative errors are acceptable. In this example, we choose the number of terms in the velocity expansion to be N_u=12 yields the relative error around 5×10^{-3}.

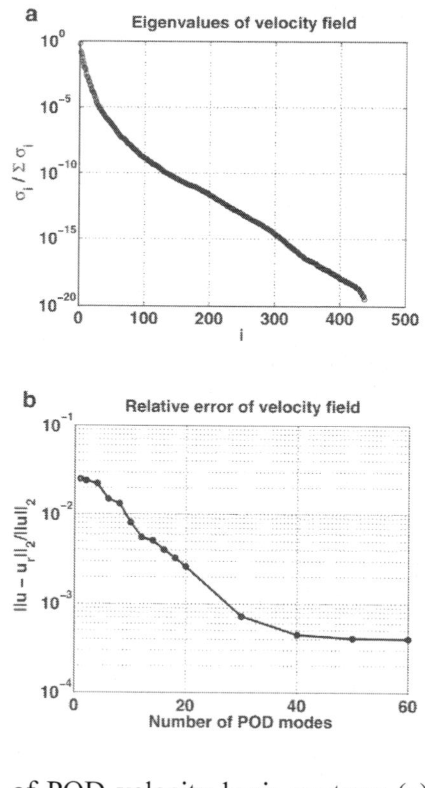

Figure 3: Properties of POD velocity basis vectors: (a) eigenvalues and (b) relative error analysis.

Figure 4 illustrates the first four POD basis vectors of the velocity field. Clearly, they represent the most important characteristics of the velocity field.

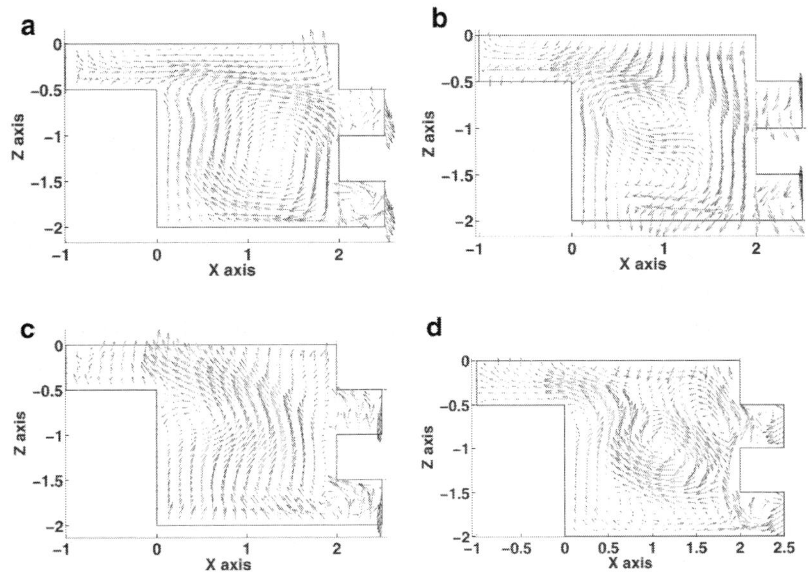

Figure 4: The first four POD basis vectors of the velocity field. (a) First POD velocity basis vector. (b) Second POD velocity basis vector. **(c)** Third POD velocity basis vector. **(d)** Fourth POD velocity basis vector.

A comparison between the original velocity and the approximate velocity representation at final simulating time is given in Figure 5. We observe that the approximate velocity profiles at x=0.5 (Figures 5a,b) and x=1.0 (Figures 5c,d) are able to represent well the characteristics of the original velocity profiles.

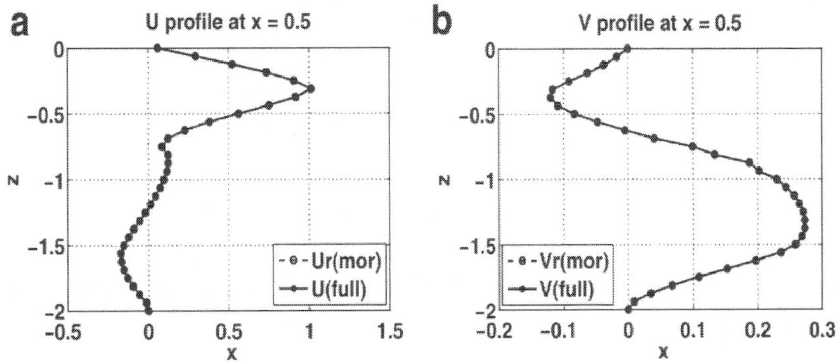

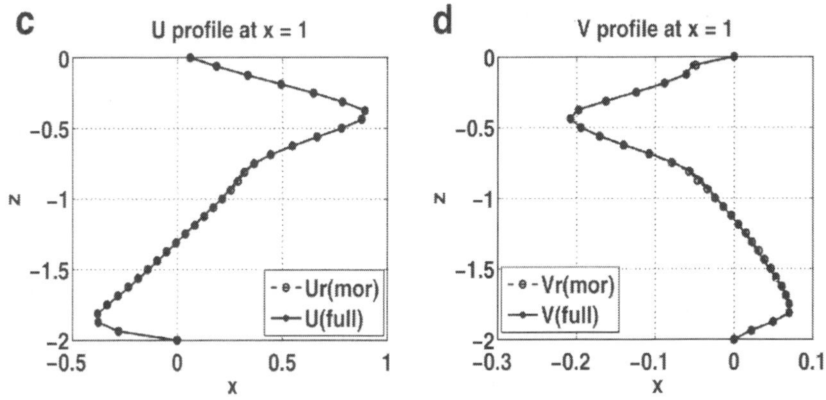

Figure 5: Velocity profile taken at $t=t_f$ at specific locations. Figures are captured (a) U profile at x=0.5, (b) V profile at x=0.5, **(c)** U profile at x=1 and **(d)** V profile at x=1. Here U(full) and V(full) are the original velocity while Ur(mor) and Vr(mor) are the approximate velocity using a POD expansion with N_u=12 modes.

To generate the snapshots needed for the POD basis to represent the contaminant concentration, we choose S location samples in the computational domain. In this example, we generate random input locations θ∈D⊂R2 for source term $f(\mathbf{x},t;\theta)$.

Table 1 shows the relative errors and computational time ratio of the reduced-order models (ROM) for different sample sizes of snapshot matrices. Here, we use the notation ROM-1 to denote the reduced model mentioned in the section 'Proper orthogonal decomposition' and ROM-2 for the one mentioned in the section 'Linear expansion techniques for velocity-dependent term'. The POD basis is chosen based on the energy as in Equation 33, with ε_E=99.999%. We observe that while the relative errors in the outputs are similar for both cases, the ratio of computational time differs significantly.

The case ROM-2 shows speedups from 15 to 111 times, while the case ROM-1 is in fact more expensive to solve than the full forward model (t_{full}≈285 seconds)[1].

This is because the first case consumes a lot of time in re-computing the velocity-dependent convection matrix and then projecting it onto the reduced space at each time step. We choose as our reduced-order model the case ROM-2 with sample size S=30.

Table 1: The properties of various reduced-order models

N_s	m	ε_{ROM-1}	ε_{ROM-2}	$\frac{t_{full}}{t_{ROM-1}}$	$\frac{t_{full}}{t_{ROM-2}}$
10	153	3.25e-1	4.32e-1	0.801	111
20	220	9.81e-2	9.74e-2	0.799	56
30	257	8.03e-3	6.75e-3	0.795	32
40	284	7.53e-3	7.08e-3	0.785	26
50	308	3.39e-3	2.12e-2	0.776	22
60	328	5.83e-3	6.08e-3	0.767	19
70	342	1.41e-2	1.21e-2	0.764	18
80	355	9.09e-3	8.21e-3	0.763	15

Figure 6a presents the eigenvalues of contaminant snapshot matrix for the sample test case S=30. We observe that POD eigenvalues of the contaminant field are decayed very fast, and they are significantly important only at the first few hundred values. The first four POD contaminant basis vectors are shown in Figure 7.

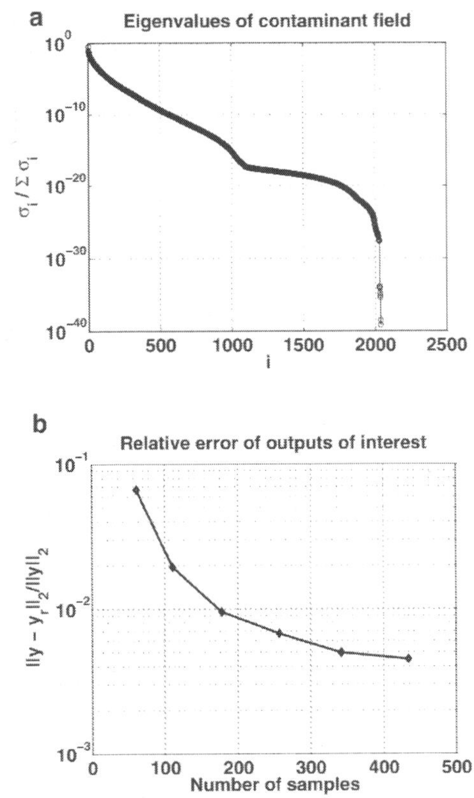

Figure 6: Properties of POD contaminant basis vectors: (a) eigenvalues and (b) relative error analysis.

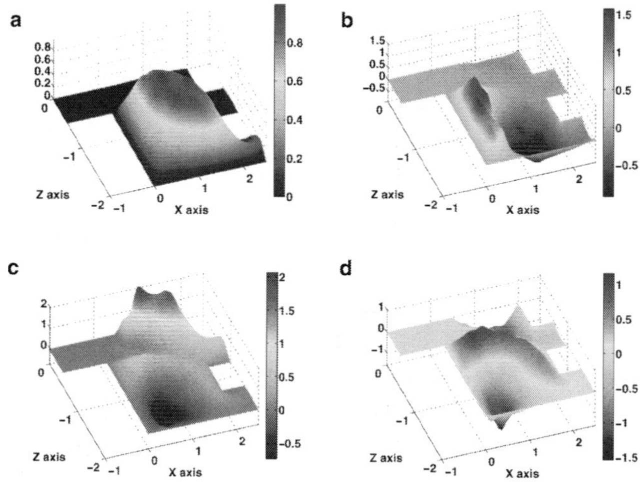

Figure 7: The first 4 POD basis vectors of the contaminant field. (a) First POD contaminant basis vector. (b) Second POD contaminant basis vector. **(c)** Third POD contaminant basis vector. **(d)** Fourth POD contaminant basis vector.

Figure 6b shows relative errors of ROMs for different numbers of POD basis vectors. With the energy taken to 99.99999%, the ROM resulted in 434 POD basis vectors with relative error around 4×10^{-3}. In the trade-off between accuracy results and computational efforts (please refer to Table 1), we can choose our ROM with 99.999% of energy captured which resulting in the size of m=257 POD basis vectors and an acceptably small error.Figure 8a shows the outputs of the full model (**y**) and ROM-2 ($\mathbf{y}_r$) with size m=257 at selected sensor locations. The selected sensors (sensor 6, 7, 10, and 11) capture more information of the contaminant than the others for the case studied. We observe that ROM-2 is able to capture well the behavior of the full model at the sensor locations. This reduced-order model will be utilized as the forward solver in solving the inverse problem.

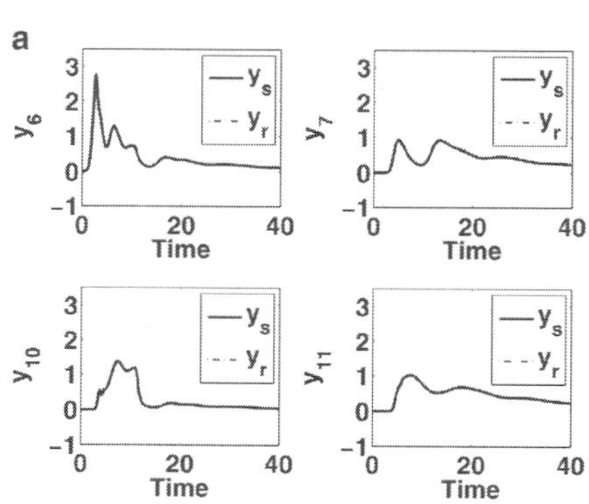

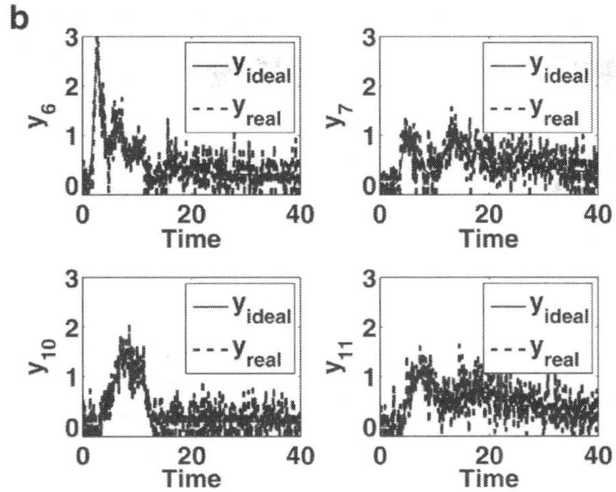

Figure 8: A comparison of the full model *(N=1377)* and ROM *(m=257)* output of interest. (a) Output comparison and **(b)** a set of generated data at sensors 6, 7, 10 and 11.

Inverse Problems

MCMC is now used to solve the inverse problem for a variety of source locations using the ROM-2 solver above. By accounting for both measured noise and uncertain information in model parameters, Bayesian inference to inverse problem leads to a well-posed problem resulting in a posterior distribution of the unknown [1, 2]. Once obtaining the posterior distribution sample, all statistical questions related to the unknown could be answered with sample averages.

To represent the behavior of uncertain variables such as wind velocity, we generate noisy data at sensor locations. The 'real' data (y_{real}) is generated synthetically by adding noise to the ideal data (y_{ideal}) as in Equation 18. The noise is assumed to be Gaussian $\eta \sim N(0,\sigma^2 I)$. Figure 8b shows the noisy data at sensor locations with σ=0.3.

We use the snapshots with S=30 samples above to generate several reduced-order models with different energies. To estimate the relative error of the inverse problem solutions (ε_θ), we choose the solution corresponding to the largest order as a 'truth' solution. We do the MCMC simulation with the starting point $\theta_{ini}=(0.821; -0.955)^T$. The total number of MCMC samples is set to N_{mcmc}=5000. The initial burn-in period is set to N_{burnin}=500. After this stage, data is saved to compute summary statistics of source locations.

Table 2 shows the results of the inverse source problems with different numbers of POD basis. For the case with m=257, the computational time is around 24 h to obtain the solution. If we use the full forward solver, with the speedup factor as given in Table 1, the computational time is estimated around 773 h or approximately 32 days.

Table 2 Estimated inverse solutions for different numbers of POD basis

$\epsilon_E(\%)$	m	θ_1	θ_2	ε_θ	$t_{mcmc}(s)$
99.0	61	0.788	-0.595	5.31e-1	1.75e+3
90.9	111	0.484	-0.491	6.95e-2	2.74e+3
99.99	178	0.451	-0.471	1.05e-2	7.55e+3
99.999	257	0.443	-0.467	2.42e-3	8.71e+4
99.9999	343	0.446	0.466	0	5.11e+5

Figure 9a shows the credible regions of the source locations θ. In this figure, both a pairwise scatter plot and 1D marginal distributions are displayed. The dash line (in θ_1 axis) and dash-dot line (in θ_2 axis) show the probability density function of each parameter while the solid blue regions show the 95% and 99% credible regions in which the parameters appeared the most (the blue dots).

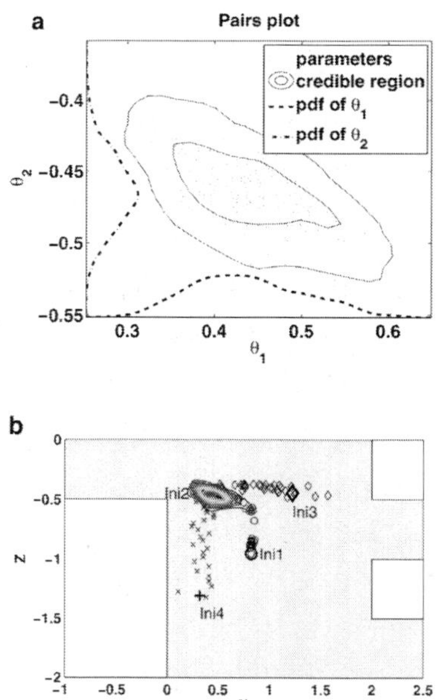

Figure 9: Credible regions and historical trajectory plots of the MCMC convergence. (a) A parameter plot showing estimated source location with credible regions. **(b)** A contour plot of the posterior pdf.

To assess the convergence of the MCMC approach, we generate four different initial points for the MH sampler. Figure 9b shows the contour plot of the posterior probability density function (the contour plot corresponds to the result of Init1) and the historical trajectories of MCMC samplers for each starting point. We observe that for all four initial points, the MCMC simulations converge to the same final solution. However, depending on the position of the initial points, the computational time varies, as given in Table 3.

Table 3 Estimated inverse solutions for different MCMC starting points

Initial	θ_{1-ini}	θ_{2-ini}	θ_{1-M}	θ_{2-M}	t_{mcmc} (s)
Ini1	0.821	-0.955	0.451	-0.467	8.71e+4
Ini2	0.322	-0.451	0.458	-0.463	4.53e+4
Ini3	1.222	-0.451	0.452	-0.464	9.80e+4
Ini4	0.323	-1.313	0.456	-0.465	1.05e+5

CONCLUSION

This study has applied successfully the combination of a model order reduction technique based on the POD and a Bayesian inference approach to solve an inverse problem that seeks to identify an uncertain contaminant location. Applying an additional POD expansion to approximate the velocity results in a reduced-order model in both state (concentration) and parameter (velocity). The resulting reduced-order model is efficient for solution of the forward problem. Solution of the Bayesian formulation of the inverse problem using the reduced-order solver is much more rapid than using the full model, yielding the probability density of the source location in reasonable computational times. The computational time of using a reduced-order model with m=257 degrees of freedom is about a factor of 32 times lower than using the full model with size N=1,377. This reduction is important in real-time water quality management applications because it reduces time cost and storage requirements.

Endnote

[1] The simulations were performed on a personal computer (PC) with processor Intel(R) Core(TM)2 Duo CPU E8200 @2.66GHz 2.66GHz, RAM 3.25GB, 32-bit Operating System.

REFERENCES

1. Tarantola A (2004) Inverse problem theory and methods for model parameter estimation. SIAM, Philadelphia
2. Sivia DS, Skilling J (2006) Data analysis: a Bayesian tutorial, Second edition. Oxford University Press Inc., New York
3. Jackson C, Sen MK, Stoffa PL (2004) An efficient stochastic Bayesian approach to optimal parameter and uncertainty estimation for climate model predictions. J Climate 17: 2828–2841
4. Snodgrass MF, Kitanidis PK (1997) A geostatistical approach to contaminant source identification. Water Resour Res 34(4): 537–546
5. Woodbury AD, Ulrych TJ (2000) A full-Bayesian approach to the groundwater inverse problem for steady state flow. Water Resour Res 36(1): 159–171
6. Marzouk YM, Najm HN, Rahn LA (2007) Stochastic spectral methods for efficient Bayesian solution of inverse problems. J Comput Phys 224(2): 560–586
7. Wang J, Zabaras N (2005) Hierarchical Bayesian models for inverse problems in heat conduction. Inverse Probl21: 183–206
8. Metropolis N, Ulam S (1949) The Monte Carlo method. J Am Stat Assoc 44: 335–341
9. Metropolis N, Rosenbluth AW, Rosenbluth MN, Teller AH, Teller E (1953) Equations of state calculations by fast computing machines. J Chem Phys 21: 1087–1092
10. Hastings WK (1970) Monte Carlo sampling methods using Markov chains and their applications. Biometrika 57: 97–109
11. Gelman A, Carlin J, Stern H, Rubin D (2003) Bayesian data analysis. Second Edition, Chapman & Hall/CRC Texts in Statistical Science
12. Zienkiewicz OC, Morgan K (1983) Finite elements and approximation. Wiley, New York
13. Antoulas AC (2005) Approximation of large-scale dynamical systems. SIAM Advances in Design and Control, Philadelphia
14. Lieberman C, Willcox K, Ghattas O (2010) Parameter and state model reduction for large-scale statistical inverse problems. SIAM J Sci Comput 32(5): 2523–2542
15. Lieberman C, Willcox K (2012) Goal-oriented inference: approach, linear theory, and application to advection-diffusion. SIAM J Sci

Comput 34(4): 1880–1904

16. Sirovich L (1987) Turbulence and the dynamics of coherent structures. Part 1: coherent structures. Q Appl Math 45(3): 561–571
17. Codina R, Blasco J (2000) Stabilized finite element method for the transient Navier-Stokes equations based on a pressure gradient projection. Comput Methods Appl Mech Engrg 182: 277–300
18. Ma X, Zabaras N (2008) A stabilized stochastic finite element second-order projection method for modeling natural convection in random porous media. J Comput Phys 227: 8448–8471
19. Jaynes ET (1957) Information theory and statistical mechanics. Phys Rev 104(4): 620–630
20. Wilcox CD (1993) Turbulence Modeling for CFD, DCW Industries, La Cañada, California
21. Crank J, Nicolson P (1996) A practical method for numerical evaluation of solutions of partial differential equations of the heat-conduction type. Adv Comput Math 6(1): 207–226

CHAPTER

14

Statistical Approaches to the Inverse Problem

A. Pascarella[1] and A. Sorrentino[2]

[1]Dipartimento di Neuroscienze, Universitá di Parma, Italy

[2]Department of Statistics, University of Warwick, United Kingdom

INTRODUCTION

Magnetoencephalography (MEG) can be considered as one of the most powerful instruments for non-invasive investigation of the brain functions (Hämäläinen et al., 1993). Its good temporal resolution (sampling frequency can reach several thousands of Hertz), only comparable to the sampling frequency of electroencephalography (EEG), allows following the neural dynamics on a millisecond time scale. This implies that, with respect to other functional imaging modalities, MEG provides an enormous amount of

Citation: Pascarella, A., & Sorrentino, A. (2011). Statistical approaches to the inverse problem (p. 93). Rijeka, Croatia: InTech.(21 pages) DOI: https://doi.org/10.5772/27932

information about the brain activity. On the other hand, localization of brain activity from MEG data requires to solve a highly ill-posed inverse problem (Sarvas, 1987): exact reconstruction of the neural current is not possible even with noise-free data, and the debate about the actual spatial resolution of MEG is still open. Despite this drawback, the appeal of MEG data is such that a large number of methods have been developed for localizing brain activity from MEG recordings. Starting from the middle Eighties, when the first Minimum Norm Estimate (Hämäläinen & Ilmoniemi, 1984) appeared, to the recent developments in spatio temporal regularization (Ou et al., 2009), state space models (Long et al., 2006; Sorrentino et al., 2009) and Markov Chain Monte Carlo (Jun et al., 2005), the attempts to insert enough a priori information to actually provide stable and reliable solutions have been rising rather than diminishing. The goal and the scope are clear: automatic and reliable source localization can be of practical use in several fields, possibly including clinical applications (pre-surgical evaluation, foci localization in epilepsy), new technology development (e.g. brain computer interfaces), and scientific research (e.g. for analysis of large datasets).

Despite having been presented in rather different frameworks and with largely different names and techniques, most methods appeared so far in the MEG inverse literature can be described rather straightforwardly in a statistical setting, which helps unifying and clarifying the assumptions and the limitations of the methods. In this chapter, we attempt a brief review of the most largely known methods, as well as of some of the most recently appeared ones, highlighting the statistical features. We first provide a brief description of the statistical approach to inverse problems in Section 2; then we describe the main features of the MEG inverse problem in Section 3; in Section 4 and 5 we review methods developed by other authors, while in Section 6 we discuss our work on Bayesian filtering for MEG.

STATISTICS AND INVERSE PROBLEMS

Inverse Problems

In applied mathematics the term inverse problem denotes a class of problems, often originating from physics equations, where the unknown is an unobserved quantity which has to be recovered from indirect measurements. Rather often, there is a causal relationship between the unknown and the data, and the inverse problem involves moving backwards from a measured effect to an unknown cause. An excellent source of inverse problems is the field of medical imaging, where one tries to recover information about quantities within the body from external measurements.

Inverse problems are usually represented in functional form as

$$g = A(f) \quad (1)$$

where $g \in H_2$, $f \in H_1$, and $A : H_1 \rightarrow H_2$ is an operator between the two possibly distinct functional spaces (e.g. Banach spaces, Hilbert spaces, etc.) H_1 and H_2. From a mathematical viewpoint, the inverse problem consists in finding the function f given the function g and the operator A. In most cases of practical interest, however, the solution of the inverse problem cannot be found exactly, because the operator A is pathological: either its kernel is non-empty - leading to non-uniqueness of the inverse solution - or the range of A is not the whole H_2 - leading to non-existence problems - or the inverse operator, when it exists, is not continuous - leading to stability problems. In applied mathematics, the typical solution to these pathologies is provided in the celebrated regularization framework for linear inverse problems (Bertero & Boccacci, 1998), where the ill-posed problem is replaced by a one-parameter family of well-posed ones. The most common regularization algorithm, Tichonov method, consists in finding the minimizer ^ f of

$$||g - A \cdot f||^2_{H_2} + \lambda ||f||^2_{H_1} \quad (2)$$

which is rather straightforwardly

$$\hat{f} = (A^* A + \lambda I)^{-1} A^* g \quad (3)$$

where A^* is the adjoint of A and λ is the regularization parameter, which tunes the balance between stability of the solution and its capability to explain the measured data. More in general, the recipe of any regularization algorithm is to minimize a functional which combines the misfit - the difference between the actual data and the model - with some type of norm of the solution - implicitely encouraging some stability property for the solution. The choice of this norm, together with a criterion for choosing the value of the regularization parameter, are key ingredients which highly depend on the specific problem.

Statistics for Inverse Problems

Given the stochastic nature of noise affecting the data, it is natural to view inverse problems as statistical inference problems. The core of this approach is to model the data as a Random Variable (RV). Then, there exist two main frameworks for dealing with the inverse problem: in the Maximum

Likelihood (ML) framework, the unknown is viewed as a set of parameters which characterize the probability distribution of the data, and the inverse problem is re-phrased as a parameter estimation problem; in the Bayesian framework, the unknown is viewed as a Random quantity as well, and the inverse problem is re-phrased as the problem of calculating the posterior distribution of the unknown, given the data.

More specifically, let y be the realization of the Random Variable Y, representing the measured data. In the ML framework, x represents the unknown parameters; the distribution of the data L(y|x) is called likelihood function and contains both the physical model - the operator A mapping the unknown onto the data - and the assumptions on the statistical distribution of noise. Typically one wants to find the x which maximizes L for the observed noisy data y; such

$$\hat{x} = \arg\max(L(y|x)) \qquad (4)$$

is naturally called Maximum Likelihood estimate of x.

In the Bayesian framework, on the other hand, x is viewed as the realization of the Random Variable X rather than as a parameter, and it is given a prior distribution; this prior is then combined with the likelihood through Bayes theorem to produce the so called posterior distribution, i.e. the probability distribution for the unknown, conditioned on the realization of the data. The solution of the inverse problem is here defined to be the posterior distribution of the unknown, which can then be used to compute point estimates and bounds. More specifically, let p(x) be the prior probability distribution for the unknown, emboding all the information which is known a priori, before the data are collected; let L(y|x) be the likelihood function, the same function which is used in the ML framework; the posterior distribution is given by Bayes theorem as

$$p(x|y) = \frac{L(y|x)p(x)}{p(y)} \qquad (5)$$

where p(y) is the normalizing constant and is given as p(y) = L(y|x)p(x) dx.

The linear Gaussian Case

Let us briefly consider a linear inverse problem with additive zero-mean Gaussian noise N, with standard deviation σ; in the ML framework, the model equation is

$$Y = Ax + N \tag{6}$$

and the likelihood function is

$$L(y|x) = \frac{1}{\sqrt{2\pi}\sigma} \exp\left(-\frac{||y - Ax||^2}{2\sigma^2}\right). \tag{7}$$

The ML estimate in this example is the solution of the least squares problem.

In Bayesian terms, the model equation becomes

$$Y = AX + N \tag{8}$$

and the posterior distribution under zero-mean Gaussian prior for the unknown turns out to be:

$$p(x|y) \propto \exp\left(-\frac{||y - Ax||^2}{2\sigma^2}\right) \exp\left(-\frac{||x||^2}{2\gamma^2}\right) \tag{9}$$

where γ is the standard deviation of the prior. Comparison of equations (9) and (2) suggests that the minimizer of the Tichonov functional is the mode of the posterior distribution, provided that the regularization parameter is identified with the ratio between the noise variance and the variance of the prior distribution.

As this simple example suggests, in the statistical framework regularization is interpreted as providing a priori information to solve the ill-posed problem. Furthermor, the statistical framework is an elegant and unifying framework where linear as well as non-linear problems can be formally described. Rather than providing computational recipes, this framework provides a theoretical background which is general enough for different kinds of problems, for different kinds of distributions of the noise, and always provides a meaningful interpretation of the variables and parameters in the game.

BASIC FACTS ABOUT THE MEG INVERSE PROBLEM

Ill Posedeness

The magnetic field b produced by a given electrical current distribution j(r) inside a volume V and measured at a point r $\in$/ V is given by the Biot-Savart equation

$$\mathbf{b}(\mathbf{r}) = \frac{\mu_0}{4\pi} \int_V \mathbf{j}(\mathbf{r}') \times \frac{\mathbf{r} - \mathbf{r}'}{|\mathbf{r} - \mathbf{r}'|^3} d\mathbf{r}' \quad . \tag{10}$$

It has been known for 150 years that the problem of estimating the electrical current from measurements of the magnetic field is an ill posed problem. Recently a few rigorous proofs have appeared in the literature, which are summarized in the following.

Let $B_1 : (C(V))^3 \rightarrow (C(D))^3$ be the operator described by (10), mapping the electrical current distribution, belonging to the space $(C(V))^3$ of continuous function within the brain volume V, onto the magnetic field, belonging to the space $(C(D))^3$ of continuous functions on a surface D which contains the boundary of V. In Kress et al. (2002) it has been proved that the kernel of B_1 contains the linear subspace

$$M = \{\mathbf{j} | \mathbf{j} = \Delta m, m \in (C_0^2(V))^3\}. \tag{11}$$

This implies that the inverse problem has no unique solution; in particular, given a solution j $*$ of the inverse problem (10), any current distribution of the form $\tilde{j} = j^* + \lambda j'$, with $j' \in M$ and $\lambda \in R$ is also a solution of the problem.

Let $B_2 : (L^2(V))^3 \rightarrow (L^2(V))^3$ be the operator described by (10), mapping the electrical current distribution in the volume V onto the magnetic field within the same volume V. In this case, all three components of both the electrical current and the magnetic field are assumed to be L^2 functions. In Cantarella et al. (2001), it has been proved that B_2 is a compact operator. This result implies that the generalized inverse is not a continuous operator.

The Inverse Model

The Biot Savart equation (10) is only a partial description of the MEG inverse problem: there are two main issues which need to be discussed in order to provide computational solutions to the inverse problem. The first issue, only mentioned here because it is not relevant in the rest of the Chapter, concerns the geometrical and physical model used to approximate the head; it is usually referred to as the forward problem, because it amounts to accurate calculations of the magnetic field produced by a known neural current distribution. Due to the non-zero conductivity of the tissues inside the head, the electrical currents elicited by brain activity are only a portion of the electrical currents flowing in the brain; this means that j(r) in (10) contains both neural activity and passively flowing currents, which are generated by the neural activity itself and contribute to the measured field as well. The forward problem accounts for the mechanisms by which the neural currents generate the passive currents, and how these passive currents flow in the brain. In the simplest model, available since the Eighties, the head is modelled as a homogeneous spherical conductor, and the effect of the passive currents can be calculated analytically (Sarvas, 1987). A more realistic model, nowadays largely used, takes into account the actual shape of the brain, while maintaining a piecewise homogeneous conductivity, and is solved with numerical integration techniques known as Boundary Element Methods (Gramfort et al., 2011). More recently the use of Finite Element Methods (Pursiainen et al., 2011), far more computationally demanding, allows implementation of completely non-homogeneous and even anisotropic conductivity models.

The second relevant point is the discretization of equation (10). The standard approach comprises discretization of the brain volume with a finite set of N_{points} points, possibly distributed along the cortical surface. The neural current is then represented as a vector field on these support points. The single current element, with support in a single point of the brain, is usually referred to as current dipole. It is characterized by a location r and a dipole moment q; mathematically speaking it is an applied vector, while from the physical viewpoint it represents the first order approximation of any current distribution. The whole neural current is then viewed as the superposition of thousands of current dipoles:

$$\mathrm{j}(\mathbf{r}) = \sum_{k=1}^{\mathcal{N}_{points}} \mathbf{q}^k \delta(\mathbf{r} - \mathbf{r}^k) \qquad (12)$$

where q^k is the dipole moment of the current dipole at location r^k and $\delta(\cdot)$ is the Dirac delta. The Biot-Savart operator is discretized in a matrix $G = [G(r^1),..., G(r^{N_{points}})]$ of size $N_{sensors} \times 3N_{points}$ where each column of the matrix $G(r^k)$ represents the magnetic field produced by a unit current dipole at a given position and oriented according to one of the three orthogonal directions; this matrix is usually referred to as the leadfield matrix. The discretized mathematical model becomes here

$$\mathrm{b} = \mathrm{G} \cdot \mathrm{j} \qquad (13)$$

where the vector b is $N_{sensors} \times 1$ and contains the collection of data measured by all the sensors, while the vector $j = [j\ 1,...,j^{\ N_{points}}]\ T$ is $3N_{points} \times 1$ and contains the dipole moments of the current dipoles located at the fixed locations contained in the leadfield G. In the remainder of the Chapter we will denote by b_t and j_t the measured data and the neural current at time t; the leadfield matrix is usually assumed to be not time dependent.

Classification of Inverse Methods

Equation (13) is the main equation of practical interest for the solution of the inverse problem. It reflects the fact that the inverse problem is linear, in the sense that the measured data are the linear superpositions of the contributions of individual current dipoles. The physical and geometrical model discussed in the previous section is embodied in the matrix G, which is assumed to be known when solving the inverse problem. However, talking about the MEG inverse problem may be misleading for the following reason. While in the general case one will model the neural current as a vector field with support in the whole brain, and the inverse problem amounts to inversion of the (non-invertible) matrix G, in an alternative approach one can assume that the brain activity is highly localized in few brain regions; if these regions are small compared to their distance from the sensors, each region can be approximated as a single current dipole; in this case, the inverse problem consists in estimating the number of sources, their location, orientation and strength. The mathematical properties of the two inverse problems are rather different. In the former case, one will face a linear equation resulting in a highly under-determined problem due to the ratio between the typical number of sensors ($\simeq 100$) and the typical number of points ($\simeq 10,000$). In the following, we will call imaging methods the methods based on a continuous current distribution model. In the second case, one will want to estimate the intensity, the orientation, as well as the number and the locations of the point sources; the dependence of the magnetic field on the two latter parameters is highly non-linear, thus resulting in a genuinely non-linear problem. In the following, we will call parametric methods the methods based on a point source model.

The classification of inverse methods in imaging and parametric methods is rather common practice in many reviews and dates back at least to Hämäläinen et al. (1993), where dipole methods are separated from regularization methods. In this chapter, we introduce a new - to the best of our knowledge - classification of inverse methods, which accounts for a recent trend in the MEG inverse problem community. We will call static methods the inverse methods which do not give a priori information on the temporal behaviour of the neural sources, and dynamic methods those which do provide a prior on it. The recent appearance of dynamic methods is indeed a consequence of the increased availability of cheap computational power. Dynamic methods are far more computationally demanding for the simple reason that they perform regularization in a much higher-dimensional space. At the same time, prior information in the temporal domain can really help in stabilizing the solution of this highly ill-posed problem.

IMAGING METHODS

Imaging methods solve the linear inverse problem defined by

$$\mathbf{B}_t = \mathbf{G}\mathbf{J}_t + \mathbf{N}_t \tag{14}$$

where: Bt ∈ $R^{Nsensors}$ is the Random Vector representing the measured data at time t; G is the leadfield matrix as in (13); Jt ∈ $R^{3Npoints}$ is the Random Vector representing the neural current distribution at time t; Nt ∈ $R^{Nsensors}$ explicitely accounts for the presence of noise in the data. The output of an imaging method is an estimate of J_t for every t; as such, it is a sequence of current distributions which can be visualized as a dynamic intensity map representing the flow of brain activity.

From a statistical perspective, all the imaging methods we will describe can be interpreted as Bayesian estimation methods; this is due to the ill-posedeness of the inverse problem which requires rather strong regularization algorithms, i.e. a priori information, to get stable solutions. Due to the fact that most of these methods have been first described in a regularization/optimization framework, the estimate they provide is always easily interpreted as the MAP of the posterior distribution. All methods share basically the same likelihood function: since noise is assumed to have a zero-mean Gaussian distribution and to be uncorrelated with the sources, the likelihood function is:

$$\pi(\mathbf{b}_t|\mathbf{j}_t) = \frac{1}{\sqrt{(2\pi)^M|\mathbf{\Sigma}|}} \exp\left(-(\mathbf{b}_t - \mathbf{G}\mathbf{j}_t)^T\mathbf{\Sigma}^{-1}(\mathbf{b}_t - \mathbf{G}\mathbf{j}_t)\right) \tag{15}$$

where b_t, j_t are realizations of the data and the unknown RV, Σ is the noise covariance matrix and $M = N_{sensors}$.

What really makes each method unique is the specific choice of the prior distribution for the spatio/temporal distribution of the neural current.

Static imaging methods - Minimum Norm Estimate

Minimum Norm Estimate (MNE) is perhaps the most popular method for solving the MEG inverse problem. It has been introduced almost thirty years ago and is the straightforward implementation of the Tichonov algorithm. As seen in Section 2, this choice corresponds to a Gaussian prior distribution

$$\pi(\mathbf{j}_t) = \frac{1}{\sqrt{2\pi\gamma}} \exp\left(-\frac{||\mathbf{j}_t||^2}{2\gamma^2}\right). \tag{16}$$

The explicit form of the MNE solution, i.e. the MAP of the posterior distribution under the given prior, is given by

$$\hat{\mathbf{j}}_t = (\mathbf{G}^T\boldsymbol{\Sigma}^{-1}\mathbf{G} + \lambda\mathbf{I})^{-1}\mathbf{G}^T\boldsymbol{\Sigma}^{-1}\mathbf{b}_t \tag{17}$$

where the regularization parameter is $\lambda \propto 1/\gamma^2$. Due to the Gaussian spatial prior, solutions provided by MNE tend to be rather smooth, a feature that makes MNE less suitable than other methods for estimating focal activity. Furthermore, another limitation of MNE is that the zero-mean Gaussian prior gives an implicit preference to superficial sources, which need lower intensity to produce the meaured field. This results in an estimation bias which is stronger for deeper sources. Despite these drawbacks, MNE is a very attractive method thanks to its simplicity and to the very low computational cost.

Static Imaging Methods - Weighted MNE, LORETA and LCMV Beamformer

The class of weighted minimum norm solutions has been developed with the aim of overcoming the depth bias introduced by the simple prior used by MNE. The main difference with respect to standard MNE is that the prior is still Gaussian but the source covariance matrix is no longer proportional to the identity matrix; instead, the source covariance matrix contains a weighting factor for each source, aimed at removing the bias towards superficial solutions. In Lin et al. (2006), for instance, the $3N_{\text{points}} \times 3N_{\text{points}}$ source covariance matrix R is diagonal with

$$\mathbf{R}_{j,j} = ||\mathbf{G}(\mathbf{r}^k)||^{-p}, j = 3k-2, 3k-1, 3k, \;\; k = 1, \ldots, \mathcal{N}_{points} \tag{18}$$

where $G(r^k)$ is the lead field corresponding to the k-th source point, $||\cdot||$ is the euclidean norm and p is the depth weighting parameter. With this choice, deeper sources have a Gaussian prior with larger variance with respect to shallower sources. The final solution of WMNE is given by the two following equivalent formulations

$$\begin{aligned}\hat{\mathbf{j}}_t = & \;\; \mathbf{R}\mathbf{G}^T(\mathbf{G}\mathbf{R}\mathbf{G}^T + \boldsymbol{\Sigma})^{-1}\mathbf{b}_t = \\ = & \;(\mathbf{G}^T\boldsymbol{\Sigma}^{-1}\mathbf{G} + \mathbf{R}^{-1})^{-1}\mathbf{G}^T\boldsymbol{\Sigma}^{-1}\mathbf{b}_t.\end{aligned} \tag{19}$$

Low Resolution Electromagnetic Tomography (LORETA) (Pascual-Marqui et al., 1994) is a WMNE with the further addition of a Laplacian operator. The prior for LORETA is given by

$$\pi(\mathbf{j}_t) \propto \exp\left(-\mathbf{j}_t^T \mathbf{R}^T \mathbf{L}^T \mathbf{L} \mathbf{R} \mathbf{j}_t\right) \tag{20}$$

where R is as in (18) and L is the discrete spatial laplacian operator (Pascual-Marqui, 1999). Such operator L produces a smoother estimate of the current distribution since the inverse matrix L^{-1} implements a discrete spatial smoothing operator. LORETA has been shown to provide accurate localization even for deep sources.

The term beamformer encompasses a broad range of algorithms which have become popular in the last fifteen years. In one way or the other, most or all of them can be proved to be particular choices of WMNE (Mosher et al., 2003). Here we only consider the Linearly Constrained Minimum Variance Beamformer (Van Veen et al., 1997). The derivation of this method is made in an engineering framework and is based on a constrained minimization. The idea behind beamforming is that of constructing a spatial filter which isolates the activity produced in a selected location, blocking the activity originated elsewhere. The main working assumption of the method described in (Van Veen et al., 1997) is that all the sources are uncorrelated. The result of the constrained minimization turns out to be

$$\hat{\mathbf{j}}_t^k = \left[(\mathbf{G}(\mathbf{r}^k))^T \mathbf{R}_b^{-1} \mathbf{G}(\mathbf{r}^k)\right]^{-1} (\mathbf{G}(\mathbf{r}^k))^T (\mathbf{G}\mathbf{R}\mathbf{G}^T + \mathbf{\Sigma})^{-1} \mathbf{b}(t) \tag{21}$$

where R_b is the data covariance matrix and $\hat{j}^k_t$ is the estimated current density at k-th location. We remark that this solution corresponds to the WMNE solution (19) if the source covariance matrix R is a block diagonal matrix, composed by the 3 × 3 matrices:

$$\mathbf{R}(\mathbf{r}^k) = \left[(\mathbf{G}(\mathbf{r}^k))^T \mathbf{R}_b^{-1} \mathbf{G}(\mathbf{r}^k)\right]^{-1} \tag{22}$$

and zero elsewhere. Importantly, the assumption that the neural sources are not correlated in time is a strong assumption, often not verified in experimental data. As a consequence of this assumption, however, beamformers face severe difficulties when the data actually contain signals produced by correlated sources. An improved version of the beamforming technique has been proposed, which tries to overcome this limitation (Sekihara et al., 2002).

Static imaging methods - dSPM and SLORETA

Dynamic Statistical Parametric Mapping (dSPM) and Standardized LORETA (sLORETA) combine Bayesian inference for the electrical current with noise normalization. The ouput of these methods is no longer an estimate of the neural current distribution, but rather a statistical measure of brain activity

Instead of imaging the estimated electrical current intensity, dSPM (Dale et al., 2000) produces maps of the following neural activity index

$$z_t^k = \frac{||\hat{\mathbf{j}}_t^k||_2^2}{||(\mathbf{W}(\mathbf{r}^k))^T \boldsymbol{\Sigma} \mathbf{W}(\mathbf{r}^k)||_F^2} \tag{23}$$

where $W(r^k)$ is the regularized inverse operator in (19)

$$\mathbf{W}(\mathbf{r}^k) = (\mathbf{GRG}^T + \boldsymbol{\Sigma})^{-1}\mathbf{G}(\mathbf{r}^k)\mathbf{R}(\mathbf{r}^k) \tag{24}$$

Σ is the noise covariance matrix, and $|| \cdot ||F$ is the Frobenius norm; in practice, the output of a weighted MNE $\hat{j}_t$ at the k-th location is divided by an estimate of the portion of the source variance, at the same location, which is due to the noise in the data. Under the null hypothesis of zero brain activity, the resulting quantity z_t has a Student's t distribution.

sLORETA (Pascual-Marqui, 2002) consists in a similar approach, but the standardization is obtained considering the variance of the estimated sources (19), instead of using just the variance due to the noise component. SLORETA produces maps of the quantity

$$z_t^k = \frac{||\hat{\mathbf{j}}_t^k||_2^2}{||(\mathbf{W}(\mathbf{r}^k))^T (\mathbf{GRG}^T + \boldsymbol{\Sigma})^{-1} \mathbf{W}(\mathbf{r}^k)||_F^2}. \tag{25}$$

The claim of the authors is that this method produces unbiased, zero-error localization of point sources. sLORETA is largely used in EEG source analysis, where source localization is made particularly difficult by the strongest impact of the passive currents on the measured data.

Static Imaging Methods - Minimum Current Estimate

The MCE algorithm (Uutela et al. (1999)) is substantially different from all the imaging methods presented so far in that it does not use a Gaussian prior for the current distribution, but an exponential prior instead:

$$\pi(\mathbf{j}_t) = \exp(-||\mathbf{j}_t||_1) \tag{26}$$

where $||\mathbf{j}_t||_1 = \sum_{k=1}^{3N_{points}} \left|\mathbf{j}_t^k\right|$.

This is usually phrased in the regularization framework as using an L^1 prior, and the main consequence of this choice is that sparsity of the solution is promoted, so that estimates provided by MCE are typically highly focused.

In contrast to the MNE, MCE is therefore well suited for estimating focal activity. As a drawback, it has been suggested that MCE is in fact not capable to recover smooth current distributions, always shrinking to few points the estimated active area. From a computational viewpoint, the minimization with the L1 norm cannot be expressed any more in a closed form. Linear programming algorithms can be used to solve the minimum problem, but the computational cost is remarkably higher compared to that of methods based on a Gaussian prior.

Dynamic Imaging Methods - L1L2 Regularization

From the conceptual point of view, the L1L2 approach (Ou et al., 2009) consists in replacing the time-varying Random Vectors in equation (14) with the full random matrices containing the whole spatio-temporal pattern of the data and of the brain activity:

$$B = GJ + N \qquad (27)$$

where $B = [B_1,\ldots,B_T]$, $J = [J_1,\ldots,J_T]$ and $N = [N_1,\ldots, N_T]$; the functional to be minimized is now

$$||\mathbf{B} - \mathbf{GJ}||_F^2 + ||\mathbf{J}||_2^1 \qquad (28)$$

with

$$||\mathbf{J}||_2^1 = \sum_{k=1}^{3N_{points}} \sqrt{\sum_{t=1}^{T} (\mathbf{J}_t^k)^2}. \qquad (29)$$

In Bayesian terms, this can be viewed as using a Gaussian prior distribution in time and a Laplacian prior distribution in space. This type of regularization promotes both sparsity in the spatial domain, in order to produce good estimates of focal sources, and smoothness in the temporal domain. From the computational viewpoint, this method is realized by means of a projection of the data and of the source space onto a subset of basis functions, to reduce the dimensionality; a gradient method is used to minimize the functional, i.e. to calculate the approximate MAP of the posterior distribution.

Dynamic Imaging Methods - Kalman Filter

In Kalman filtering, one explicitely models the neural currents and the measurements as two stochastic processes, $\{J_1,\ldots,J_T\}$, $\{B_1,\ldots,B_T\}$. Under Markovian assumptions for these processes, the posterior distribution

$\pi(j_t|b_{1:t})$ can be computed at any time-point t applying sequentially the following two equations:

$$\pi(\mathbf{j}_t|\mathbf{b}_{1:t-1}) = \int \pi(\mathbf{j}_t|\mathbf{j}_{t-1})\pi(\mathbf{j}_{t-1}|\mathbf{b}_{1:t-1})d\mathbf{j}_{t-1} \tag{30}$$

$$\pi(\mathbf{j}_t|\mathbf{b}_{1:t}) = \frac{\pi(\mathbf{b}_t|\mathbf{j}_t)\pi(\mathbf{j}_t|\mathbf{b}_{1:t-1})}{\pi(\mathbf{b}_t|\mathbf{b}_{1:t-1})} \tag{31}$$

where $\{b_{1:t}\}$ denotes the set of measurements up to time t and $\pi(j_t|j_{t-1})$ is the transition kernel wich describes the evolution of the neural currents in time. Under Gaussian noise, Gaussian prior assumptions, the posterior distribution remains Gaussian at any time point, and these two equations have closed form solution through the so-called Kalman filter: at every time point, the mean and the covariance matrix of the Gaussian distribution are updated. The main difference between this approach and all the previously described approaches is that here the data are analyzed in sequence. The solution at time t depends on all the past history of observations; the prior density in the Kalman filter is time dependent and is given by

$$\pi(\mathbf{j}_{1:t}) = \pi(\mathbf{j}_0)\prod_{n=0}^{t}\pi(\mathbf{j}_{n+1}|\mathbf{j}_n). \tag{32}$$

The main practical issue in the application of a Kalman filter to the MEG inverse problem is that the Kalman filter requires to invert the source covariance matrix at each time step. In MEG, this covariance matrix is typically very large, being $3N_{points} \times 3N_{points}$. In Galka et al. (2004), the very high dimensional problem is decomposed in a collection of low dimensional problems, exploiting the local nature of the covariance matrix; in Long et al. (2006), a network of computers is instead used to solve the full problem.

PARAMETRIC METHODS

Parametric methods solve the non-linear inverse problem defined by

$$\mathbf{B}_t = \sum_{i=1}^{\mathcal{N}_{dipoles}} \mathbf{G}(\mathbf{R}_t^i)\mathbf{Q}_t^i + \mathbf{N}_t \tag{33}$$

where $N_{dipoles}$ is the unknown number of active dipoles at time t, $R^i_{\ t}$ and $Q^i_{\ t}$ are the Random Vectors representing the location and the dipole moment of the i-th source at time t, respectively, and G(r) is the lead field of a unit dipole at location r. Remarkably, the number of active dipoles is not expected to be higher than, say, ten; this implies that the number of unknowns in the parametric approach is considerably lower than the number of unknowns in the imaging approach. However, as already observed, the data depend now non-linearly on the parameters, namely, on the number of sources and on the source locations. This non-linearity causes in fact several problems if one attempts to solve the full problem. For this reason, several methods for dipole estimation work under two important assumptions: first that the number of sources is constant in time, and second that the source locations are fixed. Only recently, Bayesian methods have appeared which try to overcome these artificial constraints. The authors have personally developed one of these methods, namely particle filtering; to this method is devoted Section 6. In the present Section, we give a brief overview of the other main parametric methods.

Static parametric methods - dipole fitting

Dipole fitting is one of the oldest and most largely used methods for solving the MEG inverse problem. In dipole fitting, current dipoles are estimated one at a time, at fixed time points; the user typically chooses the best time point, the one at which the data exhibit a dipolar field pattern, and then starts a non-linear optimization algorithm - typically a Levemberg-Marquardt algorithm - which maximizes the likelihood for a single current dipole at that particular time point

$$\exp\left(-(\mathbf{b}_t - \mathbf{G}(\mathbf{r})\cdot\mathbf{q})^T\mathbf{\Sigma}^{-1}(\mathbf{b}_t - \mathbf{G}(\mathbf{r})\cdot\mathbf{q})\right) \tag{34}$$

r and q being the location and dipole moment of one single source. The single dipole fitting procedure is typically iterated, usually at different time points, until a satisfactory set of different dipoles is found. To produce an estimate of the temporal behaviour of the estimated sources, a multiple dipole model is constructed by combining the estimated dipole locations and - possibly - orientations; the dipole strengths are then estimated linearly from the data, by optimization of the likelihood

$$\exp\left[-\left(\mathbf{b} - \sum_{i=1}^{\mathcal{N}_{dipoles}} \mathbf{G}(\mathbf{r}^i)\cdot\mathbf{q}^i\right)^T \mathbf{\Sigma}^{-1}\left(\mathbf{b} - \sum_{i=1}^{\mathcal{N}_{dipoles}} \mathbf{G}(\mathbf{r}^i)\cdot\mathbf{q}^i\right)\right] \tag{35}$$

A key problem with dipole fitting is that it strongly relies on the personal experience of the person who performs the analysis; in some sense then, despite it never formally makes use of a prior density, it may be considered as a Bayesian method due to the strong prior embodied in the human experience. A second relevant problem of dipole fitting is that, involving decision making from the user, it is usually quite time consuming.

Static parametric methods - MUSIC and RAP-MUSIC

The Recursively Applied and Projected MUltiple SIgnal Classification (RAP MUSIC) algorithm is perhaps the most largely known method for automatic estimation of current dipoles from MEG data. It was first introduced in Mosher et al. (1992) and further developed in order to deal efficiently with correlated sources (Mosher & Leahy, 1999).

MUSIC is based on a simplified version of the model described in equation (33), as it assumes that the number of dipoles and their locations are fixed in time; thus the model becomes now

$$\mathbf{B}_t = \sum_{i=1}^{\mathcal{N}_{dipoles}} \mathbf{G}(\mathbf{R}^i)\mathbf{U}^i \cdot S_t^i + \mathbf{N}_t = \mathbf{A} \cdot \mathbf{S}_t + \mathbf{N}_t \tag{36}$$

where $A = [G(R^1)U^1,..., G(R^{Ndipoles})U^{Ndipoles}]$ is the field generated by $N_{dipoles}$ dipoles with locations $\{R^i\}$ and orientations $\{U^i\}$; $S^i{}_t$ is the strength of the i-th dipole and $S_t = [S^1{}_t,..., S^{Ndipoles}{}_t]$ the amplitudes vector; the dipole moment is therefore $Q^i = U^i \cdot S^i$.

If one further assumes that the neural sources are not correlated, simple calculations yield for the data covariance matrix R_b

$$\mathbf{R}_b = \mathbf{A}^T\mathbf{R}_s\mathbf{A} + \sigma_{noise}^2\mathbf{I} \tag{37}$$

where R_s is the covariance matrix of the sources. Due to the assumption of uncorrelated sources, the source covariance matrix has full rank, and the data covariance matrix contains $N\hat{}_{dipoles}$ eigenvalues considerably larger than the others; this number represents an estimate of the number of dipoles.

Furthermore the corresponding eigenvectors span the so called signal subspace $\varphi\hat{}_s$ and an estimate of the unknown parameters $\{r^i, u^i\}$ is obtained by maximizing the subspace correlation (Golub & Loan, 1984) between the dipole field and the signal subspace:

$$\text{subcorr}(G(r)u, \varphi\hat{}s). \tag{38}$$

Once the locations and orientations are estimated, the source amplitudes are found by a least square approach. A recursive procedure deals with the problem of local minima. From a statistic view point it can be shown that the estimated parameters represent a Maximum Likelihood estimate of the Random Vector in the model (33) (Stoica & Nehorai, 1989).

Dynamic Parametric Methods - MCMC of Current Dipoles

A more recently proposed strategy for solving the parametric MEG problem is that of using Bayesian inference for approximating the posterior density in the parameter space (Jun et al., 2005). In these studies, the authors assume a fixed but unknown number of spatially stationary dipolar sources; the modeling parameters are then the Random Variable of the number of sources $N_{dipoles}$ and the Random Vectors of the source locations $R \in R3^{Ndipole}s$, of the source time courses $S \in R^{Ndipoles \times T}$ and of the dipole orientations $U \in R^{3Ndipoles}$. The authors seek an approximation of the posterior density through Bayes theorem (5):

$$\pi(n_{dipoles}, \mathbf{r}, \mathbf{s}, \mathbf{u}|\mathbf{b}) = \pi(\mathbf{b}|n_{dipoles}, \mathbf{r}, \mathbf{s}, \mathbf{u})\pi(\mathbf{u}|\mathbf{r}, n_{dipoles}) \times \times \pi(\mathbf{s}|n_{dipoles})\pi(\mathbf{r}|n_{dipoles})\pi(n_{dipoles}) \quad , \tag{39}$$

where the probability distributions for the source parameters are (necessarily) conditioned on the number of sources. Uniform distributions are used for $N_{dipoles}$, U and R, while a Gaussian distribution is assumed for S. In order to explore the high-dimensional posterior density, they resort to reversible jump Markov Chain Monte Carlo techniques after marginalization of the posterior density with respect to the noise covariance matrix and the source time courses. In Jun et al. (2006), the multi-dipole model is refined by introducing a source-dependent temporal activity window: in other words, each dipole is active only for a finite time interval, and is totally silent for the rest of the sequence, thus avoiding possible cross-talk phenomena. The implementation is again performed with a Reversible Jump Markov Chain Monte Carlo algorithm.

HIGHLY AUTOMATED DIPOLE ESTIMATION

In this section we present the formulation of the multi-dipole particle filter we have been developing in collaboration with our colleagues, and which we have published as an online open source code under the name of HADES (Highly Automated Dipole EStimation). The idea behind the method is that of using the Bayesian filtering equations (43)-(44) in a multi-dipole setting rather than in the imaging framework. In practice, the algorithm tracks in time the probability distribution of the dipole parameters; the main difficulties here are related to the non-linear dependence of the data on the dipole parameters, and on the necessity of estimating the number of dipoles as well. The particle filtering approach has been first proposed in a methodological paper, (Somersalo et al., 2003). In a subsequent set of publications, with our colleagues we have investigated the use of particle filtering for estimation of current dipoles from experimental MEG data (Sorrentino, 2010; Sorrentino et al., 2009), and compared particle filtering with other source estimation methods in Pascarella et al. (2010). Particle filtering has also been used to solve the EEG inverse problem in Antelis & Minguez (2010) and Mohseni et al. (2008); however, in these studies a simpler setting was used where the number of dipoles is fixed a priori.

The main advantages of the approach outlined below are that (i) it is a very general approach to multiple dipole modeling, as it explicitely allows the number of dipoles to be time varying and provides dynamically a posterior probability for the number of sources; (ii) it can be used online in real time, at least in principle, because data are processed in sequence and not as a whole.

In the following, we first describe the mechanism of the most basic particle filter; we then consider the practical modeling of multiple dipole with Random Finite Sets; we outline briefly the use of semi-analytic techniques, and eventually describe the open source software with Graphical User Interface for use of particle filtering.

Particle Filtering

Bayesian filtering relies on Markovian assumptions for the data and the unkown processes: if J_t and B_t represent the neural current and the measured data at time t, respectively, these assumptions can be formulated in terms of conditional distributions, and amount to

$$\pi(\mathbf{b}_t|\mathbf{j}_t, \mathbf{j}_{t-1}, \mathbf{j}_{t-2}, ..., \mathbf{j}_1) = \pi(\mathbf{b}_t|\mathbf{j}_t) \tag{40}$$

$$\pi(\mathbf{j}_t|\mathbf{j}_{t-1}, \mathbf{b}_{t-1}, ..., \mathbf{b}_1) = \pi(\mathbf{j}_t|\mathbf{j}_{t-1}) \tag{41}$$

$$\pi(\mathbf{j}_t|\mathbf{j}_{t-1}, ..., \mathbf{j}_1) = \pi(\mathbf{j}_t|\mathbf{j}_{t-1}) \ . \tag{42}$$

Given these premises, and assuming knowledge of the two conditional distributions at the right hand side of (40)-(42), the posterior distribution at time t can be obtained from the posterior distribution at time t − 1 with the two following equations

$$\pi(\mathbf{j}_t|\mathbf{b}_{1:t-1}) = \int \pi(\mathbf{j}_t|\mathbf{j}_{t-1})\pi(\mathbf{j}_{t-1}|\mathbf{b}_{1:t-1})d\mathbf{j}_{t-1} \tag{43}$$

$$\pi(\mathbf{j}_t|\mathbf{b}_{1:t}) = \frac{\pi(\mathbf{b}_t|\mathbf{j}_t)\pi(\mathbf{j}_t|\mathbf{b}_{1:t-1})}{\pi(\mathbf{b}_t|\mathbf{b}_{1:t-1})} \quad . \tag{44}$$

in (43) a prior density at time t is obtained by convolution of the posterior density at time t − 1 with the transition kernel; in (44), the posterior density at time t is obtained by combining the prior with the likelihood in the standard Bayes theorem. In addition to the two conditional distributions mentioned above, the density $\pi(j_0|b_0) = \pi(j_0)$ at time t = 0 is needed for initialization.

Exact computation of these two equations is possible under rather restrictive assumptions, such as the ones leading to Kalman filtering and described in Section 4. For more general cases, such as the one considered here which involves a non-linear relationship between the data and the unkonwns, numerical techniques have to be used to get reasonable approximations of the posterior densities. Particle filters are one of the most used numerical techniques for solving this type of problems. They are basically a sequential Monte Carlo sampling technique. In their simplest form, the SIR (Sampling Importance Resampling) particle filter, they amount to sequential application of an importance sampling - resampling mechanism.

The main steps of the SIR particle filter are:

0 [initialization] draw $N_{particles}$ sample points $\{\tilde{j}^i_0\}i=1,...,N_{particles}$ distributed according to $\pi(j_0|b_0)$; the sampled points represent therefore an approximation of the initializing density:

$$\pi(\mathbf{j}_0) \simeq \sum_{i=1}^{\mathcal{N}_{particles}} \frac{1}{\mathcal{N}_{particles}} \delta(\mathbf{j}_0 - \tilde{\mathbf{j}}_0^i) \quad ; \tag{45}$$

1 [Evolution] for $t \geq 0$, let $\{\tilde{j}^i_t\}_{i=1,...,Nparticles}$ be a sample distributed according to $\pi(j_t|b_{1:t})$; then exploiting the Chapman-Kolmogorov equation we may approximate the next prior probability density function as follows:

$$\pi(\mathbf{j}_{t+1}|\mathbf{b}_{1:t}) = \int \pi(\mathbf{j}_{t+1}|\mathbf{j}_t)\pi(\mathbf{j}_t|\mathbf{b}_{1:t})d\mathbf{j}_t \simeq$$

$$= \frac{1}{\mathcal{N}_{particles}} \sum_{i=1}^{\mathcal{N}_{particles}} \pi(\mathbf{j}_{t+1}|\tilde{\mathbf{j}}_t^i) \quad ; \tag{46}$$

sample the so obtained prior density, by drawing a single particle j^i_{t+1} from $\pi(j_{t+1}|\tilde{j}^i_t)$ for each i; we have now an approximation of the prior at time t + 1:

$$\pi(\mathbf{j}_{t+1}|\mathbf{b}_{1:t}) \simeq \sum_{i=1}^{\mathcal{N}_{particles}} \frac{1}{\mathcal{N}_{particles}} \delta(\mathbf{j}_{t+1} - \mathbf{j}_{t+1}^{i}) \quad ; \tag{47}$$

2 [Observation] apply the Bayes theorem, i.e. compute the relative weights of the particles:

$$w_{t+1}^{i} = \pi(\mathbf{b}_{t+1}|\mathbf{j}_{t+1}^{i}) \quad , \tag{48}$$

and then normalize the weights so that $\sum_{i=1}^{\mathcal{N}_{particles}} w_{t+1}^{i} = 1$; an approximation to the posterior probability density function is given by

$$\pi(\mathbf{j}_{t+1}|\mathbf{b}_{1:t+1}) \simeq \sum_{i=1}^{\mathcal{N}_{particles}} w_{t+1}^{i} \delta(\mathbf{j}_{t+1} - \mathbf{j}_{t+1}^{i}) \quad ; \tag{49}$$

3 [Resampling] resample the sample set representing the posterior density: extract a new set of particles $\{\tilde{\mathbf{j}}_{t+1}^{i}\}_{i=1}^{\mathcal{N}_{particles}}$ from the old set $\{\mathbf{j}_{t+1}^{i}\}_{i=1}^{\mathcal{N}_{particles}}$, in such a way that the probability of drawing the i-th particle is w_{t+1}^{i} (see Arulampalam et al. (2002) for the details); we get

$$\pi(\mathbf{j}_{t+1}|\mathbf{b}_{1:t+1}) \simeq \sum_{i=1}^{\mathcal{N}_{particles}} \frac{1}{\mathcal{N}_{particles}} \delta(\mathbf{j}_{t+1} - \tilde{\mathbf{j}}_{t+1}^{i}) \quad ; \tag{50}$$

go to Step 1.

Remarkably, the combination of Step 1 and 2 can be viewed as an importance sampling.

Filtering of multiple dipoles with Random Finite Sets

In order to deal with a time-varying and unknown number of sources, in Sorrentino et al. (2009) we have proposed the use of Random Finite Sets as a mathematical tool to model the neural current.

Random Finite Sets (Molchanov, 2005) are generalizations of Random Vectors to a non-fixed dimension case. As the name suggests, realizations of a Random Finite Set are sets containing a random number of random objects. The use of Random Finite Sets for multiple target tracking has been proposed in (Mahler, 2003) for the first time.

In the RFS approach to the MEG problem, the neural current at time t is assumed to be a RFS of current dipoles, denoted by $J_t = \{D,.., D^{N_t}{}_t\}$, where each D^i_t is a Random Vector containing the parameters of a single current dipole and the Random Variable N_t represents the number of sources at time t. A realization of J_t will be denoted as J_t. The model now amounts to:

$$\mathcal{J}_{t+1} = \mathcal{E}(\mathcal{J}_t) \cup \mathcal{B}_{t+1} \quad (51)$$

$$\mathbf{B}_t = \mathbf{B}(\mathcal{J}_t) + W_t \quad . \quad (52)$$

In (51), the dipole set at time t + 1 is given by the union of the set $E(J_t)$, containing the subset of dipoles active at time t which have survived and evolved, and the set of new born dipoles B_{t+1}. In (52), the Random Finite Set of dipoles J_t produces the data; since the dimension of the data vector is fixed to the number of sensors in the MEG device, the data is represented as a standard Random Vector.

Despite the formulation in terms of Random Finite Sets, single dipole probabilities can be used to describe most of the models as long as individual dipoles are assumed to behave independently on each other . In the following we describe the main features of the model in terms of single dipole probability density functions.

Initializing Density

HADES adopts an uninformative initializing density for the multi-dipole states. Let v_{max} be the maximum number of dipoles in a RFS (which we need to fix for computational reasons); then the prior probability of different models is uniform, $P(n_0) = 1/(v_{max} + 1)$ for $n_0 = 0, ..., v_{max}$. Dipoles are assumed to be independent and identically distributed with the same density used for the single-dipole case; the probability density function of each dipole $D_0 \in J_0$ is

$$\pi(\mathbf{d}_0|\mathbf{b}_0) \propto \frac{1}{V} \times \exp\left(-\mathbf{q}_0^T \Xi \mathbf{q}_0\right) \quad (53)$$

where we have defined separately the uniform density for the dipole location in the volume V and the zero-mean Gaussian density with covariance matrix Ξ for the dipole moment.

Likelihood Function

Due to the linearity of the Biot-Savart equation, the magnetic field produced by a set of sources is given by the linear combination of the magnetic fields produced by the individual dipoles contained in the set. If we denote by $b(J_t)$ the magnetic measurement produced by the dipole set J_t, under the standard assumption of additive white Gaussian noise, with covariance matrix Σ, the likelihood function is

$$\pi(\mathbf{b}_t|J_t) \propto \exp\left(-[\mathbf{b}_t - \mathbf{b}(J_t)]^T \Sigma^{-1} [\mathbf{b}_t - \mathbf{b}(J_t)]\right) . \tag{54}$$

Transition kernel

Lack of knowledge on the true dynamics of the neural sources is coded in a transition kernel based on the random walk. The transition probabilities between sets with different number of dipoles are uniform. The evolution of a dipole set containing n dipoles at time t is governed by the following rules:

- the probability that the dipole set contains n dipoles at time t + 1 is 1/3; the probability that it contains n + 1 dipoles is 1/3; the probability that it contains n − 1 dipoles is 1/3;
- if the dipole set at time t + 1 contains n − 1 dipoles, the dipole to be removed is selected randomly from the set;
- if the dipole set at time t + 1 contains n + 1 dipoles, the new dipole is drawn from the same distribution of the initializing prior, eq. (53);
- all the surviving dipoles of a dipole set evolve independently according to the single-dipole transition kernel:

$$\pi(\mathbf{d}_{t+1}|\mathbf{d}_t) \propto \mathcal{I}_{B(\mathbf{r}_t)}(\mathbf{r}_{t+1}) \times \exp - \left([\mathbf{q}_{t+1} - \mathbf{q}_t]^T \Xi_\delta^{-1} [\mathbf{q}_{t+1} - \mathbf{q}_t]\right) . \tag{55}$$

where $I_{B(r)}(\cdot)$ is the indicator function of a ball with center in r and Ξ_δ is the covariance matrix for the Gaussian random walk.

Source Estimates

Point estimates of the dipole parameters are provided in a three-step automatic procedure. First, the number of sources is estimated from the marginal distribution

$$P(|J_t| = k) = \int_{S(k)} \pi(J_t|\mathbf{b}_{1:t}) \delta J_t , \tag{56}$$

where S(k) is the set of all subsets with k elements; the maximum of the marginal distribution can be used as an estimator:

$$\hat{n}_t = \arg\max_k P(|J_t| = k) . \tag{57}$$

Second, source locations are estimated as the maxima of the marginal quantity often referred to as Probability Hypothesis Density (PHD) in the RFS framework. The PHD is defined as

$$PHD(\mathbf{d}_t) = \int \pi(\{\mathbf{d}_t\} \cup Y)\delta Y \quad , \tag{58}$$

where the integral is a set integral, $\{d_t\}$ is a single-dipole set and $\pi(\cdot)$ is the probability density function of the Random Finite Set J_t. In a particle filter implementation, the PHD can be approximated by

$$PHD(\mathbf{d}_t) = \sum_{i=1}^{N_{particles}} w_t^i \left(\sum_{\mathbf{d} \in J_t^i} \delta(\mathbf{d} - \mathbf{d}_t) \right) \quad . \tag{59}$$

The PHD can also be viewed as the intensity measure of a point process.

Third and last step, once the dipole locations have been estimated the dipole moments are estimated with a linear least squares.

Rao-Blackwellization

The SIR particle filter described in the previous subsection is very general as it allows approximation of non-Gaussian densities which arise due to the non-linear dependence between the data and the parameters. However, in the MEG case such non-linearity is limited to the parameters describing the location of the sources; on the other hand, the data depend linearly on the dipole moment; this linearity can be in fact exploited using a technique known as Rao-Blackwellization. The main idea is that the posterior density can be split (Campi et al. (2008))

$$\pi(\mathbf{r}_t, \mathbf{q}_t | \mathbf{b}_{1:t}) = \pi(\mathbf{q}_t | \mathbf{r}_t, \mathbf{b}_{1:t}) \pi(\mathbf{r}_t | \mathbf{b}_{1:t}) \tag{60}$$

At the right hand side, the density for q_t $\pi(q_t|r_t, b_{1:t})$ conditional on r_t is a Gaussian density, whose parameters can be tracked with a Kalman filter; on the other hand, the locations parameters must be sampled, which leads to an algorithm that realizes a set of Kalman filters for the dipole moment, one for each sampled location.

Clustering

Due to its dynamic nature, the particle filter described above tends to provide slightly moving estimates of static dipoles. In order to produce a compact description of the estimated dipoles, it can be useful to apply a clustering algorithm to the whole set of estimated dipoles $\{\hat{\mathbf{d}}_1^1, ..., \hat{\mathbf{d}}_1^{\hat{n}_1}, ..., \hat{\mathbf{d}}_t^1, ..., \hat{\mathbf{d}}_t^{\hat{n}_t}\}$. In Sorrentino et al. (2009) we have applied an iterative algorithm based on the k-means (Spath, 1980) to obtain a small set of neural sources. The iterations are needed because the number of clusters is unknown, while the k-means works with a fixed number of clusters: hence, we first cluster the dipoles in a large number of clusters $N_{clusters}$; then apply the Wilcoxon test (Weerahandi, 1995) and check whether all clusters are significantly diverse;

if not, we set $N_{clusters} = N_{clusters-1}$ and try again the clustering procedure; if they are all diverse, the algorithm stops. A similar idea, implemented in a more sophisticated strategy, has been proposed in Clark & Bell (2007).

HADES

The particle filter described above has been implemented in Matlab, and is currently distributed with a Graphical User Interface, named HADES (Highly Automated Dipole EStimation) (Campi et al. (2011)). The software can be downloaded from http://hades.dima.unige.it. The software is open source and is released under the standard GPL license. Through the MNE toolbox (http://www.nmr.mgh.harvard.edu/martinos/userInfo/data/MNE register/index.php) it supports input/output in Neuromag .fif, FreeSurfer .w and MNE .stc formats, further than the standard Matlab .mat format.

CONCLUSION

Localization of brain activity from MEG recordings is a challenging problem which has been solved in a variety of frameworks and using many different algorithms. In this Chapter we have attempted a short review of some of these methods, the ones which seem to be most used in the MEG community, with the aim of describing them in the unifying framework offered by statistics. Interestingly, when described in such framework some clear connections between many of these methods appear naturally. Indeed, it emerges clearly from this review that the only difference between most of these methods, except for the type of source model, relies in the choice of the prior density. It also emerges that the main difference between more recent and previously developed methods is that the recent methods tend to exploit the dynamic nature of the inverse problem by performing regularization in the whole spatio-temporal domain; in statistical terms, this corresponds to providing a priori information on the temporal behaviour of the sources in addition to the a priori information on the spatial distribution. Among these new dynamic methods, we have given particular emphasis to the description of HADES, the multi-dipole particle filter we have been developing in collaboration with our colleagues. We look forward to see how far this new generation of spatio-temporal solvers can reach, and what the next generation will be.

ACKNOWLEDGEMENTS

AS has been supported by the European Union Seventh Framework Programme (FP7/2007-2013) under the grant agreement PIEF-GA-2009-252440.

REFERENCES

1. Antelis, J. & Minguez, J. (2010). DYNAMO: dynamic multi-model source localization method for EEG and/or MEG, IEEE Engineering in Medicine and Biology Society Conference Proceedings, pp. 5141–5144.
2. Arulampalam, M., Maskell, S., Gordon, N. & Clapp, T. (2002). A tutorial on particle filters for online nonlinear/non-gaussian bayesian tracking, IEEE Transactions on Signal Processing 50(2): 174–188.
3. Bertero, M. & Boccacci, P. (1998). Introduction to inverse problems in imaging, IoP, Institute of Physics Publishing. Campi, C., Pascarella, A., Sorrentino, A. & Piana, M. (2008). A rao-blackwellized particle filter for magnetoencephalography, Inverse Problems 24: 025023.
4. Campi, C., Pascarella, A., Sorrentino, A. & Piana, M. (2011). Highly automated dipole estimation, Computational Intelligence and Neuroscience 2011: 982185.
5. Cantarella, J., Turck, D. D. & Gluck, H. (2001). The biot - savart operator for application to knot theory, fluid dynamics, and plasma physics, Journal of Mathematical Physics 42: 876–905.
6. Clark, D. & Bell, J. (2007). Multi-target state estimation and track continuity for the particle PHD filter, IEEE Transactions on Aerospace and Electronic Systems 43: 1441–1453.
7. Dale, A., Liu, A., Fischl, B., Buckner, R., Belliveau, J., Lewine, J. & Halgren, E. (2000). Dynamic statistical parametric mapping: Combining fmri and meg for high-resolution imaging of cortical activity, Neuron 26: 55–67.
8. Galka, A., Yamashita, O., Ozaki, T., Biscay, R. & Valdes-Sosa, P. (2004). A solution to the dynamical inverse problem of eeg generation using spatiotemporal kalman filtering, NeuroImage 23: 435–453.
9. Golub, G. & Loan, C. V. (1984). Matrix Computation, The Johns Hopkins University Press. Gramfort, A., Papadopoulo, T., Olivi, E. & Clerc, M. (2011). Forward field computation with openmeeg, Computational Intelligence and Neuroscience 2011: 923703.
10. Hämäläinen, M., Hari, R., Knuutila, J. & Lounasmaa, O. (1993). Magnetoencephalography: theory, instrumentation and applications to non-invasive studies of the working human brain, Reviews of Modern Physics 65: 413–498.
11. Hämäläinen, M. & Ilmoniemi, R. (1984). Interpreting measured

magnetic fields of the brain: estimates of current distributions. Jun, S., George, J., Paré-Blagoev, J., Plis, S., Ranken, D., Schmidt, D. & Wood, C. (2005). Spatiotemporal bayesian inference dipole analysis for MEG neuroimaging data, NeuroImage 28: 84–98.

12. Jun, S., George, J., Plis, S., Ranken, D., Schmidt, D. & Wood, C. (2006). Improving source detection and separation in a spatiotemporal bayesian inference dipole analysis, Physics in Medicine and Biology 51: 2395–2414.
13. Kress, R., Kuhn, L. & Potthast, R. (2002). Reconstruction of a current distribution from its magnetic field, Inverse Problems 18: 1127–1146.
14. Lin, F., Witzel, T., Ahlfors, S., Stufflebeam, S., Belliveau, J. & Hamalainen, M. (2006). Assessing and improving the spatial accuracy in MEG source localization by depth-weighted minimum-norm estimates, NeuroImage 31: 160–171.
15. Long, C., Purdon, P., Temeranca, S., Desai, N., Hämäläinen, M. & Brown, E. (2006). Large scale kalman filtering solutions to the electrophysiological source localization problem - a meg case study, Proceedings of the 28thIEEE EMBS Annual International Conference, Vol. 5, pp. 4532–4535.
16. Mahler, R. (2003). Multitarget bayes filtering via first-order multitarget moments, IEEE Transactions on Aerospace and Electronic Systems 39: 1152–1178.
17. Mohseni, H., Wilding, E. & Sanei, S. (2008). Sequential monte carlo techniques for EEG dipole placing and tracking, 2008 IEEE Sensor array and multichannel signal processing workshop, pp. 95–98.
18. Molchanov, I. (2005). Theory of Random Sets, Springer-Verlag. Mosher, J., Baillet, S. & Leahy, R. (2003). Equivalence of linear approaches in bioelectromagnetic inverse solutions, 2003 IEEE Workshop on Statistical Signal Processing pp. 294–297.
19. Mosher, J. & Leahy, R. (1999). Source localization using Recursively Applied and Projected (RAP) MUSIC, IEEE Transactions on Signal Processing 47: 332–340.
20. Mosher, J., Lewis, P. & Leahy, R. (1992). Multiple dipole modeling and localization from spatio-temporal MEG data, IEEE Transactions on Biomedical Engineering 39: 541–557.
21. Ou, W., Hämäläinen, M. & Golland, P. (2009). A distributed spatio-temporal EEG/MEG inverse solver, NeuroImage 44: 932–946.

22. Pascarella, A., Sorrentino, A., Campi, C. & Piana, M. (2010). Particle filtering, beamforming and rap-music in the analysis of magnetoencephalography time series: a comparison of algorithms, Inverse Problems and Imaging 4: 169–190.
23. Pascual-Marqui, R. (1999). Review of methods for solving the eeg inverse problem, Int. J. Bioelectromagnetism 1: 75–86.
24. Pascual-Marqui, R. (2002). Standardize low resolution electromagnetic tomography (sLORETA: technical details, Methods and Findings in Experimental and Clinical Pharmacology 24: 5–12.
25. Pascual-Marqui, R., C.Michel & D.Lehman (1994). Low resolution electromagnetic tomography: A new method for localizing electrical activity in the brain, Int. J. Psycophysiol. 18: 49–65.
26. Pursiainen, S., Sorrentino, A., Campi, C. & Piana, M. (2011). Forward simulation and inverse dipole localization with lowest order Raviart-Thomas elements for electroencephalography, Inverse Problems p. in press.
27. Sarvas, J. (1987). Basic mathematical and electromagnetic concepts of the biomagnetic inverse problem, Phys. Med. Biol. 32: 11–22.
28. Sekihara, K., Nagarajan, S., Poeppel, D. & Marantz, A. (2002). Performance of an meg adaptive beamformer technique in the presence of correlated neural activities: effects on signal intensity and time course estimates, IEEE Transactions on Biomedical Engineering 49: 1534–1546.
29. Somersalo, E., Voutilainen, A. & Kaipio, J. (2003). Non-stationary magnetoencephalography by bayesian filtering of dipole models, Inverse Problems 19: 1047–1063.
30. Sorrentino, A. (2010). Particle Filters for Magnetoencephalography, Archives of Computational Methods in Engineering 17: 213–251.
31. Sorrentino, A., Parkkonen, L., Pascarella, A., Campi, C. & Piana, M. (2009). Dynamical MEG source modeling with multi-target bayesian filtering, Human Brain Mapping 30: 1911–1921.
32. Spath, H. (1980). Cluster Analysis Algorithms for Data Reduction and Classification of Objects, Halsted Press. Stoica, P. & Nehorai, A. (1989). MUSIC, Maximum Likelihood and Cramer-Rao bound, IEEE Transactions on acoustics, speech and signal processing 37: 720–741.
33. Uutela, K., Hämäläinen, M. & Somersalo, E. (1999). Visualization of magnetoencephalographic data using minimum current estimates,

NeuroImage 10: 173–180.

34. Van Veen, B., van Drongelen, W., Yuchtman, M. & Suzuki, A. (1997). Localization of brain electrical activity via linearly constrained minimum variance spatial filtering, IEEE Transactions on Biomedical Engineering 44: 867–880.
35. Weerahandi, S. (1995). Exact Statistical Methods for Data Analysis, Springer-Verlag, New York.

INDEX

D

E

F

G

H

I